$\sigma(n)$	sum-of-divisors function, 98
$\tau(n)$	number-of-divisors function, 98
$u(n)$	unit function, 100
$z(n)$	zero function, 100
$M(n)$	identity function with respect to multiplication, 100
$I(n)$	identity function with respect to Dirichlet product, 100
$(f \cdot g)(n)$	Dirichlet product, 101
$\omega(n)$	number of distinct prime factors of n, 108
$f^{-1}(n)$	Dirichlet inverse, 110
$\mu(n)$	Moebius function, 111
$\emptyset(n)$	Euler's totient function, 112
$\text{ind}_g t$	index of t relative to g, 134
$\left(\dfrac{a}{p}\right)$	Legendre symbol, 142
(a/p)	Legendre symbol, 142
$\left(\dfrac{a}{n}\right)$	Jacobi symbol, 163
L_n	nth Lucas number, 185
$<x>$	integer nearest to x, 187
P_n	nth Pell number, 193
R_n	nth associated Pell number, 193
$[a_0, a_1, a_2, \cdots, a_n]$	finite simple continued fraction, 195
C_k	convergent to a continued fraction, 201
$[a_0, a_1, a_2, \cdots]$	infinite continued fraction, 206
$[a_0, \overline{a_1, a_2, \cdots, a_t}$	periodic continued fraction, 219
$[a_n, a_{n+1}, a_{n+2}, \cdots]$	nth complete quotient, 224
$[a_0, \cdots, a_{n-1}, \overline{g_1(n), \cdots, g_k(n)}]_{n=0}^{\infty}$	Hurwitz-type continued fraction, 231
B_n	nth Bernouilli number, 234
$O(g(x))$	big oh notation, 253
$PSP(b)$	pseudoprime to base b, 255
$SPSP(b)$	strong pseudoprime to base b, 261
$EXP(n)$	exponential function, 268
$E(n)$	enciphering key, 277
$D(n)$	deciphering key, 277

BEGINNING NUMBER THEORY

Neville Robbins

Professor of Mathematics
San Francisco State University

WCB **Wm. C. Brown Publishers**
Dubuque, Iowa • Melbourne, Australia • Oxford, England

Book Team

Editor *Earl McPeek*
Developmental Editor *Theresa Grutz*
Publishing Services Coordinator *Julie Avery Kennedy*

Wm. C. Brown Publishers
A Division of Wm. C. Brown Communications, Inc.

Vice President and General Manager *Beverly Kolz*
National Sales Manager *Vincent R. Di Blasi*
Assistant Vice President, Editor-in-Chief *Edward G. Jaffe*
Director of Marketing *John W. Calhoun*
Marketing Manager *Elizabeth Robbins*
Advertising Manager *Amy Schmitz*
Director of Production *Colleen A. Yonda*
Manager of Visuals and Design *Faye M. Schilling*
Publishing Services Manager *Karen J. Slaght*
Permissions/Records Manager *Connie Allendorf*

Wm. C. Brown Communications, Inc.

Chairman Emeritus *Wm. C. Brown*
Chairman and Chief Executive Officer *Mark C. Falb*
President and Chief Operating Officer *G. Franklin Lewis*
Corporate Vice President, President of WCB Manufacturing *Roger Meyer*

Copyediting and production by Editorial Services of New England

Library of Congress Catalog Card Number: 92–72528

ISBN 0–697–11614–X

Printed in the United States of America by Wm. C. Brown Communications, Inc.,
2460 Kerper Boulevard, Dubuque, IA 52001

10 9 8 7 6 5 4 3 2 1

THIS BOOK IS DEDICATED TO

the memories of those great mathematicians of ages past
who advanced the cause of number theory

the students in my own recent number theory classes
at San Francisco State University.

CONTENTS

5   Arithmetic Functions   97

6   Primitive Roots and Indices   119

7   Quadratic Congruences   141

8   Sums of Squares   169

PREFACE

Beginning Number Theory is intended for use in an upper-division undergraduate course in elementary number theory. The usual audience for such a course consists of math majors who have had three or more semesters of calculus, plus some exposure to set theory. (In reality, calculus is used quite sparingly in elementary number theory. It suffices to be able to compute derivatives of polynomials with integer coefficients.)

Numbers such as 1, 2, 3, etc. are the first objects that we encounter in our mathematical education. The theory of numbers is that branch of mathematics which is concerned with the study of numbers for their own sake. For this reason, number theory, which has a nearly 4000-year history, has traditionally been considered as pure mathematics. Number theory has fascinated some of the best minds of every era. In recent years, as high-speed computers have become more powerful and more available, number theory has developed some significant applications, especially to cryptology.

ABOUT THIS BOOK

This text attempts to strike a balance between the traditional and algorithmic approaches to elementary number theory. The traditional subject matter—namely, divisibility, primes, congruences, number theoretic functions, primitive roots, quadratic reciprocity, and sums of squares—is covered logically and thoroughly in the first eight chapters, which might serve as the basis of a one-semester course. For a one-quarter course, one could use the first seven chapters, omitting Sections 2.4, 4.3, 4.6, 4.9, 6.4, 7.5, and 7.6. The algorithmic viewpoint, which has strongly influenced number theory in recent years, appears within the traditional subject matter. For example, in Chapter 7, not only is the existence of solutions to quadratic congruences (mod p) discussed, but also an algorithm is presented whereby such solutions can be obtained, when they exist.

In recognition of recent developments, the two final chapters are devoted to computational number theory and its most important application, cryptology. The instructor who wishes to treat these exciting new topics could base a course on Chapters 1 through 4 plus Chapters 12 and 13. (It would be necessary to refer briefly to that part of Chapter 5 where the definition and properties of Euler's function are given.)

Fibonacci numbers, which often play a supporting role in number theory, are first introduced in Chapter 1 and then are taken up in greater detail in Chapter 9. Chapter 10 is devoted to continued fractions, which are essential in the solution of Pell's equation in Chapter 11. In addition, continued fractions lead to a method of factoring large integers in Chapter 12.

FEATURES

1. *Classical number theory* is presented in a thoroughly logical, well-organized manner. Particular attention has been devoted to making proofs of major theorems more lucid. This is accomplished (a) by presenting necessary preliminary results beforehand, (b) by avoiding needlessly elaborate notation, and (c) in certain instances by providing original (or partly original) proofs that are simpler, more explicit, or more algorithmic than the conventional proofs.

2. *Computational number theory* is treated in Chapter 12. This chapter begins with an overview of this new branch of number theory and includes definitions of key concepts such as *computational feasibility* and *polynomial time*. After discussing tests for compositeness and primality, this chapter offers a thorough treatment of factoring, including the *continued fraction* method, the *quadratic sieve* method, and the *Pollard p-1* method, all with examples worked out in detail.

3. *Cryptology* is the subject of Chapter 13. After giving some details of the history of cryptology and its influence on world affairs, this chapter discusses various methods of cryptography, from Caesar ciphers to public-key cryptology.

4. *Computer exercises*, as well as regular exercises, appear at the end of nearly every section. The computer exercises include both (a) implementation of an algorithm that has already been described and (b) the "discovery" type, where the student must compute certain quantities and then infer an appropriate conclusion.

5. *Conjectures, theorems, and proofs* are discussed in Section 1.2. This section is intended to help the student make the transition from lower-division mathematics, where proofs are not greatly stressed, to

upper-division mathematics, where the understanding of proofs takes on a greater importance.

6. *The Dirichlet product* is used in Chapter 5 to derive the multiplicative properties of four standard number theoretic functions: number-of-divisors function (tau), sum-of-divisors function (sigma), Moebius function (mu), and Euler's function (phi). This leads to a better-motivated definition of the Moebius function, as well as an easy proof of the Moebius inversion formula.

7. *Full biographies* of four important number theorists of earlier ages: Fermat, Euler, Gauss, and Lagrange. In addition, some biographic material is given regarding Eratosthenes, Fibonacci, Wallis, and Pell. As a result, the student may better appreciate number theory from a historical perspective.

8. *Equivalence relations* and the related notion of equivalence classes are taken up in Section 4.3. This helps prepare the student for the study of a particular equivalence relation: congruence (mod m).

9. *Starred exercises* offer challenges to those readers whose enthusiasm for the subject surpasses the ordinary.

ACKNOWLEDGMENTS

I wish to express my gratitude to the following individuals, whose suggestions helped improve this book: H.W. Lenstra (University of California, Berkeley), S.Y. Tang and R. Pasker (San Francisco State University), and my number theory students at San Francisco State University during the fall semester 1991.

CHAPTER ONE

PRELIMINARIES

1.1 INTRODUCTION

Most of us begin our mathematical education by learning how to count. One becomes familiar with the natural numbers, that is, 1, 2, 3, 4, 5, etc., and learns to add, subtract, multiply, and divide. Later, one is exposed to more general number systems, namely, (1) the rational numbers, or fractions, (2) the integers, which include negative numbers and zero, (3) the real numbers, and (4) the complex numbers.

The branch of mathematics that is concerned with the properties of numbers (integers and others) for their own sake is called *number theory*.

As an example of a problem in number theory, consider the Pythagorean equation:

1.1 $$x^2 + y^2 = z^2$$

This equation has geometric significance, since it is satisfied by the lengths of the three sides of a right triangle. Suppose that we wish to find all solutions of equation 1.1 such that $0 < x < y < z$ and each of x, y, and z is an *integer*. It is easily seen that the smallest such solution is $x = 3$, $y = 4$, and $z = 5$. The problem of finding integer solutions of equation 1.1 was of interest in ancient times; a Babylonian tablet from the period 1900–1600 B.C. contains a list of such solutions. The complete solution of equation 1.1, which has been known for some time, is given in Chapter 4.

Let us modify equation 1.1 by replacing the exponent 2 with n to obtain

1.2 $$x^n + y^n = z^n$$

In 1638 the French mathematician Fermat claimed that equation 1.2 has no solution in positive integers x, y, and z for any integer $n > 2$. This statement, known as *Fermat's last theorem*, or *Fermat's conjecture*, appears to be true and indeed is known to be true for all n such that $3 \leq n \leq 10^6$. Nevertheless, it has yet to be conclusively proved or disproved. Fermat's conjecture is the best-known example of an open question in number theory. We discuss this further in Chapter 11.

The first systematic treatment of number theory, which is sometimes called "higher arithmetic," was given by Euclid in Books 7, 8, and 9 of his *Elements* about 300 B.C. In Book VII of his *Republic*, Plato advises: "We must endeavor to persuade those who are to be the principal men of the state to go and learn arithmetic."

Part of number theory's charm lies in its accessibility to the layperson. Many of its problems can be stated in terms that are comprehensible to anyone familiar with grade-school arithmetic and elementary algebra. One such problem, which we have already mentioned, was to find all integer solutions of the Pythagorean equation. As a second example, we briefly discuss perfect numbers.

According to Euclid, a positive integer is called *perfect* if it equals the sum of its own proper divisors. (Any divisor of an integer other than the integer itself is a *proper divisor*.) At present, 32 perfect numbers are known, all of them even, and the least of them is $6 = 1 + 2 + 3$. (Perfect numbers are discussed in detail in Chapter 5.) Can an odd number be perfect? Although 23 centuries have elapsed since Euclid's time, no one knows.

There are several kinds of number theory: (1) elementary or classical number theory, (2) algebraic number theory, (3) analytic number theory, and (4) computational number theory. The origins of elementary number theory are found in antiquity. Algebraic number theory uses the concepts of abstract algebra: groups, rings, fields, etc. Analytic number theory uses the tools of real and complex analysis: inequalities, derivatives, integrals, series, etc. Algebraic number theory and analytic number theory have produced significant results since their inception about two centuries ago. Computational number theory, which is a blend of number theory and computer science, is the newest branch of number theory; it has developed rapidly since 1970. Chapter 12 is devoted to computational number theory.

This textbook is mostly about elementary number theory, but computational number theory and one of its major applications, cryptology, are the subjects of the two final chapters. Three semesters of calculus plus some exposure to set theory are an adequate prerequisite for most of the subject matter that will be presented. When one encounters modular arithmetic in Section 4.7, some familiarity with abstract algebra is helpful, although not absolutely necessary. Two topics in Chapters 11 and 12 require familiarity with linear algebra. Specifically, the concise formulation of the solution to Pell's equation (Chap. 11, Theorem 11.12) involves matrix multiplication. In addition, the quadratic sieve method of factoring large integers (Sec. 12.3B) uses the concept of linearly dependent sets of vectors.

1.2   CONJECTURES, THEOREMS, AND PROOFS

In elementary number theory we are interested mostly in natural numbers, but we must occasionally deal with integers, rational numbers, and real numbers. Note that *positive integer* is a synonym for *natural number*. We assume the usual ordering on N, the set of natural numbers, namely, $1 < 2 < 3 < \cdots$

If we add together the first n natural numbers, what do we get?

For $n = 2$, we have $\qquad\qquad 1 + 2 = 3$
For $n = 3$, we have $\qquad\qquad 1 + 2 + 3 = 6$
For $n = 4$, we have $\qquad\qquad 1 + 2 + 3 + 4 = 10$

At first, no pattern is apparent. Suppose, however, that we rewrite these three equations as

$$1 + 2 = \tfrac{1}{2}(2 \cdot 3)$$
$$1 + 2 + 3 = \tfrac{1}{2}(3 \cdot 4)$$
$$1 + 2 + 3 + 4 = \tfrac{1}{2}(4 \cdot 5)$$

Generalizing, it appears that

1.3 $$1 + 2 + 3 + \cdots + n = \frac{1}{2}n(n + 1)$$

This last equation seems to be valid for any natural number n. For the moment, call this equation a *conjecture*, that is, a guess. Our conjecture may be stated as follows: "If n is any natural number, then the sum of all natural numbers from 1 to n is $\frac{1}{2}n(n + 1)$."

If we manage to *prove* our conjecture, then we call it a *theorem*. (A proof of equation 1.3 is supplied in the following section.) A *theorem* is a statement that can be proved by means of prior assumptions, prior knowledge, and logical reasoning. Usually, a theorem says: "If A, then B." A is called the *hypothesis* of the theorem, and B is called the *conclusion*. The *proof* of a theorem consists of a logical sequence of assertions that link the hypothesis to the conclusion. Methods of proof include (1) *direct substitution*, (2) *case analysis*, (3) *indirect proof*, that is, proof by contradiction, and (4) *inductive proof*. We now present examples of proof by direct substitution, proof by case analysis, and indirect proof. Inductive proof is discussed at length in Section 1.3.

Suppose that two interrelated sequences of integers are defined as follows. Let $u_n = (a^n - b^n)/(a - b)$ and $v_n = a^n + b^n$, where $n \geq 0$ and a and b are the distinct roots of a certain quadratic equation. The two following theorems are proved by *direct substitution*.

Theorem 1.1 $u_{2n} = u_n v_n$

PROOF:
$$u_n v_n = \left(\frac{a^n - b^n}{a - b}\right)(a^n + b^n)$$
$$= \frac{(a^n - b^n)(a^n + b^n)}{a - b}$$
$$= \frac{a^{2n} - b^{2n}}{a - b}$$
$$= u_{2n} \qquad \blacksquare$$

Theorem 1.2   *If $D = (a - b)^2$ and $Q = ab$, then*
$$v_n^2 - Du_n^2 = 4Q^n$$

PROOF: $v_n^2 - Du_n^2 = (a^n + b^n)^2 - (a - b)^2 \left(\frac{a^n - b^n}{a - b}\right)^2$
$$= (a^n + b^n)^2 - (a^n - b^n)^2$$
$$= (a^{2n} + 2a^n b^n + b^{2n}) - (a^{2n} - 2a^n b^n + b^{2n})$$
$$= 4a^n b^n = 4(ab)^n$$
$$= 4Q^n \qquad \blacksquare$$

Let n be an integer, not necessarily positive. If n is even, then $n = 2m$ for some integer m. For example, $10 = 2(5)$ and $-12 = 2(-6)$. If n is odd, then $n = 2j - 1$ for some integer j. For example, $13 = 2(7) - 1$ and $-7 = 2(-3) - 1$. It is also true that if n is odd, then $n = 2k + 1$ for some integer k. For example, $13 = 2(6) + 1$ and $-7 = 2(-4) + 1$.

Next, we prove by case analysis that every odd number has the form $4j - 1$ or $4k + 1$.

Theorem 1.3   *If n is odd, then n = 4j − 1 for some integer j or n = 4k + 1 for some integer k.*

PROOF: Since n is odd by hypothesis, $n = 2m - 1$ for some integer m.
Case 1:   If m is even, then $m = 2j$ for some j, so $n = 2(2j) - 1 = 4j - 1$.
Case 2:   If m is odd, then $m = 2k + 1$ for some k. Therefore,

$$n = 2(2k + 1) - 1 = 4k + 1$$ ∎

For example, $7 = 4(2) - 1$ and $13 = 4(3) + 1$. Can an odd number have both the form $4j - 1$ and the form $4k + 1$? We shall use indirect proof to see that this cannot occur.

In the *indirect method* of proof, we assume the contrary of what we wish to prove and then show that this assumption leads to a contradiction.

Theorem 1.4   *If n is odd, then there are no integers j and k such that n = 4j − 1 = 4k + 1.*

PROOF: Assume that $4j - 1 = 4k + 1$ for some integers j and k. Then $4j - 4k = 2$, so $2j - 2k = 1$, that is, $2(j - k) = 1$. The left side of this last equation is an even integer, while the right side is an odd integer, an impossibility. ∎

As a second example of indirect proof, consider the following:

Theorem 1.5   *If ab = 4m − 1, then a or b = 4k − 1 for some k.*

PROOF: If a or b is even, then ab is even. Since ab is odd by hypothesis, it follows that both a and b are odd. Therefore, Theorem 1.3 implies that $a = 4h \pm 1$ for some h and $b = 4j \pm 1$ for some j. If Theorem 1.5 is false, then $a = 4h + 1$ and $b = 4j + 1$. However, then

$$ab = (4h + 1)(4j + 1) = 16hj + 4h + 4j + 1 = 4(4hj + h + j) + 1$$

contrary to hypothesis and Theorem 1.4. We therefore conclude that Theorem 1.5 is true. ∎

For example, $91 = 4(23) - 1$. Now $91 = 7 \cdot 13$, and $7 = 4(2) - 1$.

1.3 WELL-ORDERING AND MATHEMATICAL INDUCTION

Consider the usual ordering on the set of natural numbers: $1 < 2 < 3 < \cdots$ This order has the property that if a, b, and c are distinct natural numbers, then (1) either $a < b$ or $b < a$, but not both and (2) if $a < b$ and $b < c$, then $a < c$. We also assume the

Well-Ordering Principle: The set of natural numbers is well-ordered with respect to the usual order; that is, every nonempty subset of natural numbers has a least element.

For example, let set S be defined by $S = \{x : x$ is the product of distinct even integers$\}$. Listing the elements of S in order, we see that $S = \{8, 12, 16, 20, 24, \ldots\}$. Note that S does indeed have a least element, namely, 8.

Although the well-ordering principle may appear to be trivially true for N, the set of natural numbers, it is *not* true for Z, the set of integers, with the usual order on Z, namely

$$\cdots < -3 < -2 < -1 < 0 < 1 < 2 < 3 < \cdots$$

since the nonempty subset of negative integers has no least element.

Using indirect proof and the well-ordering principle, we prove the following:

Theorem 1.6   *Let $S = \{x_1, x_2, x_3, \ldots\}$ be a set of natural numbers such that $x_1 > x_2 > x_3 > \cdots$. Then S is finite.*

PROOF: By definition of S, S is nonempty. Suppose that S is infinite. Since $x_1 > x_2 > x_3 > \cdots$ by hypothesis, it follows that S has no least element. However, this contradicts the well-ordering principle. Therefore, S is finite.   ∎

An important consequence of the well-ordering principle is the following:

Theorem 1.7   *(Principle of Mathematical Induction)*
Let S be a set of natural numbers with the two following properties: (i) 1 belongs to S and (ii) if n belongs to S, then so does $n + 1$. Then every natural number belongs to S; that is, $S = N$.

PROOF: Let S^C be the complement of S, that is, the set of all natural numbers that do not belong to S. If S^C is nonempty, then, by the well-ordering principle, S^C has a least element, say n. Now (i) implies that 1 does not belong to S^C, so $n \neq 1$. Since 1 is the least natural number, it follows that $n > 1$, so $n - 1$ is a natural number. Since n is the least element of S^C and $n - 1$ is a natural number, it follows that $n - 1$ belongs to S. However, then (ii) implies that $(n - 1) + 1$ belongs to S; that is, n belongs to S. This contradicts the definition of n. Therefore, S^C is empty, so every natural number belongs to S; that is, $S = N$. It is sometimes more convenient to use the following: ∎

Theorem 1.7a   *(Alternate Form of the Principle of Mathematical Induction)*
Let S be a set of natural numbers with the two following properties: (i) 1 belongs to S and (ii) if $n > 1$ and if m belongs to S for each m such that $1 \leq m < n$, then n belongs to S. Then every natural number belongs to S; that is, $S = N$.

PROOF: Exercise

Referring to the set of natural numbers, we have shown that the well-ordering principle implies the principle of mathematical induction. The converse also holds, namely:

Theorem 1.8   *The principle of mathematical induction implies the well-ordering principle.*

PROOF: We wish to show that if T is a nonempty set of natural numbers, then T has a least element. It suffices to show that if T has no least element, then T is empty. If 1 belongs to T, then 1 is the least element of T, which contradicts the definition of T. Therefore, (i) 1 belongs to T^C, the complement of T. Now suppose that (ii) $n > 1$ and $1, 2, 3, \ldots, n - 1$ belong to T^C. If n belongs to T, then n is the least element of T, which again contradicts the definition of T. Therefore, n belongs to T^C. Now the alternate form of the principle of mathematical induction implies that all natural numbers belong to T^C, so T is empty. ∎

We may therefore say that on the set of natural numbers the well-ordering principle and the principle of mathematical induction are logically equivalent.

Suppose we wish to prove a statement about natural numbers, such as an equation or an inequality. Let S be the set of natural numbers for which the statement is true. We wish to prove that the statement is true for all natural numbers. Proof by induction requires two steps: (i) showing that 1 belongs to S, that is, the statement is true for $n = 1$, and (ii) showing that if n belongs to S, then so does $n + 1$; that is, if the statement is true for n, then it is also true for $n + 1$. The assertion "n belongs to S" is called the *induction hypothesis*.

We are now ready to prove equation 1.3 from page 3, which we had previously called a conjecture.

Theorem 1.9 *The sum of the first n natural numbers is $\frac{1}{2}n(n + 1)$.*

PROOF: (Induction on n) Let S_n denote the sum of the first n natural numbers, that is, $S_n = 1 + 2 + 3 + \cdots + n$. We must show that $S_n = \frac{1}{2}n(n + 1)$. To verify (i) of the principle of mathematical induction (Theorem 1.7), we note that $\frac{1}{2}(1)(1 + 1) = \frac{1}{2}(2) = 1 = S_1$. To verify (ii), we must show that if $S_n = \frac{1}{2}n(n + 1)$, then $S_{n+1} = \frac{1}{2}(n + 1)(n + 2)$. Now,

$$S_{n+1} = 1 + 2 + 3 + \cdots + n + (n + 1) = S_n + (n + 1)$$

By induction hypothesis, $S_n = \frac{1}{2}n(n + 1)$. Therefore,

$$S_{n+1} = \frac{1}{2}n(n + 1) + (n + 1)$$
$$= (n + 1)\left(\frac{1}{2}n + 1\right)$$
$$= \frac{1}{2}(n + 1)(n + 2) \qquad \blacksquare$$

Next, we use induction to prove an inequality:

Theorem 1.10 $2^n > n$ *for every natural number n*

PROOF: (Induction on n) To verify (i), we note that $2^1 = 2$ and $2 > 1$, so $2^1 > 1$. To verify (ii), we must show that if $2^n > n$, then $2^{n+1} > n + 1$. By induction hypothesis, $2^n > n$. Multiplying both sides of this inequality by 2, we get $2^{n+1} > 2n$. Since n is a natural number, certainly $n \geq 1$. Adding n to both sides of this inequality, we get $2n \geq n + 1$. If $a > b$ and $b \geq c$, then $a > c$. Therefore, $2^{n+1} > n + 1$. $\qquad \blacksquare$

Mathematical induction also may be used to prove statements that are true not for all n but for all sufficiently large n, that is, for all $n \geq n_0$ for some n_0. In such instances, step (i) in the proof consists of verifying that the statement is true for $n = n_0$.

An example of this follows:

Theorem 1.11 $2^n > 3n$ *if $n \geq 4$*

PROOF: (Induction on n) (i) $2^4 = 16$, $3(4) = 12$, and $16 > 12$, so $2^4 > 3(4)$. (ii) Given that $2^n > 3n$, we must show that $2^{n+1} > 3(n + 1)$. By induction hypothesis, $2^n > 3n$. Multiplying this inequality by 2, we get $2^{n+1} > 6n$. Since $n \geq 4$ by hypothesis, certainly $n > 1$. Adding n to both sides of this inequality, we get $2n > n + 1$. Multiplying by 3, we get $6n > 3(n + 1)$. Therefore, $2^{n+1} > 3(n + 1)$.   ■

Mathematical induction also can be used to define certain sequences of real numbers. Suppose that an initial term a_1 is defined. Suppose furthermore that a_{n+1} is defined whenever a_n is defined. Then, by the principle of mathematical induction (Theorem 1.7), a_n is defined for all n. We present two examples of definition by mathematical induction.

Definition 1.1   *Let $a_1 = 1$, and let $a_n = na_{n-1}$ for all $n \geq 2$. Then*

$$a_2 = 2a_1 = 2 \cdot 1 = 2$$
$$a_3 = 3a_2 = 3 \cdot 2 \cdot 1 = 6$$
$$a_4 = 4a_3 = 4 \cdot 3 \cdot 2 \cdot 1 = 24$$
$$a_5 = 5a_4 = 5 \cdot 4 \cdot 3 \cdot 2 \cdot 1 = 120$$
$$\vdots$$

The usual notation for a_n is $n!$ (called "n factorial"). Note that if n is a natural number, then $n!$ is the product of all natural numbers from 1 to n and that $n! = n[(n - 1)!]$. Our calculations above show that $1! = 1$, $2! = 2$, $3! = 6$, $4! = 24$, and $5! = 120$. Note that $n!$ increases rapidly with n. For example, $12! = 479{,}001{,}600$. In general, $n!$ is the number of possible permutations, that is, rearrangements, of a set of n distinct objects. If you have n books, you can arrange them on a bookshelf in $n!$ distinct ways.

The following inductively defined sequence was discovered by Leonardo Fibonacci, an important mathematician of the Middle Ages.

Definition 1.2   ***Fibonacci Numbers***
Let $F_1 = 1$, $F_2 = 1$, $F_n = F_{n-1} + F_{n-2}$ for $n \geq 3$. F_n is called the nth *Fibonacci number*. Note that

$$F_3 = F_2 + F_1 = 1 + 1 = 2$$
$$F_4 = F_3 + F_2 = 2 + 1 = 3$$
$$F_5 = F_4 + F_3 = 3 + 2 = 5$$
$$F_6 = F_5 + F_4 = 5 + 3 = 8$$
$$\vdots$$

The Fibonacci numbers, which occur in nature and have many interesting properties and applications, will be treated in greater detail in Chapter 9. An important constant related to the Fibonacci numbers is the *golden ratio*, namely, $\alpha = \frac{1}{2}(1 + \sqrt{5})$. The approximate value of α is 1.618. One can verify by direct computation that the golden ratio satisfies the equation

1.4
$$\alpha^2 = \alpha + 1$$

The following inequality will be useful in Chapters 9 and 10:

Theorem 1.12 $F_n \geq \alpha^{n-2}$ *for all $n \geq 1$*

PROOF: (Induction on n) Since the Fibonacci sequence has two initial terms, step (i) consists of verifying that Theorem 1.12 holds for $n = 1$ and $n = 2$. Since $\alpha > 1$, we have $1 > \alpha^{-1}$. Therefore, $F_1 = 1 > \alpha^{-1} = \alpha^{1-2}$. Also, $F_2 = 1 = \alpha^0 = \alpha^{2-2}$. In step (ii) we use the alternate form of the principle of mathematical induction (Theorem 1.7a). Given that $F_m \geq \alpha^{m-2}$ for all m such that $1 \leq m < n$, we must show that $F_n \geq \alpha^{n-2}$. If $n \geq 3$, then by induction hypothesis, we have

$$F_{n-1} \geq \alpha^{n-3}$$
and
$$F_{n-2} \geq \alpha^{n-4}$$
Therefore,
$$F_{n-1} + F_{n-2} \geq \alpha^{n-3} + \alpha^{n-4}$$
that is,
$$F_n \geq (\alpha + 1)\alpha^{n-4} = \alpha^2\alpha^{n-4}$$
so
$$F_n \geq \alpha^{n-2} \qquad \blacksquare$$

SECTION 1.3 EXERCISES

1. Use indirect proof to show that the least element of any nonempty set of natural numbers is unique.

2. Compute $n!$ for $n = 6, 7, 8, 9, 10$.

3. Compute F_n for $n = 7, 8, 9, \ldots, 20$.

Use mathematical induction to prove each of 4 through 16.

4. The sum of the first n odd natural numbers is n^2.

5. The sum of the first n squares is $n(n + 1)(2n + 1)/6$.

6. The sum of the first n odd squares is $n(4n^2 - 1)/3$.

7. The sum of the first n cubes is $n^2(n + 1)^2/4$.

8. The sum of the first n odd cubes is $n^2(2n^2 - 1)$.

9. $3^n > 3n - 1$

10. $5^n > n^2$

11. $F_n \leq \alpha^{n-1}$

12. $F_{n-1}F_{n+1} - F_n^2 = (-1)^n$

13. $F_1 + F_2 + F_3 + \cdots + F_n = F_{n+2} - 1$

14. $F_2 + F_4 + F_6 + \cdots + F_{2n} = F_{2n+1} - 1$

15. $F_1 + F_3 + F_5 + \cdots + F_{2n-1} = F_{2n}$

16. Let a "good" word have the following properties: (i) each letter is A or B and (ii) no two A's are adjacent. Let w_n be the number of "good" words with n letters. Show that $w_n = F_{n+2}$.

SECTION 1.3 COMPUTER EXERCISES

Write a computer program to do each of the following.

17. Find the least n such that $n! > 10^{100}$.

18. Evaluate the first 100 Fibonacci numbers.

19. Compute the ratio F_{n+1}/F_n for $1 \le n \le 50$. What seems to be happening?

20. Compute $\alpha^n/\sqrt{5}$ rounded to the nearest integer for $1 \le n \le 50$. What seems to be happening?

21. Compute the sum $a_2 + a_3 + a_4 + \cdots + a_n$ for $5 \le n \le 20$, where each $a_k = F_{k-1}/2^k$. What seems to be happening?

1.4   SIGMA NOTATION AND PRODUCT NOTATION

In number theory, as well as in other branches of mathematics, one frequently encounters sums of similar terms. A convenient abbreviation for the sum $x_1 + x_2 + x_3 + \cdots + x_n$ is

$$\sum_{k=1}^{n} x_k$$

This symbol is read "the sum of all the x_k as k runs from 1 to n." k is called a *running index*, while 1 and n are the *lower* and *upper limits of summation*, respectively. Often x_k is a function of k. For example, the sum of the first n odd natural numbers may be represented as

$$\sum_{k=1}^{n} (2k - 1)$$

Products of n terms also arise in number theory and elsewhere. A convenient abbreviaton for the product $x_1 x_2 x_3 \cdots x_n$ is

$$\prod_{k=1}^{n} x_k$$

This symbol is read "the product of all the x_k as k runs from 1 to n." For example, note that if n is a natural number, then

$$\prod_{k=1}^{n} k = n!$$

1.5   BINOMIAL COEFFICIENTS

In elementary or intermediate algebra, one learns that

$$
\begin{aligned}
(x + y)^1 &= 1x + 1y \\
(x + y)^2 &= 1x^2 + 2xy + 1y^2 \\
(x + y)^3 &= 1x^3 + 3x^2y + 3xy^2 + 1y^3 \\
(x + y)^4 &= 1x^4 + 4x^3y + 6x^2y^2 + 4xy^3 + 1y^4 \\
(x + y)^5 &= 1x^5 + 5x^4y + 10x^3y^2 + 10x^2y^3 + 5xy^4 + 1y^5 \\
(x + y)^6 &= 1x^6 + 6x^5y + 15x^4y^2 + 20x^3y^3 + 15x^2y^4 + 6xy^5 + 1y^6 \\
&\ \ \vdots
\end{aligned}
$$

If we copy just the coefficients, we obtain what is called *Pascal's triangle:*

$$
\begin{array}{ccccccccccccc}
&&&&&& 1 && 1 \\
&&&&& 1 && 2 && 1 \\
&&&& 1 && 3 && 3 && 1 \\
&&& 1 && 4 && 6 && 4 && 1 \\
&& 1 && 5 && 10 && 10 && 5 && 1 \\
& 1 && 6 && 15 && 20 && 15 && 6 && 1
\end{array}
$$

. . . .

Blaise Pascal (1623–1662) was a French mathematician, one of the founders of probability theory. There is evidence that Pascal's triangle was known earlier in Asia. The numbers that appear in Pascal's triangle are called *binomial coefficients* because they are coefficients that arise in the expansion of $(x + y)^n$. (Recall that $x + y$ is called a *binomial.*)

In Section 1.3, $n!$ was defined for natural numbers n. Now we extend this concept by defining $0! = 1$. If n is a natural number and k is an integer such that $0 \le k \le n$, then we define the *binomial coefficient* as follows:

$$\binom{n}{k} = \frac{n!}{k!(n-k)!}$$

$\binom{n}{k}$, which may be read "n choose k," counts the number of distinct ways in which a subset of k elements may be chosen, without regard to order, from a set of n distinct elements. Binomial coefficients occur in probability, statistics, and combinatorics, as well as in number theory. Note that

$$\binom{n}{k} = \frac{n(n-1)(n-2)\cdots(n-k+1)}{k(k-1)(k-2)\cdots 1} \qquad \text{if } 1 \leq k \leq n, \text{ while } \binom{n}{0} = 1$$

For example,

$$\binom{7}{3} = \frac{7\cdot 6\cdot 5}{3\cdot 2\cdot 1} = 35$$

$$\binom{12}{2} = \frac{12\cdot 11}{2\cdot 1} = 66$$

The following theorem gives several properties of binomial coefficients.

Theorem 1.13   *(Properties of Binomial Coefficients)*
Let n be a natural number. Let k be an integer such that $0 \leq k \leq n$. Then

B_1: $$\binom{n}{n-k} = \binom{n}{k}$$ *(Symmetry property)*

B_2: $$\binom{n}{k-1} + \binom{n}{k} = \binom{n+1}{k}$$ *if $1 \leq k \leq n$ (Pascal's identity)*

B_3: $$(x + y)^n = \sum_{k=0}^{n}\binom{n}{k}x^{n-k}y^k$$ *(Binomial theorem)*

B_4: $$\sum_{k=0}^{n}\binom{n}{k} = 2^n$$

B_5: $$\sum_{k=0}^{n}(-1)^k\binom{n}{k} = 0$$

B_6: $$\sum_{k \text{ odd}}\binom{n}{k} = \sum_{k \text{ even}}\binom{n}{k} = 2^{n-1}$$

B_7: $$\sum_{k=0}^{n}\binom{n}{k}^2 = \binom{2n}{n}$$

REMARKS: B_1, the *symmetry property*, says that Pascal's triangle is symmetrical with respect to a vertical line drawn through the middle. B_2, known as *Pascal's identity*, can be used to generate the $(n + 1)$th row of Pascal's triangle from the nth row. B_4 says that the sum of the entries in the nth row is 2^n.

We prove B_2, B_3, and B_6, leaving the proofs of the other properties as exercises.

PROOF OF B_2: $$\frac{1}{n+1-k} + \frac{1}{k} = \frac{n+1}{k(n+1-k)}$$

Multiply by
$$\frac{n!}{(k-1)!(n-k)!}$$

and simplify to obtain
$$\frac{n!}{(k-1)!(n-k+1)!} + \frac{n!}{k!(n-k)!} = \frac{(n+1)!}{k!(n+1-k)!}$$

that is,
$$\binom{n}{k-1} + \binom{n}{k} = \binom{n+1}{k}$$

∎

PROOF OF B_3:   (Induction on n)
$$(x+y)^1 = x + y = \binom{1}{0}x^1y^0 + \binom{1}{1}x^0y^1$$

Thus the theorem holds for $n = 1$. By induction hypothesis,

$$(x+y)^n = \sum_{k=0}^{n}\binom{n}{k}x^{n-k}y^k$$

$$= x^n + \sum_{k=1}^{n}\binom{n}{k}x^{n-k}y^k$$

$$= \sum_{k=0}^{n-1}\binom{n}{k}x^{n-k}y^k + y^n$$

Now

$$(x+y)^{n+1} = (x+y)(x+y)^n = x(x+y)^n + y(x+y)^n$$

$$= x\left\{x^n + \sum_{k=1}^{n}\binom{n}{k}x^{n-k}y^k\right\} + y\left\{\sum_{k=0}^{n-1}\binom{n}{k}x^{n-k}y^k + y^n\right\}$$

$$= x^{n+1} + \sum_{k=1}^{n}\binom{n}{k}x^{n+1-k}y^k + \sum_{k=0}^{n-1}\binom{n}{k}x^{n-k}y^{k+1} + y^{n+1}$$

$$= x^{n+1} + \sum_{k=1}^{n}\binom{n}{k}x^{n+1-k}y^k + \sum_{k=1}^{n}\binom{n}{k-1}x^{n+1-k}y^k + y^{n+1}$$

$$= x^{n+1} + \sum_{k=1}^{n}\left\{\binom{n}{k} + \binom{n}{k-1}\right\}x^{n+1-k}y^k + y^{n+1}$$

$$= x^{n+1} + \sum_{k=1}^{n}\binom{n+1}{k}x^{n+1-k}y^k + y^{n+1}$$

$$= \sum_{k=0}^{n+1}\binom{n+1}{k}x^{n+1-k}y^k$$

∎

Note that the proof of the binomial theorem (B_3) made use of Pascal's identity (B_2).

PROOF OF B_6: $\displaystyle\sum_{k\text{ even}}\binom{n}{k} - \sum_{k\text{ odd}}\binom{n}{k} = \sum_{k=0}^{n}(-1)^k\binom{n}{k} = 0$ by B_5.

Also $\displaystyle\sum_{k\text{ even}}\binom{n}{k} + \sum_{k\text{ odd}}\binom{n}{k} = \sum_{k=0}^{n}\binom{n}{k} = 2^n$ by B_4.

Therefore, $\displaystyle\sum_{k\text{ even}}\binom{n}{k} = \sum_{k\text{ odd}}\binom{n}{k} = \frac{1}{2}(2^n) = 2^{n-1}$ ∎

For example, in row 4 of Pascal's triangle, we have $1 + 6 + 1 = 4 + 4 = 8 = 2^{4-1}$.

SECTION 1.5 EXERCISES

22. Write the statement of each of the following previous exercises from Section 1.3 using sigma notation.

(a) 4	(b) 5
(c) 6	(d) 7
(e) 8	(f) 13
(g) 14	(h) 15

23. Write the product of the first n even natural numbers in product notation, and then simplify your result.

24. Write the product of the first n odd natural numbers in product notation. Use factorials to give a formula for this product.

25. Evaluate the following:

(a) $\binom{5}{2}$ (b) $\binom{8}{3}$

(c) $\binom{9}{4}$ (d) $\binom{10}{8}$

(e) $\binom{12}{9}$

26. Prove Theorem 1.13, part B_1.

27. Prove Theorem 1.13, part B_4. (*Hint:* Use B_3.)

28. Prove Theorem 1.13, part B_5. (*Hint:* Use B_3.)

29. Prove Theorem 1.13, part B_7. {*Hint:* $(x + y)^{2n} = [(x + y)^n]^2$.}

30. Use Pascal's identity (Theorem 1.13, part B_2) to generate the seventh and eighth rows of Pascal's triangle.

31. In Pascal's triangle, any entry other than a 1 has six neighbors: the two adjacent entries in the same row, the two nearest entries in the row above, and the two nearest entries in the row below. For example, if we look at the first 10 in row 5, we see

$$\begin{array}{ccc} & 4 & \quad 6 \\ 5 & 10 & 10 \\ & 15 & 20 \end{array}$$

Computing products of alternate neighbors of 10, we obtain $5 \cdot 6 \cdot 20 = 600$ and $4 \cdot 10 \cdot 15 = 600$. Prove that if $0 < k < n$, then the two products of alternate neighbors of $\binom{n}{k}$ are equal.

32. Use Theorem 1.13, part B_2 and induction on n to prove that $\binom{n}{k}$ is an integer for all $n \geq 1$ and all k such that $0 \leq k \leq n$.

33. Simplify $\frac{n+k}{n}\binom{n}{k}$. (Your answer should contain no fractions.)

34. Prove that if $n \geq 2$, then $2^n < \binom{2n}{n} < 2^{2n}$.

35. Prove that if $n \geq 3$ and $0 \leq k \leq n$, then $\binom{n}{k} < 2^{n-1}$.

36. Use induction on n to prove that if $A = \prod_{i=1}^{n} a_i$, where each $a_i = 4k_i + 1$ for some k_i, then $A = 4K + 1$ for some K.

37. Let $u_n = (a^n - b^n)/(a - b)$ and $v_n = a^n + b^n$, where a and b are distinct. Use Theorem 1.1 to simplify the expression

$$u_n \prod_{i=0}^{r} v_{2^i n}$$

Use induction on r to prove your result. (*Hint:* Write out the product longhand for $r = 1, 2, 3$, etc.)

SECTION 1.5 COMPUTER EXERCISES

Write a computer program to do each of the following.

38. For each n such that $1 \le n \le 15$ and all k such that $0 \le 2k \le n$, find the sum of all binomial coefficients $\binom{n-k}{k}$. Is there a pattern?

39. For each n such that $1 \le n \le 10$, evaluate $\sum_{k=1}^{n} \binom{n}{k} F_k$. Is there a pattern?

1.6 GREATEST INTEGER FUNCTION

Let us take a second look at Theorem 1.13, part B$_6$. This property says, in part, that

1.5
$$\sum_{k \text{ even}} \binom{n}{k} = 2^{n-1}$$

Notice that we used a modified, somewhat vaguer form of sigma notation here; that is, we did not specify the limits of summation. Why was this necessary? If k is even, then $k = 2j$ for some $j \ge 0$. Therefore, if $0 \le k \le n$, then $0 \le 2j \le n$, so $0 \le j \le n/2$. We might be tempted to rewrite equation 1.5 as

1.6
$$\sum_{j=0}^{n/2} \binom{n}{2j} = 2^{n-1}$$

but this would be valid only when $n/2$ is an integer, that is, when n is even. Upon reflection, we see that j, the index of summation in equation 1.6, can be as large as the *greatest integer* that is less than or equal to $\frac{1}{2}n$. This leads us to define what is known as the *greatest integer function*.

Definition 1.3   ***Greatest Integer Function***
If x is a real number, let n be the unique integer such that $n \le x < n + 1$. Then we say that n is the integer part of x, and we write

$$[x] = n$$

For example, $[4.6] = 4$, $[\sqrt{5}] = 2$, $[\frac{1}{2}] = 0$, $[7] = 7$, and $[-2.7] = -3$.

Theorem 1.14   *(Properties of the Greatest Integer Function)*

I_1: $x - 1 < [x] \leq x < [x] + 1$ *for all real x.*
I_2: $[x] = x$ *if and only if x is an integer.*
I_3: $[x + n] = [x] + n$ *for all real x and all integers n.*
I_4:   *If* $x \leq y$, *then* $[x] \leq [y]$.
I_5: $[x] + [y] \leq [x + y] \leq [x] + [y] + 1$ *for all real x and y.*
I_6:   *If* $x \geq 0$ *and* $y \geq 0$, *then* $[x][y] \leq [xy]$.
I_7:   *If* $a = qb + r$, *with* $0 \leq r < b$, *then* $q = [a/b]$ *and* $r = a - b[a/b]$.

We will prove I_1, I_4, I_5, and I_6. The proofs of I_2, I_3, and I_7 are left as exercises.

PROOF OF I_1:   By Definition 1.3, we have $x = [x] + b$, where $0 \leq b < 1$. ($b = x - [x]$ is called the *fractional part* of x.) Therefore, $[x] \leq x$. Also, $[x] + 1 = x + 1 - b > x$, since $b < 1$.   ∎

PROOF OF I_4:   It suffices to prove that if $[x] > [y]$, then $x > y$. Let $x = [x] + b$ and $y = [y] + k$, where $0 \leq b < 1$ and $0 \leq k < 1$. Since $[x]$ and $[y]$ are integers and $[x] > [y]$ by hypothesis, it follows that $[x] \geq [y] + 1 = y + 1 - k$. Now $x \geq [x]$ by I_1, so $x \geq y + 1 - k$. Since $k < 1$, it follows that $x > y$.   ∎

PROOF OF I_5:   Using the notation used in the proof of I_4, we have $x + y = [x] + [y] + (b + k)$, with $0 \leq b + k < 2$. Therefore, by I_3 and I_2, $[x + y] = [x] + [y] + [b + k]$. If $0 \leq b + k < 1$, then $[b + k] = 0$, so $[x + y] = [x] + [y]$. If $1 \leq b + k < 2$, then $[b + k] = 1$, so $[x + y] = [x] + [y] + 1$.   ∎

PROOF OF I_6:   Using the same notation as that above, we have $xy = ([x] + b)([y] + k) = [x][y] + b[y] + k[x] + bk$. Now $b \geq 0$ and $k \geq 0$. Since $x \geq 0$ and $y \geq 0$ by hypothesis, it follows that $[x] \geq 0$ and $[y] \geq 0$. Therefore, $xy \geq [x][y]$. Now I_4 implies that $[xy] \geq [[x][y]]$. However, I_2 implies that $[[x][y]] = [x][y]$. Therefore, $[xy] \geq [x][y]$.   ∎

As an example of I_3, $[3 + \sqrt{2}] = 3 + [\sqrt{2}] = 3 + 1 = 4$. As an example of I_6, if $x = 2$ and $y = \sqrt{3}$, then $[x][y] = [2][\sqrt{3}] = 2 \cdot 1 = 2$, while $[xy] = [2\sqrt{3}] = 3$.

Now that the greatest integer function is at our disposal, we may correct equation 1.6 by replacing $n/2$ as the upper limit of summation by $[n/2]$. We thus obtain

1.7
$$\sum_{j=0}^{[n/2]} \binom{n}{2j} = 2^{n-1}$$

The other part of Theorem 1.13, part B_6 said that

1.8
$$\sum_{k \text{ odd}} \binom{n}{k} = 2^{n-1}$$

If k is odd, then $k = 2j + 1$ for some $j \geq 0$. Since $0 \leq k \leq n$, we have $2j + 1 \leq n$, so $j \leq \frac{1}{2}(n - 1)$; hence $j \leq [\frac{1}{2}(n - 1)]$. Therefore, we rewrite equation 1.8 as

1.9
$$\sum_{j=0}^{\left[\frac{n-1}{2}\right]} \binom{n}{2j+1} = 2^{n-1}$$

We conclude this section with a theorem regarding the greatest integer function that will be useful in Chapter 10.

Theorem 1.15   *If x is real and n is a natural number, then*
$$\left[\frac{[x]}{n}\right] = \left[\frac{x}{n}\right].$$

PROOF: Exercise.

For example, $\left[\dfrac{\sqrt{10}}{2}\right] = \left[\dfrac{[\sqrt{10}]}{2}\right] = \left[\dfrac{3}{2}\right] = 1$.

SECTION 1.6 EXERCISES

40. Evaluate the following:

    (a) $[23/3]$　　　　(b) $[\sqrt{24}]$

    (c) $[10\pi]$　　　　(d) $[-10\pi]$

41. Prove Theorem 1.14, part I_2.

42. Prove Theorem 1.14, part I_3.

43. Prove Theorem 1.14, part I_7.

44. Prove Theorem 1.15.

SECTION 1.6 COMPUTER EXERCISES

Write a computer program to do each of the following.

45. For each n such that $1 \le n \le 10$, evaluate

$$\frac{1}{2^{n-1}} \sum_{k=0}^{\left[\frac{n-1}{2}\right]} \binom{n}{2k+1} 5^k$$

Is there a pattern?

46. For each n such that $1 \le n \le 10$, evaluate

$$\frac{1}{2^{n-1}} \sum_{k=0}^{\left[\frac{n-1}{2}\right]} \binom{n}{2k+1} 3^{n-(2k+1)}$$

Is there a pattern?

47. For each n such that $1 \le n \le 10$, evaluate

$$\sum_{k=-\left[\frac{n+1}{5}\right]}^{\left[\frac{n}{5}\right]} (-1)^k \binom{n}{\left[\frac{1}{2}(n-5k)\right]}$$

Is there a pattern?

CHAPTER TWO

DIVISIBILITY

2.1 INTRODUCTION

In classical number theory, certain questions arise concerning the divisors of a given natural number. Recall that a natural number is called *perfect* if it equals the sum of its proper divisors. (The two smallest perfect numbers are $6 = 1 + 2 + 3$ and $28 = 1 + 2 + 4 + 7 + 14$.) Can one find a formula that yields all perfect numbers? If F_m and F_n are Fibonacci numbers (see page 9), what conditions on m and n guarantee that F_m divides F_n? Do the divisors of $2^n - 1$ have any special form?

In this chapter we develop the divisibility properties of natural numbers, thereby taking the first steps toward answering the questions just raised. Of particular importance are the concepts of the greatest common divisor and the least common multiple of a pair of natural numbers. Lowercase letters refer to natural numbers, unless otherwise stated.

2.2 DIVISIBILITY, GREATEST COMMON DIVISOR, EUCLID'S ALGORITHM

Definition 2.1   *Divisibility*

We say that *a divides b* and write $a \mid b$ if there exists k such that $b = ka$.

For example, $3 \mid 12$, since $12 = 4 \cdot 3$; also $8 \mid 40$, since $40 = 5 \cdot 8$.

If $a \mid b$, we might also say "*a* is a divisor of *b*," "*a* is a factor of *b*," or "*b* is a multiple of *a*."

If a does *not* divide b, we write $a \nmid b$. For example, $3 \nmid 14$; also $8 \nmid 42$.

Definition 2.2   *Proper Divisor*

We say that a is a *proper divisor* of b if $a \mid b$ and $a < b$.

For example, 3 is a proper divisor of 12, but 12 is not a proper divisor of 12. Note that for any b, all divisors of b are proper, with the exception of b itself.

Definition 2.3   *Nontrivial Divisor*

We say that a is a *nontrivial divisor* of b if $a \mid b$ and $1 < a < b$.

For example, the nontrivial divisors of 6 are 2 and 3. Note that 1 has no proper divisors itself, but 1 is a proper divisor of every larger integer.

The following theorem states some properties of divisibility.

Theorem 2.1   *(Properties of Divisibility)*

D_1: $1 \mid a$ *and* $a \mid a$ *for all* a.
D_2: *If* $a \mid b$, *then* $a \leq b$.
D_3: *If* $a \mid b$ *and* $b \mid a$, *then* $a = b$.
D_4: *If* $a \mid b$ *and* $b \mid c$, *then* $a \mid c$.
D_5: $a \mid b$ *if and only if* $ac \mid bc$ *for all* c.
D_6: *If* $a \mid b$ *and* $c \mid d$, *then* $ac \mid bd$.
D_7: *If* $a \mid b$ *and* $a \mid c$, *then* $a \mid (bx + cy)$ *for all* x *and* y.
D_8: *If* $ab \mid c$, *then* $a \mid c$.

We prove D_2, D_4, and D_7, leaving the proofs of the other divisibility properties as exercises.

PROOF OF D_2:   If $a \mid b$, then $b = ka$, where $1 \leq k$, so $a \leq ka$, that is, $a \leq b$.   ∎

PROOF OF D_4:   If $a \mid b$ and $b \mid c$, then there exist j and k such that $b = ja$ and $c = kb$. Therefore, $c = k(ja) = (kj)a$, so $a \mid c$.   ∎

PROOF OF D_7:   If $a \mid b$ and $a \mid c$, then there exist j and k such that $b = ja$ and $c = ka$. Therefore, $bx + cy = (ja)x + (ka)y = (jx + ky)a$, so $a \mid (bx + cy)$.   ∎

Because of D_3, one way to prove that $a = b$ is to prove that $a \mid b$ and $b \mid a$. As an example of D_4, $3 \mid 6$ and $6 \mid 24$, so $3 \mid 24$. As an example of D_5, $3 \mid 6$, so $300 \mid 600$. As an example of D_8, $(3 \cdot 6) \mid 54$, so $3 \mid 54$.

REMARK: If a, b, c, x, and y refer to nonzero integers rather than to natural numbers, then the properties of divisibility given above remain valid except for D_2 and D_3, which must be adjusted as follows:

D_2:   *If $a \mid b$, then $|a| \le |b|$.*
D_3:   *If $a \mid b$ and $b \mid a$, then $a = \pm b$.*

When 23 is divided by 5, the quotient is 4 and the remainder is 3. Since we prefer to deal with integers, instead of writing $\frac{23}{5} = 4\frac{3}{5}$, we write: $23 = 4(5) + 3$. Whenever one natural number is divided by another, the remainder is strictly less than the divisor. The following theorem explains why this is so.

Theorem 2.2   *(Division Algorithm)*
For all a and b there exist unique q and r such that $b = qa + r$ and $0 \le r < a$.

PROOF: If $b < a$, then $q = 0$ and $r = b$; if $a \mid b$, then $q = b/a$ and $r = 0$. Having disposed of these cases, let us assume that $a < b$ and $a \nmid b$. Let S be the set of all natural numbers of the form $x_n = b - na$. By hypothesis, we know that $x_1 = b - a > 0$, so x_1 belongs to S; that is, S is nonempty. By the well-ordering principle, S has a least element x_q. Let $x_q = b - qa = r$. If $r = a$, then $b - qa = a$, so $b = qa + a = (q + 1)a$. However, then $a \mid b$, contrary to hypothesis. If $r > a$, then $x_{q+1} = b - (q + 1)a = (b - qa) - a = r - a > 0$, so x_{q+1} belongs to S, yet $x_{q+1} < x_q$. This contradicts the definition of x_q, however. Therefore, $r < a$. Since $r > 0$, we have $0 < r < a$. Since r is the least element of S, we know by Exercise 1 in Chapter 1 (page 10) that r is unique. Since $q = (b - r)/a$, q is also unique.   ∎

REMARK: Note that q is the quotient and r is the remainder when b is divided by a. For example, if $a = 13$ and $b = 50$, then $q = 3$ and $r = 11$, since $50 = 3(13) + 11$.

Definition 2.4

Common Divisor

If $c \mid a$ and $c \mid b$, then we say c is a *common divisor* of a and b.

For example, the common divisors of 36 and 60 are 1, 2, 3, 4, 6, and 12. Also, the common divisors of 20 and 30 are 1, 2, 5, and 10. Since $1 \mid a$ and $1 \mid b$ for all a and b, the set of common divisors of a and b is always nonempty. Note also that if $c \mid a$ and $c \mid b$, then $c \leq$ Min$\{a, b\}$. An especially important common divisor of a and b is their *greatest* common divisor. For example, the greatest common divisor of 36 and 60 is 12; the greatest common divisor of 20 and 30 is 10.

Definition 2.5

Greatest Common Divisor

Suppose that (i) $d \mid a$ and $d \mid b$, and also (ii) if $c \mid a$ and $c \mid b$, then $c \mid d$. Then we say that d is the *greatest common divisor* of a and b. We write $d = $ GCD(a, b) or, simply, $d = (a, b)$.

Using this notation, we write $(36, 60) = 12$ and $(20, 30) = 10$. In other words, the greatest common divisor of a and b is that common divisor which is divisible by all other common divisors.

The following theorem states some properties of the greatest common divisor of a pair of natural numbers.

Theorem 2.3

(Properties of the Greatest Common Divisor)
Let $d = (a,b)$

G_1: *If $c \mid a$ and $c \mid b$, then $c \leq d$.*
G_2: *$(a, b) = a$ if and only if $a \mid b$.*
G_3: *$1 \leq d \leq $ Min$\{a, b\}$.*
G_4: *If $c \mid a$ and $c \mid b$, then $\left(\frac{a}{c}, \frac{b}{c}\right) = \frac{d}{c}$.*
G_5: *$\left(\frac{a}{d}, \frac{b}{d}\right) = 1$.*
G_6: *$(an, bn) = (a, b)n$ for all n.*

We prove G_1, G_4, and G_5, leaving the proofs of G_2, G_3, and G_6 as exercises.

PROOF OF G_1:   By hypothesis and Definition 2.5, $c \mid d$; by Theorem 2.1, part D_2, $c \leq d$.   ∎

PROOF OF G_4:   Since $c \mid a$ and $c \mid b$ by hypothesis, and since $d = (a, b)$, we have $c \mid d$, so d/c is a natural number. Now $d \mid a$, that is, $\left(\frac{d}{c}\right)c \mid \left(\frac{a}{c}\right)c$. Therefore, Theorem 2.1, part D_5 implies $\frac{d}{c} \mid \frac{a}{c}$. Similarly, $\frac{d}{c} \mid \frac{b}{c}$. Having shown that $\frac{d}{c}$ is a common divisor of $\frac{a}{c}$ and $\frac{b}{c}$, we must show that $\frac{d}{c}$ is their greatest common divisor. If $k \mid \frac{a}{c}$ and $k \mid \frac{b}{c}$, then Theorem 2.1, part D_5 implies that $kc \mid a$ and $kc \mid b$. Therefore, $kc \mid d$. Again applying Theorem 2.1, part D_5, we have $k \mid \frac{d}{c}$. Therefore, $\frac{d}{c} = \left(\frac{a}{c}, \frac{b}{c} \right)$.   ∎

PROOF OF G_5:   Apply G_4 with $c = d$.   ∎

As an example of G_1, $(30, 54) = 6$. $3 \mid 30$ and $3 \mid 54$, so $3 < 6$. As an example of G_2, $(30, 60) = 30$, since $30 \mid 60$. As an example of G_4, $\left(\frac{30}{3}, \frac{54}{3} \right) = \frac{6}{3}$, that is, $(10, 18) = 2$. As an example of G_5, $\left(\frac{30}{6}, \frac{54}{6} \right) = \frac{6}{6}$, that is, $(5, 9) = 1$. As an example of G_6, $(30 \cdot 5, 54 \cdot 5) = 6 \cdot 5$, that is, $(150, 270) = 30$.

Given a pair of natural numbers, how does one compute their greatest common divisor? The following method, due to Euclid, consists of a finite sequence of divisions.

Euclid's Algorithm for Finding the Greatest Common Divisor

Suppose we are given natural numbers a and b with $a \leq b$, and we wish to compute $(a, b) = d$. We start by letting a be the divisor t, and letting b be the dividend D. Next, perform the *key step:* Divide D by t, obtaining a quotient q and a remainder r such that $0 \leq r < t$. If $r > 0$, then replace t by r, replace D by t, and repeat the key step. If $r = 0$, then $d = t$ (the last divisor).

Example 1:   Find $(27, 87)$.

Solution:

$$\begin{array}{r} 3 \\ 27\overline{)87} \\ 81 \\ \hline 6 \end{array} \qquad 87 = 3(27) + 6$$

$$\begin{array}{r} 4 \\ 6\overline{)27} \\ 24 \\ \hline 3 \end{array} \qquad 27 = 4(6) + 3$$

$$\begin{array}{r} 2 \\ 3\overline{)6} \\ 6 \\ \hline 0 \end{array} \qquad 6 = 2(3) + 0$$

Result: $(27, 87) = 3$

Example 2:   Find (165, 418).

Solution:

$$
\begin{array}{r}
2 \\
165\overline{)418} \\
330 \\
\hline
88
\end{array}
\qquad 418 = 2(165) + 88
$$

$$
\begin{array}{r}
1 \\
88\overline{)165} \\
88 \\
\hline
77
\end{array}
\qquad 165 = 1(88) + 77
$$

$$
\begin{array}{r}
1 \\
77\overline{)88} \\
77 \\
\hline
11
\end{array}
\qquad 88 = 1(77) + 11
$$

$$
\begin{array}{r}
7 \\
11\overline{)77} \\
77 \\
\hline
0
\end{array}
\qquad 77 = 7(11) + 0
$$

Result: (165, 418) = 11

Why does this work, and will it always work? That is, are we sure to obtain $r = 0$ after sufficiently many iterations? To verify that this is indeed the case, note that each iteration produces an equation. The first iteration yields $b = q_1 a + r_1$, where $0 \le r_1 < a$. If $r_1 > 0$, then the second iteration yields $a = q_2 r_1 + r_2$, where $0 \le r_2 < r_1$. If $r_2 > 0$, then the third iteration yields $r_1 = q_3 r_2 + r_3$, where $0 \le r_3 < r_2$. Our notation is simplified if we choose more convenient names for a and b. For example, let $b = r_{-1}$ and $a = r_0$. Then our iterations yield the following sequence of equations:

2.1 $\qquad r_{-1} = q_1 r_0 + r_1 \qquad 0 < r_1 < r_0$

2.2 $\qquad r_0 = q_2 r_1 + r_2 \qquad 0 < r_2 < r_1$

2.3 $\qquad r_1 = q_3 r_2 + r_3 \qquad 0 < r_3 < r_2$

$$\vdots$$

2.4 $\qquad r_{j-2} = q_j r_{j-1} + r_j \qquad 0 < r_j < r_{j-1}$

$$\vdots$$

The sequence of positive r_i, namely, $\{r_{-1}, r_0, r_1, r_2, \cdots\}$ is strictly decreasing. By Theorem 1.4, this sequence must be finite. Therefore, there exists $k \geq 1$ such that

(k) $r_{k-2} = q_k r_{k-1} + r_k \qquad 0 < r_k < r_{k-1}$

($k + 1$) $r_{k-1} = q_{k+1} r_k \qquad$ (that is, $r_{k+1} = 0$)

We claim that $d = r_k$. We will prove this by showing that $d \mid r_k$ and $r_k \mid d$. Now $d = (a, b) = (r_{-1}, r_0)$, so $d \mid r_{-1}$ and $d \mid r_0$. Therefore, equation 2.1 implies that $d \mid r_1$. Since $d \mid r_0$ and $d \mid r_1$, equation 2.2 implies that $d \mid r_2$. Continuing in a like manner, equation (k) implies $d \mid r_k$.

Conversely, since $r_k \mid r_k$ and $r_k \mid r_{k-1}$, equation (k) implies that $r_k \mid r_{k-2}$. Look at the following equation (not previously shown):

($k - 1$) $r_{k-3} = q_{k-1} r_{k-2} + r_{k-1} \qquad r_{k-1} < r_{k-2}$

Since $r_k \mid r_{k-1}$ and $r_k \mid r_{k-2}$, it follows that $r_k \mid r_{k-3}$. Similarly, $r_k \mid r_i$, where $i = k - 4, k - 5$, etc., $0, -1$. We have $r_k \mid r_0$ and $r_k \mid r_{-1}$, that is, $r_k \mid a$ and $r_k \mid b$. Therefore, $r_k \mid d$.

Definition 2.6   ***Relatively Prime***

If a and b are natural numbers such that $(a, b) = 1$, then we say that a and b are *relatively prime*.

For example, 6 and 35 are relatively prime; so are 20 and 27.

SECTION 2.2 EXERCISES

1. Prove the following parts of Theorem 2.1.

   (a) D_1
   (b) D_3
   (c) D_5
   (d) D_6
   (e) D_8

2. Prove that if $a \mid 1$, then $a = 1$.

3. Prove that if $a \mid b$, then $a^n \mid b^n$ for any natural number n.

4. Prove that if $n \mid (a - 1)$ and $n \mid (b - 1)$, then $n \mid (ab - 1)$.

5. Prove that if $a \neq 1$, then $(a - 1) \mid (a^n - 1)$ for all n.

6. Prove that if $a \neq -1$ and n is odd, then $(a + 1) \mid (a^n + 1)$.

7. Find all the divisors of every integer from 20 to 30.

8. Find all the common divisors of each of the following pairs of integers.

   (a) 12, 30
   (b) 12, 16
   (c) 5, 8
   (d) 10, 20
   (e) 28, 33
   (f) 36, 63

9. For each pair of integers, use Euclid's algorithm to find the greatest common divisor.

    (a) 44, 102   (b) 102, 402

    (c) 75, 95   (d) 231, 273

    (e) 29, 123   (f) 104, 299

    (g) 27, 32   (h) 221, 273

    (i) 126, 621   (j) 18, 54

10. Prove that $(1, n) = 1$ for all n.

11. Reduce each of the following fractions to lowest terms.

    (a) 91/140   (b) 176/1001

    (c) 63/171   (d) 155/217

    (e) 185/222

12. Prove that $(n, n + 1) = 1$ for all n.

13. Prove that $(2n - 1, 2n + 1) = 1$ for all n.

14. Let F_n be the nth Fibonacci number (see page 10). Prove that (a) $(F_n, F_{n+1}) = 1$ and (b) $(F_n, F_{n+2}) = 1$.

15. Prove that if b is odd, then $8 \mid (b^2 - 1)$. Draw the conclusion that every odd square has the form $8k + 1$.

16. Prove that the $n!$ divides the product of any n consecutive integers.

*17. Prove that if $a \neq 1$, then
$$(a^m - 1, a^n - 1) = a^{(m,n)} - 1.$$

18. Prove the following parts of Theorem 2.3.
    (a) G_2 (b) G_3 (c) G_6

*19. Prove that if $2^n + 1 = a^2$, then $n = a = 3$.

20. Prove that if $2^n + 1 = a^b$ and $b \geq 2$, then $b = 2$.

SECTION 2.2 COMPUTER EXERCISES

21. Write a computer program to find the greatest common divisor of a pair of given natural numbers using Euclid's algorithm.

22. Let $L_n = F_{2n}/F_n$. Compute L_n for each n from 1 to 30. For which k does it seem to be true that $L_n \mid L_{kn}$?

23. Compute $(F_m, F_n) - F_{(m,n)}$ for all m and n such that $1 \leq m \leq n \leq 30$. Is there a pattern?

2.3   LINEAR FORM OF GREATEST COMMON DIVISOR, EUCLID'S LEMMA, LEAST COMMON MULTIPLE

We saw earlier that $(36, 60) = 12$. Now $12 = 36(2) + 60(-1)$; that is, the greatest common divisor of 36 and 60 is a linear combination of 36 and 60, with integer coefficients. The following theorem states that the greatest common divisor of any two natural numbers can be represented as a linear combination of the given natural numbers.

Theorem 2.4   *(Linear Form of the Greatest Common Divisor)*
*Given natural numbers a and b, then (a, b) is the least natural number
of the form ax + by, where x and y are integers.*

PROOF: Let S be the set of all natural numbers of the form $ax + by$, where
x and y are integers. Since $a = a(1) + b(0)$, a belongs to S, so S is
nonempty. By the well-ordering principle, S has a least element $d =
am + bn$ for some m, n. By the division algorithm (Theorem 2.2), there
exist integers q and r such that $a = qd + r$ and $0 \le r < d$. Now $r = a -
qd = a - q(am + bn) = a(1 - qm) + b(-qn)$. If $r \ne 0$, then r belongs
to S, yet $r < d$, which contradicts the definition of d. Therefore, $r = 0$, so
$d \mid a$. Similarly, $d \mid b$. Having shown that d is a common divisor of a and
b, we must still show that d is their greatest common divisor. If $c \mid a$ and
$c \mid b$, then by Theorem 2.1, D_7, we have $c \mid (am + bn)$; that is, $c \mid d$.
Therefore, $d = (a, b)$.   ∎

REMARK: The integers x and y such that $(a, b) = ax + by$ are *not* unique,
since for any m we have $ax + by = a(x + bm) + b(y - am)$.

In Chapter 4, given a and b, we will need to compute x and y such that
$(a, b) = ax + by$. This requires computing (a, b) and then doing some
backtracking, as is shown in the following examples.

Example 1:   Find x and y such that $(27, 87) = 27x + 87y$.

Solution: Recall that

$$87 = 3(27) + 6$$
$$27 = 4(6) + 3$$
$$6 = 2(3)$$

So $(27, 87) = 3$. Ignoring the last of these three equations, we rewrite the
first two equations as

$$3 = 27 - 4(6)$$
$$6 = 87 - 3(27)$$

Now we eliminate the middleman, namely 6, between these last two
equations to obtain

$$3 = 27 - 4[87 - 3(27)]$$
$$= 27 - 4(87) + 12(27)$$
$$= 27(13) + 87(-4)$$

Therefore, $x = 13$ and $y = -4$. Although 27 and 87 are constants, in this computation we must treat them as if they were unknowns. That is to say, we must refrain from carrying out some of the indicated multiplication, such as $-3(27)$.

Example 2:   Find x and y such that $(165, 418) = 165x + 418y$.

Solution: Recall that

$$418 = 2(165) + 88$$
$$165 = 1(88) + 77$$
$$88 = 1(77) + 11$$
$$77 = 7(11)$$

So $(165, 418) = 11$. Ignoring the last of these four equations, we rewrite the first three equations as

$$11 = 88 - 1(77)$$
$$77 = 165 - 1(88)$$
$$88 = 418 - 2(165)$$

Now we eliminate the middlemen, namely 77 and 88, as follows: First, eliminate 77 between the first two of the last three equations to obtain

$$11 = 88 - (165 - 88)$$
$$= 88 - 165 + 88$$
$$= 2(88) - 1(165)$$

Now look at the two equations

$$11 = 2(88) - 165$$
$$88 = 418 - 2(165)$$

We now eliminate 88 to obtain

$$11 = 2[418 - 2(165)] - 165$$
$$= 2(418) - 4(165) - 165$$
$$= 165(-5) + 418(2)$$

Therefore, $x = -5$ and $y = 2$.

Euclid's lemma, which follows, is an important consequence of Theorem 2.4. In Chapter 3, which deals with prime numbers, Euclid's lemma is used to prove a key property of primes, namely, Theorem 3.2.

Theorem 2.5   *(Euclid's Lemma)*
If a | bc and (a, b) = 1, then a | c.

PROOF: Since $(a, b) = 1$ by hypothesis, Theorem 2.4 implies that there exist integers x and y such that $ax + by = 1$, so $cax + cby = c$. Now $a \mid cax$, and $a \mid bc$ by hypothesis, so $a \mid cby$. Therefore, $a \mid (cax + cby)$; that is, $a \mid c$.   ∎

As a corollary, we obtain the following:

Theorem 2.6   *Let (a, b) = 1. Then ab | m if and only if a | m and b | m.*

PROOF: If $ab \mid m$, then Theorem 2.1, part D_7 implies that $a \mid m$ and $b \mid m$. Conversely, assume that $a \mid m$ and $b \mid m$. Therefore, m/b is an integer, so $a \mid b(m/b)$. Now $(a, b) = 1$ by hypothesis, so Theorem 2.5 implies that $a \mid (m/b)$. Finally, Theorem 2.1, part D_5 implies that $ab \mid m$.   ∎

Another consequence of Euclid's Lemma is:

Theorem 2.7   *If (a, n) = (b, n) = 1, then (ab, n) = 1.*

PROOF: By hypothesis and Theorem 2.5, there exist integers $u, v, x,$ and y such that $1 = au + nv$ and $1 = bx + ny$. Multiplying these equations, we get

$$1 = (au + nv)(bx + ny)$$
$$= ab(ux) + n(auy + bvx + nvy)$$

Let $d = (ab, n)$. Since 1 is the least natural number of any kind, 1 is certainly the least natural number that is a linear combination of ab and n. Therefore, Theorem 2.4 implies that $d = 1$.   ∎

The notion of greatest common divisor can be extended to sets of three or more natural numbers. Suppose that $a_1, a_2, a_3, \cdots, a_n$ are natural numbers such that $c \mid a_i$ for each index i. Then we say that c is a *common divisor* of the a_i. For example, 5 is a common divisor of 20, 30, and 40; also, 6 is a common divisor of 36, 54, and 72.

If d is a common divisor of the a_i, and if for every common divisor c of the a_i we have $c \mid d$, then we say that d is the *greatest common divisor* of the a_i, and we write $d = (a_1, a_2, a_3, \cdots, a_n)$. For example, 10 is the greatest common divisor of 20, 30, and 40, so we write $(20, 30, 40) = 10$; also, 18 is the greatest common divisor of 36, 54, and 72, so we write $(36, 54, 72) = 18$.

We use the following theorem to compute (a_1, a_2, a_3).

Theorem 2.8 $(a_1, a_2, a_3) = ((a_1, a_2), a_3)$

PROOF: Exercise

REMARK: Theorem 2.8 can be extended to sets of arbitrarily many natural numbers. (See Exercise 2.8 below). For example, $(12, 18, 32) = ((12, 18), 32) = (6, 32) = 2$; also, $(36, 84, 105) = ((36, 84), 105) = (12, 105) = 3$.

It may occur that $(a_1, a_2, a_3) = 1$ even though each of (a_1, a_2), (a_1, a_3), and (a_2, a_3) is greater than 1. For example, $(6, 10, 15) = 1$ even though $(6, 10) = 2$, $(6, 15) = 3$, $(10, 15) = 5$. This leads us to make the following definition:

Definition 2.7   ***Pairwise Relatively Prime***
Let $a_1, a_2, a_3, \cdots, a_n$ be natural numbers such that $(a_i, a_j) = 1$ whenever $i \neq j$. Then we say that the a_i are *pairwise relatively prime*.

For example, 6, 55, and 91 are pairwise relatively prime, since $(6, 55) = (6, 91) = (55, 91) = 1$.

So far we have considered divisors, common divisors, and greatest common divisors. Recall that if a divides b, then b is a multiple of a. We now consider the analogous concepts of common multiple and least common multiple.

Definition 2.8   ***Common Multiple***
If $a \mid n$ and $b \mid n$, then we say that n is a *common multiple* of a and b.

For example, the common multiples of 8 and 12 are 24, 48, 72, 96, etc. Also, the common multiples of 9 and 15 are 45, 90, 135, 180, etc. Since $a \mid ab$ and $b \mid ab$ for all a and b, the set of common multiples of a and b always includes ab and each multiple thereof. Note also that if $a \mid n$ and $b \mid n$, then $\text{Max}\{a, b\} \leq n$. An especially important common multiple of a and b is their *least* common multiple. For example, the least common multiple of 8 and 12 is 24; the least common multiple of 9 and 15 is 45.

Definition 2.9   ***Least Common Multiple***
Suppose that (i) $a \mid m$ and $b \mid m$, and also (ii) if $a \mid n$ and $b \mid n$, then $m \mid n$. Then we say that m is the *least common multiple* of a and b. We write $m = \text{LCM}(a, b)$ or, simply, $m = [a, b]$.

Using this notation, we write $[8, 12] = 24$ and $[9, 15] = 45$.

Theorem 2.9   *(Properties of the Least Common Multiple)*
Let $m = [a, b]$

$$L_1 = \quad \textit{If } a \mid n \textit{ and } b \mid n, \textit{ then } m \leq n.$$
$$L_2 = \quad [a, b] = b \textit{ if and only if } a \mid b.$$
$$L_3 = \quad \textit{Max}\{a, b\} \leq m \leq ab.$$
$$L_4 = \quad \textit{If } c \mid a \textit{ and } c \mid b, \textit{ then } \left[\frac{a}{c}, \frac{b}{c}\right] = \frac{m}{c}.$$
$$L_5 = \quad [an, bn] = [a, b]n \textit{ for all } n.$$
$$L_6 = \quad \textit{If } (a, b) = 1, \textit{ then } [a, b] = ab.$$
$$L_7 = \quad [a, b] = ab/(a, b).$$

We prove L_4 through L_7, leaving the proofs of L_1, L_2, and L_3 as exercises.

PROOF OF L_4:   Let $j = \left[\frac{a}{c}, \frac{b}{c}\right]$, so $\frac{a}{c} \mid j$ and $\frac{b}{c} \mid j$. Now Theorem 2.1, part D_5 implies that $a \mid cj$ and $b \mid cj$. Therefore, $m \mid cj$, so Theorem 2.1, part D_5 implies that $\frac{m}{c} \mid j$. Since $a \mid m$ and $b \mid m$, Theorem 2.1, part D_5 implies that $\frac{a}{c} \mid \frac{m}{c}$ and $\frac{b}{c} \mid \frac{m}{c}$, so $j \mid \frac{m}{c}$. Therefore, $j = \frac{m}{c}$; that is, $\left[\frac{a}{c}, \frac{b}{c}\right] = \frac{m}{c}$.   ∎

PROOF OF L_5:   Let $f = [an, bn]$. We must show that $f = mn$. Now $a \mid m$ and $b \mid m$, so Theorem 2.1, part D_5 implies that $an \mid mn$ and $bn \mid mn$. Therefore, $f \mid mn$. Let $mn = hf = h[an, bn]$. Now $[an, bn] = mn/h$. L_4 implies that $[a, b] = m/h$, so $h = 1$; hence, $f = mn$.   ∎

PROOF OF L_6:   Since $m = [a, b]$, there exist u and v such that $m = au = vb$. L_3 implies that $m \leq ab$, so $v \leq a$. Now $a \mid m$; that is, $a \mid vb$. Since $(a, b) = 1$ by hypothesis, Theorem 2.5 implies that $a \mid v$. Now Theorem 2.1, part D_2 implies that $a \leq v$, so $a = v$; hence $m = ab$.   ∎

PROOF OF L_7:   Let $d = (a, b)$. L_4 implies that $\left[\frac{a}{d}, \frac{b}{d}\right] = \frac{m}{d}$. Theorem 2.3, part G_5 implies that $\left(\frac{a}{d}, \frac{b}{d}\right) = 1$. Therefore, L_6 implies that $\left[\frac{a}{d}, \frac{b}{d}\right] = \frac{ab}{dd}$. Finally, L_5 implies that $[a, b] = ab/d$.   ∎

To compute $[a, b]$, we first use the Euclidean algorithm to find (a, b). Then, by L_7, we have $[a, b] = ab/(a, b)$. For example,

$$[27, 87] = \frac{27 \cdot 87}{(27, 87)}$$
$$= \frac{27 \cdot 87}{3}$$
$$= 9 \cdot 87$$
$$= 783$$

Also,

$$[165, 418] = \frac{165 \cdot 418}{(165, 418)}$$
$$= \frac{165 \cdot 418}{11}$$
$$= 15 \cdot 418$$
$$= 6270$$

The notion of the least common multiple can be extended to sets of three or more natural numbers. Suppose that $a_1, a_2, a_3, \cdots, a_n$ are natural numbers such that $a_i \mid k$ for each index i. Then we say that k is a *common multiple* of the a_i. For example, 144 is a common multiple of 6, 8, and 9; also, 60 is a common multiple of 6, 10, and 15.

If m is a common multiple of the a_i, and if for every common multiple k of the a_i we have $m \mid k$, then we say that m is the *least common multiple* of the a_i, and we write $m = [a_1, a_2, a_3, \cdots, a_n]$. For example, 72 is the least common multiple of 6, 8, and 9, so we write $[6, 8, 9] = 72$. Also, 30 is the least common multiple of 6, 10, and 15, so we write $[6, 10, 15] = 30$.

We use the following theorem to compute $[a_1, a_2, a_3]$.

Theorem 2.10 $[a_1, a_2, a_3] = [[a_1, a_2], a_3]$

PROOF: Exercise.

For example, $[12, 16, 30] = [[12, 16], 30] = [48, 30] = 240$. Also, $[63, 91, 104] = [[63, 91], 104] = [819, 104] = 6552$.

SECTION 2.3 EXERCISES

24. Find all pairs of a, b such that $1 \le a < b \le$ 10 and $(a, b) = 1$.

25. For each pair of integers m, n in Exercise 9, find (i) x, y such that $(m, n) = mx + ny$ and (ii) $[m, n]$.

26. Prove that if m and n are natural numbers and $(m, n) = mx + ny$, then $xy \le 0$.

27. Prove Theorem 2.8.

28. Generalize Theorem 2.8 by proving that if $n \ge 3$, then $(a_1, a_2, a_3, \cdots, a_n) = ((a_1, a_2, a_3, \cdots, a_{n-1}), a_n)$.

29. Prove that if a, b, and c are pairwise relatively prime, then $(a, b, c) = 1$.

30. Generalize Theorem 2.7 by proving that if $(a, b_i) = 1$ for all i such that $1 \le i \le n$ and if $B = \prod_{i=1}^{n} b_i$, then $(a, B) = 1$.

In each of 31 through 39, find the greatest common divisor and the least common multiple of the three given integers.

31. 30, 42, 70

32. 51, 153, 240

33. 296, 444, 555

34. 21, 28, 32

35. 20, 30, 60

36. 32, 72, 84

37. 90, 120, 135

38. 114, 133, 171

39. 45, 81, 87

40. Prove Theorem 2.9, parts L_1, L_2, and L_3.

41. Prove Theorem 2.10.

42. Prove that if $a \mid c$ and $b \mid c$, then $[a, b] \mid c$. Note in particular that if $a \mid c, b \mid c$, and $(a, b) = 1$, then $ab \mid c$. This provides an alternative proof of Theorem 2.6.

43. Generalize Theorem 2.9, part L_6 by proving that given any finite set of natural numbers that are pairwise relatively prime, their least common multiple is their product.

44. One might be tempted to generalize Theorem 2.9, part L_7 as follows: $[a, b, c] = abc/(a, b, c)$. Is this valid? (Explain.)

45. When discussing Euclid's algorithm, we showed that $(a, b) = r_k$ for some $k \ge 0$. (Recall that r_k is the remainder that results from the kth iteration.) Using the notation of pages 26 and 27, solve explicitly for (a, b) in terms of a and b (i) when $k = 2$ and (ii) when $k = 3$.

*46. Prove that if $(a, b) = 1$, then $(a + b, ab) = 1$.

SECTION 2.3 COMPUTER EXERCISES

47. Write a computer program that, given natural numbers a and b, finds integers x and y such that $(a, b) = xa + yb$.

48. Let F_n be the nth Fibonacci number. For $2 \le n \le 12$, compute x_n and y_n such that $(F_n, F_{n+1}) = x_n F_n + y_n F_{n+1}$. Is there a pattern?

2.4   DECIMAL AND BINARY
REPRESENTATIONS OF INTEGERS

We usually represent natural numbers in the decimal, or base 10, system. We now examine the decimal system and obtain several divisibility tests therefrom. We also discuss the binary, or base 2, system, which is used in computers, as well as the interrelationship between the decimal and binary systems.

Recall that

$$4372 = 4000 + 300 + 70 + 2$$
$$= 4(10^3) + 3(10^2) + 7(10^1) + 2(10^0)$$

in the decimal, or base 10, system; each natural number is represented as a sum of multiples of powers of 10. Specifically, if n is the natural number whose decimal representation is

$$a_r a_{r-1} a_{r-2} \cdots a_2 a_1 a_0$$

where $0 \leq a_i \leq 9$ for all a_i and $a_r \neq 0$, then

$$n = \sum_{i=0}^{r} a_i 10^i$$

The use of 10 as the base of our most common number system is arbitrary. It may be related to the use of fingers for counting in times past.

Familiarity with decimal representations of natural numbers can be used to provide several divisibility tests, which we now present. We prove the first two of these divisibility tests, leaving the proofs of the others as exercises.

Theorem 2.11   *n is even if and only if its last decimal digit is even.*

PROOF: Let $n = \sum_{i=0}^{r} a_i 10^i = a_0 + \sum_{i=1}^{r} a_i 10^i$. Since $2 \mid 10^i$ for all $i \geq 1$, we have $2 \mid \sum_{i=1}^{r} a_i 10^i$. Therefore, $2 \mid n$ if and only if $2 \mid a_0$.   ∎

Theorem 2.12   *If $n = \sum_{i=0}^{r} a_i 10^i$, then $3 \mid n$ if and only if $3 \mid \sum_{i=0}^{r} a_i$.*

PROOF: $$n = \sum_{i=0}^{r} a_i(1 + 10^i - 1) = \sum_{i=0}^{r} a_i + \sum_{i=0}^{r} a_i(10^i - 1)$$

$$= \sum_{i=0}^{r} a_i + \sum_{i=1}^{r} a_i(10^i - 1)$$

Now $3 \mid (10 - 1)$, and according to Exercise 9, $(10 - 1) \mid (10^i - 1)$ if $i \geq 1$. Therefore, $3 \mid \sum_{i=1}^{r} a_i(10^i - 1)$. It follows that $3 \mid n$ if and only if $3 \mid \sum_{i=0}^{r} a_i$. ∎

REMARK: In plain language, Theorem 2.12 says that an integer is divisible by 3 if and only if the sum of its decimal digits is divisible by 3. For example, looking at 78,162, we see that $7 + 8 + 1 + 6 + 2 = 24$. Now $3 \mid 24$, so $3 \mid 78,162$. Similarly, looking at 617, we see that $6 + 1 + 7 = 14$. Now $3 \nmid 14$, so $3 \nmid 617$.

Theorem 2.13   *For any $m \geq 1$, $2^m \mid n$ if and only if $2^m \mid k$, where k is the integer consisting of the last m decimal digits of n.*

PROOF: Exercise.

For example, looking at 7148, we see that $2^1 \mid 8$ and $2^2 \mid 48$, but $2^3 \nmid 148$. Therefore, $2^2 \mid 7148$ but $2^3 \nmid 7148$. Similarly, looking at 9,753,432, we see that $2^1 \mid 2$, $2^2 \mid 32$, and $2^3 \mid 432$, but $2^4 \nmid 3432$. Therefore, $2^3 \mid 9,753,432$ but $2^4 \nmid 9,753,432$.

Note that Theorem 2.10 generalizes Theorem 2.8.

Theorem 2.14   *For any $m \geq 1$, $5^m \mid n$ if and only if $5^m \mid k$, where k is the integer consisting of the last m decimal digits of n.*

PROOF: Exercise.

For example, looking at 97,450, we see that $5^1 \mid 0$ and $5^2 \mid 50$, but $5^3 \nmid 450$. Therefore, $5^2 \mid 97,450$ but $5^3 \nmid 97,450$. Similarly, looking at 447,375, we see that $5^1 \mid 5$, $5^2 \mid 75$, and $5^3 \mid 375$, but $5^4 \nmid 7375$. Therefore, $5^3 \mid 447,375$ but $5^4 \nmid 447,375$.

Theorem 2.15   *If $n = \sum_{i=0}^{r} a_i 10^i$, then $9 \mid n$ if and only if $9 \mid \sum_{i=0}^{r} a_i$.*

PROOF: Exercise.

REMARK: In plain language, Theorem 2.15 says that an integer is divisible by 9 if and only if the sum of its decimal digits is divisible by 9. For example, looking at 62,541, we see that $6 + 2 + 5 + 4 + 1 = 18$. Now $9 \mid 18$, so $9 \mid 62,541$. Similarly, looking at 80,437, we see that $8 + 0 + 4 + 3 + 7 = 22$. Now $9 \nmid 22$, so $9 \nmid 80,437$.

Theorem 2.16 *If $n = \sum_{i=0}^{r} a_i 10^i$, then $11 \mid n$ if and only if*

$$11 \mid \left(\sum_{i=0}^{[r/2]} a_{2i} - \sum_{i=0}^{\left[\frac{r-1}{2} \right]} a_{2i+1} \right)$$

PROOF: Exercise.

In plain language, Theorem 2.16 says that an integer is divisible by 11 if and only if the difference between the sums of its alternate decimal digits is divisible by 11. For example, looking at 90,816, we see that $(9 + 8 + 6) - (0 + 1) = 23 - 1 = 22$. Now $11 \mid 22$, so $11 \mid 90,816$. Similarly, looking at 2584, we see that $(2 + 8) - (5 + 4) = 10 - 9 = 1$. Now $11 \nmid 1$, so $11 \nmid 2584$.

In the binary, or base 2, system, each natural number n is represented as a sum of powers of 2. If n is the natural number whose *binary* representation is

$$a_r a_{r-1} a_{r-2} \cdots a_2 a_1 a_0$$

then $n = \sum_{i=0}^{r} a_i 2^i$, where the lead digit $a_r = 1$ and $a_i = 0$ or 1 for each i such that $0 \le i \le r - 1$. The a_i are called the *binary digits*, or *bits*, of n. For example,

$$372 = 256 + 64 + 32 + 16 + 4$$
$$= 2^8 + 2^6 + 2^5 + 2^4 + 2^2$$

so the binary representation of 372 is 101110100. When more than one base is used to represent integers, it is customary to indicate the appropriate base at the lower right. We therefore write

$$372_{10} = 101110100_2$$

One can convert from decimal to binary notation by performing successive divisions by 2 on the decimal representation of an integer and preserving the remainders. For example, to convert 372_{10} to binary, proceed as follows:

$$
\begin{array}{lll}
2 & 0 & r = 1 \\
2 & \overline{)1} & r = 0 \\
2 & \overline{)2} & r = 1 \\
2 & \overline{)5} & r = 1 \\
2 & \overline{)11} & r = 1 \\
2 & \overline{)23} & r = 0 \\
2 & \overline{)46} & r = 1 \\
2 & \overline{)93} & r = 0 \\
2 & \overline{)186} & r = 0 \\
2 & \overline{)372} &
\end{array}
$$

These divisions are performed upwards from the bottom, but then the resulting binary bits are read from the top down, so

$$372_{10} = 101110100_2$$

To convert from binary to decimal notation, one multiplies each 1 in the binary representation of an integer by the appropriate power of 2 and then adds those powers of 2. For example,

$$
\begin{aligned}
11001_2 &= 1(2^4) + 1(2^3) + 0(2^2) + 0(2^1) + 1(2^0) \\
&= 16 + 8 + 1 \\
&= 25_{10}
\end{aligned}
$$

Note that the number of bits in the binary representation of n is $1 + [\log_2 n]$, whereas the number of digits in the decimal representation of n is $1 + [\log_{10} n]$.

SECTION 2.4 EXERCISES

49. Prove the following theorems.

    (a) 2.13   (b) 2.14

    (c) 2.15   (d) 2.16

50. Test the integer 12,167,320 for divisibility by

    (a) powers of 2   (b) powers of 5

    (c) 9   (d) 11

51. Test the integer 1,415,926,535 for divisibility by (a) 9 and (b) 11.

52. Test 814,375 for divisibility by (a) powers of 5 and (b) 11.

53. Convert each of the following decimal integers to binary.

    (a) 13   (b) 36

    (c) 127   (d) 100

    (e) 1001

54. Convert each of the following binary integers to decimal.

    (a) 1010   (b) 10000

    (c) 1100001   (d) 10101

    (e) 1110

SECTION 2.4 COMPUTER EXERCISES

55. Write computer programs to convert the representation of an integer (a) from decimal to binary and (b) from binary to decimal.

56. Let $r(n)$ be the number of 0's at the end of the binary representation of $\binom{2n}{n}$. Let $s(n)$ be the number of 1's in the binary representation of n. Computer $r(n) - s(n)$ for each n such that $1 \leq n \leq 10$. Care to conjecture?

CHAPTER THREE

PRIMES

3.1 INTRODUCTION

In the scientific quest for understanding, it is often helpful to break a complex whole into basic components. In biology, one learns that complex life forms consist of many cells. In chemistry and physics, one learns that a drop of water consists of many molecules and that each molecule consists of atoms. The very word *atom* means indivisible.

Returning to number theory, consider the problem of representing a given natural number as the product of two or more smaller factors. For example,

$$350 = 10 \cdot 35$$

or

$$350 = 14 \cdot 25$$

We could go further:

$$10 = 2 \cdot 5$$
$$35 = 5 \cdot 7$$

so that

$$350 = 2 \cdot 5 \cdot 5 \cdot 7$$

This seems to be as far as we can go in the decomposition of 350 into smaller factors. To see why this is so, we must distinguish between those natural numbers other than 1 which can be factored and those which cannot be factored. The former are called *composite*, while the latter are called *prime*.

Prime numbers are important because, as we shall see, every natural number that is greater than 1 can be represented as a product of primes in a way that is essentially unique. As a consequence, many questions concerning the natural number n can be reduced to questions concerning the prime factors of n. We therefore devote this chapter to the study of primes.

3.2   PRIMES, PRIME COUNTING FUNCTION, PRIME NUMBER THEOREM

Definition 3.1   *Composite Number*

A natural number n is said to be *composite* if it has a nontrivial divisor, that is, if it has a divisor d such that $1 < d < n$. If so, then $n = dc$, where $c = n/d$.

For example, the computations above show that 350, 10, and 35 are composite (as are 14 and 25).

Definition 3.2   *Prime Number*

A natural number p is said to be *prime* if $p > 1$ and p has no nontrivial divisor. (In other words, a prime has no divisors except 1 and itself.)

The primes below 100 are 2, 3, 5, 7, 11, 13, 17, 19, 23, 29, 31, 37, 41, 43, 47, 53, 59, 61, 67, 71, 73, 79, 83, 89, and 97. Table 1 in Appendix C lists the primes below 10,000.

Note that the number 1 is neither prime nor composite; we call it a *unit*. Every natural number other than 1 is either prime or composite, but not both.

Theorem 3.1   *If p is prime and m is arbitrary, then*

$$(m, p) = \begin{cases} p & \text{if } p \mid m \\ 1 & \text{if } p \nmid m \end{cases}$$

PROOF: Let $d = (m, p)$. Since $d \mid p$ and p is prime, we know that either $d = p$ or $d = 1$. If $p \mid m$, then Theorem 2.3, part G_2 implies that $d = p$. If $p \nmid m$, then Theorem 2.3, part G_2 implies that $d \neq p$, so $d = 1$.   ■

For example, $(31, 124) = 31$, since 31 is prime and $31 \mid 124$. Also, $(31, 200) = 1$ since 31 is prime and $31 \nmid 200$.

The following theorem establishes one of the most important properties of primes.

Theorem 3.2   *If p is prime and $p \mid ab$, then $p \mid a$ or $p \mid b$.*

PROOF: It suffices to show that if $p \mid ab$ and $p \nmid a$, then $p \mid b$. If $p \nmid a$, then Theorem 3.1 implies that $(a, p) = 1$. Since $p \mid ab$ by hypothesis, Euclid's lemma (Theorem 2.5) implies that $p \mid b$.   ■

For example, 3 is prime and $3 \mid (6 \cdot 7)$, so $3 \mid 6$; also, 5 is prime and $5 \mid (11 \cdot 15)$, so $5 \mid 15$.

Note that if n is composite and $n \mid ab$, it need not follow that $n \mid a$ or $n \mid b$. For example, 6 is composite and $6 \mid (3 \cdot 4)$, yet $6 \nmid 3$ and $6 \nmid 4$. In fact, an alternate way to *define* primes would be to say that p is prime if whenever $p \mid ab$ it is also true that $p \mid a$ or $p \mid b$.

In the next theorem, the result of Theorem 3.2 is extended to products of more than two factors.

Theorem 3.3   *If p is prime and $p \mid \prod_{i=1}^{r} a_i$, then $p \mid a_i$ for some i.*

PROOF: Exercise.

For example, 7 is prime and $7 \mid (33 \cdot 56 \cdot 65)$, so $7 \mid 56$. Also, 13 is prime and $13 \mid (33 \cdot 56 \cdot 65)$, so $13 \mid 65$.

Theorem 3.4   *If $n > 1$, then n has a prime divisor p.*

PROOF: If n is prime, then $p = n$. If n is composite, let p be the least nontrivial divisor of n. We must show that p is prime. If p is composite, then p has a nontrivial divisor d. Now $d \mid p$ and $p \mid n$, so $d \mid n$, yet $d < p$. This contradicts the definition of p. Therefore, p is prime.   ■

For example, the least prime divisor of 111 is 3, the least prime divisor of 323 is 17, and the least prime divisor of 1001 is 7. the following theorem establishes an upper bound for the least prime factor of a composite number; as a by-product, we obtain a test for primality.

Theorem 3.5   *If n is composite and p is the least prime factor of n, then $p \leq \sqrt{n}$.*

PROOF: Let $n = pm$. Since n is composite, $m > 1$. Therefore, by Theorem 3.4, there is a prime q such that $q \mid m$, so $m = qk$ for some $k \geq 1$. Since $m \mid n$, we have $q \mid n$. Since p is the least prime factor of n by hypothesis, we must have $p \leq q$. Therefore, $p^2 \leq pq \leq pqk$; that is, $p^2 \leq n$, so $p \leq \sqrt{n}$.   ∎

Suppose we wish to know whether 10,001 is prime or composite. If 10,001 is composite, then according to Theorem 3.5, 10,001 has a prime factor p such that $p \leq \sqrt{10,001}$; that is, $p < 100$. Therefore, we try to divide 10,001 by each of the primes below 100, in order. (These primes were listed on page 42.) Eventually, we see that $73 \mid 10,001$, and $10,001 = 73 \cdot 137$. This approach can be generalized as follows.

Test of Primality by Trial Division

Let $p_1 = 2, p_2 = 3, p_3 = 5$, etc. Given that $n > 1$, let k be the largest integer such that $p_k^2 \leq n$. If $p_i \mid n$ for some i such that $1 \leq i \leq k$, then n is composite. If no such p_i divides n, then n is prime.

For example, suppose we wish to determine whether 103 is prime or composite without using a table of primes. Now $p_4 = 7, p_5 = 11$, and $7^2 < 103 < 11^2$, so $k = 4$. We attempt to divide 103 by each of the first four primes, namely, by 2, 3, 5, and 7. Since $2 \nmid 103$, $3 \nmid 103$, $5 \nmid 103$, $7 \nmid 103$, we conclude that 103 is prime.

Testing a natural number n for primality by trial division requires that we know all the primes p such that $p^2 \leq n$. We shall see that if n is sufficiently large, then there are too many such primes, so trial division would require too much time and space. More efficient methods of testing large integers for primality are discussed in Chapter 12.

Although primes grow more scarce among larger integers, they never run out, as Euclid proved about 300 B.C.

Theorem 3.6   *(Euclid)*
There exist infinitely many primes.

PROOF: Suppose that $p_1, p_2, p_3, \cdots, p_n$ are the only primes. Let $m = 1 + \prod_{i=1}^{n} p_i$. Since $m > 1$, Theorem 3.4 implies that there exists a prime q such that $q \mid m$. However, $(m, p_i) = 1$ for all i, so $q \neq p_i$ for all i, contrary to hypothesis.   ∎

REMARK: The proof of Theorem 3.6 indicates a constructive method such that given a finite set of primes, one can find at least one additional prime. For example, given 2, 3, and 5, one computes $1 + 2 \cdot 3 \cdot 5 = 31$, which is prime; given 2, 7, and 11, one computes $2 \cdot 7 \cdot 11 + 1 = 155 = 5 \cdot 31$.

The distribution of primes seems highly irregular. There is no convenient formula that yields the value of p_n, the nth prime, in terms of n. The prime that follows 197 is 199, but then the next prime is 211. If p and q are odd primes with $p < q$, then $q - p \geq 2$. If $q - p = 2$, then we say that p and q are a pair of *twin primes*. Some examples of twin prime pairs are 5 and 7, 197 and 199, and 1,000,000,009,649 and 1,000,000,009,651. It is not known whether infinitely many pairs of twin primes exist. On the other hand, the following is known:

Theorem 3.7   *There exist arbitrarily large gaps between consecutive primes.*

PROOF: We will show that for any $n \geq 2$, there exist consecutive primes p and q such that $q - p \geq n$. If $2 \leq k \leq n$, then $k \mid (n! + k)$. Therefore, $n! + 2, n! + 3, n! + 4, \cdots, n! + n$ is a sequence of $n - 1$ consecutive composite numbers. Let p be the greatest prime such that $p \leq n! + 1$. We know that such a prime exists, since 2 is a prime and $2 < 2! + 1 \leq n! + 1$. Let q be the smallest prime such that $q \geq n! + n + 1$. Theorem 3.6 implies that such a q exists. Now p and q are consecutive primes, and $q - p \geq n$.   ∎

We might also note that if $2 \leq k \leq n$, then $k \mid (n! - k)$, so $n! - 2, n! - 3, n! - 4, \cdots, n! - n$ are consecutive composite numbers. Therefore, the primes closest to $n! - 1$ and $n! - (n + 1)$ differ by at least n. For example, for $n = 4$, we have $4! - 1 = 23$ (a prime) and $4! - 5 = 19$ (a prime).

As a consequence of the prime number theorem, which follows shortly, it can be shown that the average distance between consecutive primes in the vicinity of n is approximately $\log n$. (Note that $\log n$ denotes the natural logarithm of n.) For example, choose an interval of length 200 centered at $n = 1000$. The reader can verify by referring to Table 1 in Appendix C that there are 28 primes in this interval and that the average distance between consecutive primes is 6.8. On the other hand, $\log 1000$ is approximately equal to 6.9.

Let us consider the problem of estimating the number of primes up to a given bound.

Definition 3.3

Prime Counting Function

Let $\pi(x)$ denote the number of primes p such that $p \leq x$, where x is a positive real number.

For example, $\pi(10) = 4$, since the primes not exceeding 10 are 2, 3, 5, and 7. Also, $\pi(30) = 10$, since the primes not exceeding 30 are 2, 3, 5, 7, 11, 13, 17, 19, 23, and 29. Theorem 3.5 implies that $\pi(x)$ tends toward infinity as x tends toward infinity. Table 3.1 shows how $\pi(10^n)$ grows with n.

One of the most important theorems in number theory follows:

Table 3.1. Table of the Prime Counting Function

n	$\pi(10^n)$	n	$\pi(10^n)$
1	4	9	50847534
2	25	10	455051511
3	168	11	4118054813
4	1229	12	37607912018
5	9592	13	346065536839
6	78498	14	3204941750802
7	664579	15	29844570422669
8	5761455	16	279238341033925

Source: Riesel.

Prime Number Theorem $\pi(x)$ *is asymptotic to* x/log x, *that is,*

$$\lim_{x \to \infty} \frac{\pi(x)}{x/\log x} = 1$$

The prime number theorem, which suggests that we estimate $\pi(x)$ by $x/\log x$, was first conjectured by Gauss and by Legendre in the 1790s. The first complete proofs, based on analytic methods, were given by Hadamard and De la Vallee Poussin in 1896. An elementary (but not simple) proof was given by Selberg and Erdos in 1949. We omit the proof, which is beyond the scope of this text. (See Apostol (1976).)

Let $f(x) = \frac{\pi(x)}{x/\log x}$. The convergence of $f(x)$ to its limiting value of 1 is slow, since $f(10^{16}) = 1.029$. There are better approximations to $\pi(x)$, such as the so-called *logarithmic integral:*

$$\text{li}(x) = \int_2^x \frac{dt}{\log t}$$

Using the prime number theorem, however, one can easily estimate $\pi(x)$ with a pocket calculator, even for very large x. For example, $\pi(10^{20})$ is about $10^{20}/\log(10^{20}) = 10^{20}/20 \log 10 = 2.17 \cdot 10^{18}$.

The prime number theorem implies that the proportion of natural numbers below x that are prime is about $1/\log x$. Therefore, the expected number of primes in an interval of length b centered at x is about $b/\log x$. For example, if $x = 1000$ and $b = 200$, our expected number of primes is $200/\log 1000$ or approximately 29. (One can verify by consulting Table 1 in Appendix C that the actual number of primes in this interval is 28.)

Special methods exist to verify the primality of natural numbers t such that $t = m - 1$, where m is an easily factored number. An example is the *Mersenne numbers* $M_n = 2^n - 1$. These are discussed in Chapter 5. At present, the largest verified prime is $2^{756839} - 1$.

Consider the problem of finding a function from the nonnegative integers to the natural numbers, all of whose values are primes. For example, if $f(x) = 2x^2 + 29$, then $f(0) = 29$ (a prime), $f(1) = 31$ (a prime), $f(2) = 37$ (a prime), ..., $f(28) = 1597$ (a prime). However $f(29) = 1711 = 29 \cdot 59$ (a composite). In fact, $f(29k)$ is divisible by and greater than 29 for all $k \geq 1$ and hence is composite.

Similarly, if $g(x) = x^2 + x + 41$, then $g(0) = 41$ (a prime), $g(1) = 43$ (a prime), $g(2) = 47$ (a prime), ..., $g(39) = 1601$ (a prime). However, $g(40) = 1681 = 41^2$ (a composite). In fact, $g(41k)$ is divisible by and greater than 41 for all $k \geq 1$ and hence is composite.

The following theorem shows that no polynomial of a single variable can evaluate primes only. More specifically, every polynomial of a single variable with integer coefficients evaluates infinitely many composite numbers.

Theorem 3.8   *Let $f(x)$ be a polynomial of degree $r \geq 1$ with integer coefficients and positive lead coefficient, that is,*

$$f(x) = \sum_{k=0}^{r} a_k x^k \quad \text{with } a_r > 0$$

Then there are infinitely many n such that $f(n)$ is composite.

PROOF: Since $a_r > 0$, there exists c such that $f(x) \geq 2$ for all $x \geq c$. If $f(x)$ is composite for all $x \geq c$, then we are done. Otherwise, there exists $t \geq c$ such that $f(t) = p$, a prime. Let $m \geq 1$. Now

$$f(t + mp) = \sum_{k=0}^{r} a_k (t + mp)^k = \sum_{k=0}^{r} a_k \left[\sum_{j=0}^{k} \binom{k}{j} t^{k-j} (mp)^j \right]$$

$$= \sum_{k=0}^{r} a_k \left[t^k + \sum_{j=1}^{k} \binom{k}{j} t^{k-j} (mp)^j \right]$$

$$= \sum_{k=0}^{r} a_k t^k + \sum_{k=0}^{r} a_k \left[\sum_{j=1}^{k} \binom{k}{j} t^{k-j} (mp)^j \right]$$

$$= f(t) + \sum_{k=0}^{r} a_k \left[\sum_{j=1}^{k} \binom{k}{j} t^{k-j} (mp)^j \right]$$

Therefore, $p \mid f(t + mp)$.

Since a polynomial of degree r has at most r real roots, there are at most r values of m such that $f(t + mp) = p$. Therefore, there are infinitely many values of m such that $f(t + mp)$ is divisible by and greater than p (hence composite).   ∎

SECTION 3.2 EXERCISES

1. Prove that if $a \geq 2$, $n \geq 2$, and $a^n - 1$ is prime, then $a = 2$ and n is prime.

2. Prove that if $a \geq 2$, $n \geq 2$, and $a^n + 1$ is prime, then $2 \mid a$ and $n = 2^k$.

3. Prove that if n is the product of k prime factors, of which p is the least, then $p \leq n^{1/k}$. (Note that this generalizes Theorem 3.5.)

4. Prove Theorem 3.3. (Use induction on n.)

5. With very little computation, find a prime factor of $10^{10} + 1$.

6. Using Table 3.1, find the five smallest primes of the form $x^2 + 1$.

7. Using Table 3.1, find the five smallest primes of the form $x^2 + x + 1$.

8. Using Table 3.1, find the five smallest primes p such that $2p + 1$ is also prime.

9. If p is prime, $p^k \mid a^n$, and $p^{k+1} \nmid a^n$, prove that $n \mid k$.

10. Prove that there exist arbitrarily long sequences of consecutive odd composite numbers.

11. Use the prime number theorem to estimate the number of primes below 10^{10}. Now look up the actual number from Table 3.1 and compute the relative error in the estimate.

12. In an application to cryptology (see the last section of Chap. 13), we need a prime greater than 10^{100}. If odd numbers greater than 10^{100} are chosen at random until a prime is found, how many odd numbers should one expect to test for primality?

SECTION 3.2 COMPUTER EXERCISES

13. For a given natural number n, let $S_n = \{2, 3, 5, \cdots, q\}$ be the set of primes p such that $p \leq \sqrt{n}$. Suppose that S_n has k elements. A counting argument shows that $\pi(n)$ may be computed from the formula

$$\pi(n) = k + n - 1 - \Sigma\left[\frac{n}{p}\right] + \Sigma\left[\frac{n}{p_1 p_2}\right]$$

$$- \Sigma\left[\frac{n}{p_1 p_2 p_3}\right] + \cdots$$

(Each sum is taken over all products of one or more distinct primes from S_n.) For example, if $n = 10$, then $S_{10} = \{2, 3\}$, so $k = 2$ and

$$\pi(10) = 2 + 10 - 1 - \left(\left[\frac{10}{2}\right] + \left[\frac{10}{3}\right]\right)$$

$$+ \left[\frac{10}{6}\right]$$

$$= 11 - (5 + 3) + 1$$

$$= 4$$

Write a computer program to compute $\pi(10^4)$, making use of the preceding formula.

14. Write a computer program to find the least odd number n such that n is the least of 10 consecutive odd composite numbers.

15. Write a computer program to test natural numbers for primality by trial division.

16. Write a computer program to determine which of the first 30 Fibonacci numbers are prime. If F_n is prime, does this tell us something about n?

3.3   SIEVE OF ERATOSTHENES, CANONICAL FACTORIZATION, FUNDAMENTAL THEOREM OF ARITHMETIC

Let us now consider the problem of finding all the primes in a given interval. A method for doing so was given by Eratosthenes (276–195 B.C.). This method is known as the *Sieve of Eratosthenes*.

Sieve of Eratosthenes

Given $a < b$, we wish to find all primes p such that $a \leq p \leq b$. By Theorem 3.5, if n is a composite number in this interval, then n has a prime factor not exceeding $\sqrt{b}$. Suppose that the primes not exceeding $\sqrt{b}$ are $p_1 = 2$, $p_2 = 3$, $p_3 = 5$, $\cdots$, p_k. First, we write down all the integers from a to b. Then we eliminate all nontrivial multiples of each of the p_i. That is to say, we cross out all nontrivial multiples of p_1, then cross out all remaining nontrivial multiples of p_2, then cross out all remaining nontrivial multiples of p_3, etc. After we have eliminated all remaining nontrivial multiples of p_k, the remaining integers, if any, are all prime, since we have "sieved out" all the composites.

For example, let us use the sieve of Eratosthenes to find all primes, if any, between 110 and 130. We can simplify our task by omitting the even integers in the given interval. We write

$$111, \ 113, \ 115, \ 117, \ 119, \ 121, \ 123, \ 125, \ 127, \ 129$$

Since $11 < \sqrt{130} < 13$, any odd composite numbers in our interval must be divisible by an odd prime not exceeding 11, that is, by 3, 5, 7, or 11. Now $3 \mid 111$, $3 \mid 117$, $3 \mid 123$, and $3 \mid 129$; $5 \mid 115$ and $5 \mid 125$; $7 \mid 119$; and $11 \mid 121$. The remaining integers, namely, 113 and 127, are prime.

Eratosthenes was born in Cyrene, in what is now Libya, and he became head of the library at Alexandria, Egypt. One of the foremost scholars of his time, he wrote works on geography, mathematics, literary criticism, grammar, and poetry. He devised an experiment to measure the circumference of the earth. Eratosthenes also served as tutor to the sons of the Pharaoh, Ptolemy II.

According to Theorem 3.4, if $n > 1$, then n has a prime divisor p. If n is composite, then $n/p > 1$, so Theorem 3.4 can be applied to n/p. This leads to the following theorem.

Theorem 3.9   *If $n > 1$, then n can be represented as a product of one or more primes.*

PROOF: (Induction) If n is prime, then n is the product of a single prime, namely, itself. Such is the case for $n = 2$. If n is composite, then Theorem 3.4 implies that $n = pm$, where p is prime and $2 \le m < n$. By induction hypothesis, m is a product of primes. Therefore, so is n.   ■

For example, $2646 = 2 \cdot 1323 = 2 \cdot 3 \cdot 441 = 2 \cdot 3 \cdot 3 \cdot 147 = 2 \cdot 3 \cdot 3 \cdot 3 \cdot 49 = 2 \cdot 3 \cdot 3 \cdot 3 \cdot 7 \cdot 7$. It is more convenient to write

$$2646 = 2^1 3^3 7^2$$

This last equation is the *canonical factorization* of 2646.

Definition 3.4   **Canonical Factorization**

If $n > 1$, we write n as a product of primes:

$$n = p_1^{e_1} p_2^{e_2} p_3^{e_3} \cdots p_k^{e_k}$$

That is,

$$n = \prod_{i=1}^{k} p_i^{e_i}$$

where the p_i are distinct primes such that $p_1 < p_2 < p_3 < \cdots < p_k$ and each exponent $e_i \geq 1$. This is called the *canonical factorization* of n.

For example, the canonical factorization of 202,400 is

$$202,400 = 2^5 5^2 11^1 23^1$$

Before discussing how to obtain the canonical factorization of a given natural number, we need to introduce some more notation. We are concerned with the highest power of a prime that divides a given integer. If p is prime, $p^k \mid n$, and $p^{k+1} \nmid n$, then we write

$$p^k \parallel n$$

For example, $3^4 \mid 162$ but $3^5 \nmid 162$, so we write: $3^4 \parallel 162$. Also, $2^3 \mid 40$ but $2^4 \nmid 40$, so we write $2^3 \parallel 40$. If $p^k \parallel n$, we also write $o_p(n) = k$. The two preceding computations show that $o_3(162) = 4$ and that $o_2(40) = 3$.

Determining the Canonical Factorization of a Natural Number

Given $n > 1$, divide n by each of the primes in increasing order, that is, by 2, 3, 5, 7, etc., until p_1, the smallest prime factor of n, is found.

Determine the exponent e_1 such that $p_1^{e_1} \parallel n$. Let $n_1 = n/p_1^{e_1}$. Find p_2, the smallest prime factor of n_1. Determine the exponent e_2 such that $p_2^{e_2} \parallel n_1$. Let $n_2 = n_1/p_2^{e_2}$. Find p_3, the smallest prime factor of n_2. Determine the exponent e_3 such that $p_3^{e_3} \parallel n_2$. Let $n_3 = n_2/p_3^{e_3}$, etc. Ultimately, for some $k \geq 1$, $n_{k+1} = 1$. Then our result is

$$n = \prod_{i=1}^{k} p_i^{e_i}$$

For example, let us obtain the canonical factorization of 1386. $1386 = 2 \cdot 693$. Now $2 \nmid 693$, but $693 = 3 \cdot 231 = 3^2 \cdot 77$. Now $3 \nmid 77$ and $5 \nmid 77$, but $77 = 7 \cdot 11$. Therefore, $1386 = 2^1 3^2 7^1 11^1$.

As a second example, let us obtain the canonical factorization of 2125. $2 \nmid 2125$ and $3 \nmid 2125$, but $2125 = 5 \cdot 425 = 5^2 \cdot 85 = 5^3 \cdot 17$. Since 17 is prime, we are done. Our result is $2125 = 5^3 \cdot 17^1$.

The fundamental theorem of arithmetic, which follows, states that the canonical factorization of the natural number n is unique if $n > 1$.

Theorem 3.10   *(Fundamental Theorem of Arithmetic)*
If $n > 1$, then the canonical factorization of n is unique.

PROOF: Let n have a canonical factorization:

$$n = \prod_{i=1}^{k} p_i^{e_i}$$

First, we show that the set of prime factors of n, namely $\{p_1, p_2, p_3, \cdots , p_k\}$ is uniquely determined. Suppose that q is prime and $q \mid n$. Applying Theorem 3.3, we see that $q \mid p_i^{e_i}$ for some index i. Applying Theorem 3.3 again, we see that $q \mid p_i$. However, this implies that $q = p_i$.

Next, we show that the set of exponents, namely, $\{e_1, e_2, e_3, \cdots , e_k\}$ is uniquely determined. Suppose that $n = p_i^{e_i}c = p_i^{f_i}d$, where $(p_i, c) = (p_i, d) = 1$ and, without loss of generality, $e_i \leq f_i$. Therefore, $c = p_i^{f_i - e_i}d$. If $e_i < f_i$, then $p_i \mid c$, an impossibility. Therefore, $e_i = f_i$. Since i is arbitrary, we are done.   ∎

REMARK: Although the fundamental theorem of arithmetic may seem obvious, at least for natural numbers, it is noteworthy that number systems exist in which the fundamental theorem fails to hold. In such number systems, the factorization of certain numbers as products of primes may fail to be unique. For example, consider the set of all numbers of the form $a + b\sqrt{-5}$, where a and b are ordinary integers. We define multiplication by

$$(a + b\sqrt{-5})(c + d\sqrt{-5}) = (ac - 5bd) + (ad + bc)\sqrt{-5}$$

In this system, we have

$$6 = 2 \cdot 3$$

Also,

$$6 = (1 + \sqrt{-5})(1 - \sqrt{-5})$$

It can be shown that in this number system, each of the factors 2, 3, and $1 \pm \sqrt{-5}$ is prime. Therefore, *in this number system*, the factorization of 6 as a product of primes is *not* unique.

We saw earlier (Theorem 3.6) that there exist infinitely many primes. By adapting the argument used to prove Theorem 3.6, we now prove the existence of infinitely many primes of a certain special type.

Theorem 3.11   *There exist infinitely many primes of the form $4k - 1$.*

PROOF: Suppose that $q_1, q_2, q_3, \cdots, q_n$ are the only primes of the form $4k - 1$. Let $m = -1 + 4\prod_{i=1}^{n} q_i$, so that m is odd. If all the prime factors of m were of the form $4k + 1$, then by the result of Exercise 36 in Chapter 1, m also would have the form $4k + 1$. Since m has the form $4k - 1$, it follows that m must have a prime factor p of the form $4k - 1$. By hypothesis, $p = q_i$ for some i. Since $p \mid m$, we have $p \mid (-1)$, an impossibility.   ■

REMARK: The first few primes of the form $4k - 1$ are 3, 7, 11, 19, and 23. It is also true that there are infinitely many primes of the form $4k + 1$, such as 5, 13, 17, 29, 37, etc., but this cannot be proved in the same manner as Theorem 3.11. (Why not?) We furnish a proof of this assertion at the end of Chapter 4.

Next we state without proof two important theorems regarding primes. The first theorem, which is due to Dirichlet, greatly generalizes Theorem 3.11.

Theorem 3.12   *(Dirichlet's Theorem)*
If $(a, b) = 1$, then there are infinitely many primes of the form $an + b$.

For example, there are infinitely many primes whose decimal representations end in 3, such as 3, 13, 23, 43, 53, 73, 83, 103, 113, etc., since these are the primes of the form $10n + 3$ and $(10, 3) = 1$. The proof of Theorem 3.12 uses analytic methods that are beyond the scope of this text. [See Apostol (1976).]

Theorem 3.13   *(Bertrand's Postulate)*
For all $n \geq 2$, there is a prime p such that $n < p < 2n$.

For example, if $n = 5$, then $p = 7$. Note that in terms of the prime counting function, Theorem 3.13 says that $\pi(2n) - \pi(n) \geq 1$ if $n \geq 2$. The proof of Theorem 3.13 is elementary but quite lengthy. (See Hardy & Wright.)

We conclude this chapter by presenting a theorem that allows us to obtain the canonical factorization of $n!$ without calculating $n!$. First, a preliminary result is needed.

Theorem 3.14   *If d and n are natural numbers, then*

$$\left[\frac{n}{d}\right] - \left[\frac{n-1}{d}\right] = \begin{cases} 1 & \text{if } d \mid n \\ 0 & \text{if } d \nmid n \end{cases}$$

PROOF: Exercise.

Theorem 3.15   *For any natural number n, we have*

$$n! = \prod p^{\sum_{k=1}^{\infty}[n/p^k]}$$

the product being taken over all primes.

REMARKS: Although the product is taken over all primes, only those finitely many primes p such that $p \leq n$ will carry a positive exponent. Therefore, we could write

$$n! = \prod_{p \leq n} p^{\sum_{k=1}^{\infty}[n/p_k]}$$

Furthermore, if $p^j \parallel n$, then $[n/p^k] = 0$ for all $k > j$. Therefore, $\sum_{k=1}^{\infty}[n/p^k] = \sum_{k=1}^{j}[n/p^k]$; that is, all sums that occur as exponents in the statement of Theorem 3.15 are actually finite.

PROOF: (Induction on n)   Theorem 3.15 is trivially true for $n = 1$. By induction hypothesis,

$$(n-1)! = \prod p^{\sum_{k=1}^{\infty}[(n-1)/p^k]}$$

We wish to prove that

$$n! = \prod p^{\sum_{k=1}^{\infty}[n/p^k]}$$

Since $n = (n!)/(n-1)!$, it suffices to show that

$$n = \prod p^{\sum_{k=1}^{\infty}\{[n/p^k] - [(n-1)/p^k]\}}$$

In other words, it suffices to show that if $p^j \parallel n$, then $j = \sum_{k=1}^{\infty}\{[n/p^k] - [(n-1)/p^k]\}$, that is, $j = \sum_{k=1}^{j}\{[n/p^k] - [(n-1)/p^k]\}$. This last equation is valid, because $p^k \mid n$ for $k = 1, 2, 3, \cdots, j$, so each of the j summands on the right side of the equation is 1, by Theorem 3.14.   ∎

For example, $10! = 2^a 3^b 5^c 7^d$, where $a = [10/2] + [10/4] + [10/8] = 5 + 2 + 1 = 8$, $b = [10/3] + [10/9] = 3 + 1 = 4$, $c = [10/5] = 2$, and $d = [10/7] = 1$. That is, $10! = 2^8 3^4 5^2 7^1$.

Much remains to be learned about primes. For example, one may ask whether there exist infinitely many primes of the forms (1) $n^2 + 1$, (2) $n^2 + n + 1$, (3) $2^n - 1$, (4) $2^n + 1$, and (5) Fibonacci numbers. If there are infinitely many primes of a certain type, one would like to estimate the number of such primes that are less than or equal to x, where x is a positive real number.

A celebrated open problem regarding primes is the following:

Goldbach's Conjecture

If $n \geq 4$, then there are distinct primes p and q such that $p + q = 2n$.

SECTION 3.3 EXERCISES

17. Using the sieve of Eratosthenes, find all primes between 220 and 250.

18. Without using Dirichlet's theorem, prove that there are infinitely many primes of the form $6k - 1$.

19. Find the canonical factorization of each of the following.

    (a) 299   (b) 480

    (c) 777   (d) 399

    (e) 145   (f) 221

    (g) 72   (h) 2450

    (i) 5005   (j) 191

    (k) 896   (l) 529

    (m) 104   (n) 171

20. Prove that if $n \geq 9$ and $n - 2$ and $n + 2$ are both prime, then $3 \mid n$.

21. Prove that if $(a, b) = 1$ and $ab = c^n$, then there exist x and y such that $a = x^n$, $b = y^n$, $xy = c$, and $(x, y) = 1$.

22. Prove that if p and q are distinct primes such that $pq \mid n^2$, then $pq \mid n$.

23. Prove that if p is prime and $0 < k < p$, then $p \mid \binom{p}{k}$.

24. Find all n with the property that for every prime p such that $p \leq \sqrt{n}$, it is true that $p \mid n$.

*25. Prove that if $1 < m < n$ and $m^n = n^m$, then $m = 2$ and $n = 4$.

26. Let $f(x)$ be a monic polynomial, that is, a polynomial whose lead coefficient is 1. Prove that if r is a rational root of $f(x)$, that is, $f(r) = 0$, then r is an integer.

27. Use the result of Exercise 26 to prove that $\sqrt{2}$ is irrational.

28. If p_n is the nth prime and $n \geq 5$, prove that $p_n^2 < p_1 p_2 p_3 \cdots p_{n-1}$.

29. If $m! = a^n$ and $a > 1$, prove that $n = 1$.

30. Prove Theorem 3.14.

31. Theorem 3.15 could be stated as follows. Let $p^j \| n!$, where p is prime. Then $j = \sum_{k=1}^{\infty} [n/p^k]$. Prove the following alternative formula. Let $n = \sum_{i=0}^{r} a_i p^i$, where $0 \leq a_i \leq p - 1$ for all i and $a_r > 0$. (This is the representation of n to the base p.) Let

$t_p(n) = \sum_{i=0}^{r} a_i$. In other words, $t_p(n)$ is the sum of the digits of n to the base p. Then $j = [n - t_p(n)]/(p - 1)$. For example, if $n = 22$, $p = 3$, and $3^j \| 22!$, then according to Theorem 3.15, $j = [22/3] + [22/9] = 7 + 2 = 9$. According to the alternative formula, $22_{10} = 221_3$ [since $22 = 2(3^2) + 1(3^1) + 1(3^0)$], so $j = \frac{1}{2}[22 - (2 + 1 + 1)] = \frac{1}{2}(22 - 4) = \frac{1}{2}(18) = 9$.

32. Using the formula from Exercise 31, prove that if $p^j \| \binom{n}{k}$, then
$$j = \frac{t_p(k) + t_p(n-k) - t_p(n)}{p - 1}$$

33. Prove that if $2^j \| \binom{2n}{n}$, then j is the sum of the binary digits of n. Conclude that for all n, $\binom{2n}{n}$ is even.

34. Find k such that $3^k \| 100!$.

35. How many zeroes does the decimal representation of 1000! end with?

36. Prove that $[2, 3, 4, \cdots, n] = \prod_{p \leq n} p^{[\log_p n]}$.

*37. Prove that if $n \geq 2$, then $\frac{1}{2} + \frac{1}{3} + \frac{1}{4} + \cdots + \frac{1}{n}$ is not an integer.

38. Let $p_1 = 2, p_2 = 3, p_3 = 5$, etc. Find all natural numbers m such that there exists $n \geq 2$ so that m! = $p_1^{2^{n-1}} p_2^{2^{n-2}} p_3^{2^{n-3}} \cdots p_{n-1}^2 p_n$.

SECTION 3.3 COMPUTER EXERCISES

39. Write a computer program that, given a large integer n, will attempt to at least partially factor n by using trial division to find all prime factors of n that are less than 100.

40. Write a computer program that will use the sieve of Eratosthenes to find all primes between A and B, where $A \leq B \leq 10,000$.

41. Write a computer program that, given an integer n such that $4 \leq n \leq 100$, will find all representations of $2n$ as a sum of two distinct odd primes.

42. It has been conjectured by Ronald Graham that $\binom{2n}{n}$ is divisible by each of the primes 3, 5, and 7 for infinitely many n. Write a computer program to find all n such that $1 \leq n \leq 100$ for which $3 \cdot 5 \cdot 7 \mid \binom{2n}{n}$. (Use the formula from Exercises 28 in order to avoid needless computation.)

43. The following problem is related to Goldbach's conjecture. If $n \geq 4$, let $g(n)$ be the number of pairs of noncomposite integers p and q such that $1 \leq p < q \leq 2n - 1$ and $p + q = 2n$. (a) Prove that $g(n) \leq \pi(2n) - \pi(n)$. (b) Prove that if $g(n) = \pi(2n) - \pi(n)$, then $3 \mid n$. (c) Find all $n \leq 200$ such that $g(n) = \pi(2n) - \pi(n)$.

CHAPTER FOUR

CONGRUENCES

4.1 INTRODUCTION

Many questions in number theory concern the nature of the remainder when one natural number is divided by another. An integer is *odd* if division by 2 leaves a remainder of 1; it is *even* if division by 2 leaves a remainder of 0. We saw by way of Theorem 3.11 that there exist infinitely many primes of the form $4k - 1$. Now $4k - 1 = 4(k - 1) + 3$, so these are the primes which when divided by 4 leave a remainder of 3. Toward the end of this chapter we will prove that there are also infinitely many primes of the form $4k + 1$.

If m is a natural number, and if a and b are integers that leave the same remainder when divided by m, then we say that a is *congruent* to b (modulo m), and we write

$$a \equiv b \ (\text{mod } m)$$

Such a statement is called a *congruence*. For example, $11 \equiv 3 \ (\text{mod } 2)$, since 11 and 3 both leave remainder 1 when divided by 2.

The theory of congruences was invented by the great German mathematician Carl Friedrich Gauss (1777–1855). We shall see that many questions involving divisibility become more tractable when phrased in terms of congruences.

4.2   DEFINITION AND PROPERTIES OF CONGRUENCE

Let a, b, c, and d denote integers, while m and n denote natural numbers.

Definition 4.1   *Congruence   (mod m)*

If $m \mid (a - b)$, then we say a is *congruent* to b (modulo m), and we write

$$a \equiv b \quad (\text{mod } m)$$

We call m the *modulus of the congruence*.

For example, $17 \equiv 1$ (mod 8), since $17 - 1 = 16 = 2(8)$; $21 \equiv 0$ (mod 7), since $21 - 0 = 21 = 3(7)$; and $13 \equiv -2$ (mod 5), since $13 - (-2) = 15 = 3(5)$. If $m \nmid (a - b)$, then we say a is *not congruent* to b (modulo m) and we write

$$a \not\equiv b \; (\text{mod } m)$$

For example, $17 \not\equiv 1$ (mod 6), since $17 - 1 = 16$ and $6 \nmid 16$; $21 \not\equiv 0$ (mod 8), since $21 - 0 = 21$ and $8 \nmid 21$; and $13 \not\equiv -2$ (mod 7), since $(13 - -2) = 15$ and $7 \nmid 15$.

Theorem 4.1   *(Properties of Congruence)*

C_1 *If $a \equiv b$ (mod m) and $c \equiv d$ (mod m),*
    then $a \pm c \equiv b \pm d$ (mod m).

C_2 *If $a \equiv b$ (mod m), then $ac \equiv bc$ (mod m).*

C_3 *If $a \equiv b$ (mod m) and $c \geq 1$, then $ac \equiv bc$ (mod mc)*

C_4 *If $a \equiv b$ (mod m) and $c \equiv d$ (mod m), then $ac \equiv bd$ (mod m)*

C_5 *If $a \equiv b$ (mod m), then $a^n \equiv b^n$ (mod m) for all $n \geq 1$.*

C_6 *$a \equiv 0$ (mod m) if and only if $m \mid a$.*

C_7 *If $ac \equiv bc$ (mod m) and $(c, m) = 1$, then $a \equiv b$ (mod m).*

C_8 *If $ac \equiv bc$ (mod m) and $(c, m) = n$, then $a \equiv b$ (mod m/n).*

C_9 *If $a \equiv b$ (mod m) and $n \mid m$, then $a \equiv b$ (mod n).*

We prove several of these properties, leaving the proofs of the remaining properties as exercises.

PROOF OF C$_2$:   By hypothesis and Definition 4.1, $m \mid (a - b)$. Now $(a - b) \mid c(a - b)$, so Theorem 2.1, part D$_4$ implies that $m \mid c(a - b)$; that is, $m \mid (ac - bc)$. Therefore, $ac \equiv bc \pmod{m}$.   ∎

PROOF OF C$_7$:   If $ac \equiv bc \pmod{m}$, then $m \mid (ac - bc)$; that is, $m \mid c(a - b)$. By hypothesis, $(c, m) = 1$. Therefore, Euclid's lemma (Theorem 2.5) implies that $m \mid (a - b)$, so $a \equiv b \pmod{m}$.   ∎

PROOF OF C$_8$:   As in the proof of C$_7$, $m \mid c(a - b)$. Since $(c, m) = n$ by hypothesis, we know that $n \mid c$ and $n \mid m$, so m/n and c/n are integers. Now $n(m/n) \mid n(c/n)(a - b)$. Theorem 2.1, part D$_5$ implies that $m/n \mid c/n(a - b)$, but Theorem 2.3, part G$_5$ implies that $(m/n, c/n) = 1$. Therefore, Euclid's lemma (Theorem 2.5) implies that $m/n \mid (a - b)$; that is, $a \equiv b \pmod{m/n}$.   ∎

Note that arithmetic operations on congruences are similar to, but not identical with, arithmetic operations on equations. Parts C$_1$ and C$_4$ of Theorem 4.1 say that congruences (mod m) may be added, subtracted, and multiplied. Part C$_8$, which generalized part C$_7$, tells what happens when a common factor is canceled from a congruence (mod m); namely, the resulting congruence may have a reduced modulus. For example,

$$110 \equiv 50 \pmod{15}$$

That is,

$$10 \cdot 11 \equiv 10 \cdot 5 \pmod{15}.$$

Now $(10, 15) = 5$. Therefore, if we cancel 10 from both sides of the last congruence, we get

$$11 \equiv 5 \pmod{15/5}$$

That is,

$$11 \equiv 5 \pmod{3}$$

Congruence (mod m) is a special case of a more general mathematical concept known as an *equivalence relation*. We therefore digress from our discussion of congruences to study equivalence relations and the related concept of equivalence classes.

4.3 EQUIVALENCE RELATIONS

Suppose that S is a nonempty set and R is a set of ordered pairs of elements from S. If the ordered pair (a, b) belongs to R, then we write aRb, and we say a is *related to b* via R.

Definition 4.2   *Equivalence Relation*
If R and S are as above, we say that R is an *equivalence relation* on S if all three of the following properties hold:

(1) (Reflexive Property) For all a in S, aRa.
(2) (Symmetrical property) For all a and b in S, if aRb, then bRa.
(3) (Transitive property) For all a, b, and c in S, if aRb and bRc, then aRc.

Example 1: S is the set of all plane triangles. R is the relation of similarity. (The two triangles are said to be similar if all three pairs of corresponding sides are in the same proportion.) In less precise language, two triangles are said to be similar if they have the same shape but not necessarily the same size.

Example 2: S' is the set of all English words. R' is the relation of beginning with the same letter.

Example 3: S'' is the set of all polynomials in a single variable of degree at least 1 with integer coefficients. R'' is the relation of having the same degree.

We now give some examples of relations that are not equivalence relations. Let a relation R be defined on N by aRb if $a = b^n$ for some $n \geq 1$. Then R is reflexive and transitive but not symmetrical. Therefore, R is not an equivalence relation on N.

As a second example, let S be the set of all plane triangles. Define a relation R on S by $t_1 R t_2$ if t_1 and t_2 are both right triangles. Then R is symmetrical and transitive but not reflexive; hence R not an equivalence relation.

Associated with any equivalence relation on a set are certain subsets known as *equivalence classes*, which are defined below.

Definition 4.3   *Equivalence Class*

If R is an equivalence relation on set S, and if a belongs to S, we define the *equivalence class* of a under R, denoted $[a]_R$, as the set of all b in S such that aRb.

Referring to Example 1 above, if t is an equilateral triangle, then $[t]_R$ is the set of all equilateral triangles. Referring to Example 2, [mathematics]$_{R'}$ is the set of all English words that begin with m. Referring to Example 3, $[3x^2 + 4x - 5]_{R''}$ is the set of all polynomials of degree 2 with integer coefficients.

An equivalence class may have many different names, as the following theorem shows.

Theorem 4.2   *Let a and b belong to set S. Let R be an equivalence relation on S. Then aRb if and only if $[a]_R = [b]_R$.*

PROOF: First we show that if aRb, then $[a]_R = [b]_R$. Suppose that c belongs to $[a]_R$. Then aRc. Since R is symmetrical, it follows that cRa. Since aRb by hypothesis and R is transitive, it follows that cRb, so c belongs to $[b]_R$. Therefore, $[a]_R$ is a subset of $[b]_R$. Similarly, $[b]_R$ is a subset of $[a]_R$, so $[a]_R = [b]_R$. Conversely, suppose that $[a]_R = [b]_R$. Since R is reflexive, bRb, so b belongs to $[b]_R$. Therefore, b belongs to $[a]_R$, so aRb. ∎

For example, referring to Example 2, we have

$$[\text{mathematics}]_{R'} = [\text{milk}]_{R'} = [\text{monarch}]_{R'} = \text{etc.}$$

Definition 4.4   *Partition*

A *partition* of a nonempty set is a collection of subsets that are mutually exclusive and collectively exhaustive. That is to say, (i) the subsets are pairwise disjoint (no two distinct subsets share a common element), and (ii) the union of all these subsets is the original set. In other words, a partition of set S is a collection of one or more subsets of S such that each element of S belongs to precisely one subset.

For example, if $S = \{1, 2, 3, 4, 5, 6, 7, 8, 9, 10\}$, one partition of S consists of the subsets $S_1 = \{1, 4, 7, 10\}$, $S_2 = \{2, 5, 8\}$, and $S_3 = \{3, 6, 9\}$.

One reason that equivalence relations are important is that if R is an equivalence class on set S, then the set of all equivalence classes under R forms a partition of S. [In the example above, the subsets that form the partition are equivalence classes under the equivalence relation aRb if $3 \mid (a - b)$.]

SECTION 4.3 EXERCISES

1. Prove each of the following parts of Theorem 4.1.

    (a) C_1 (b) C_3

    (c) C_4 (d) C_5

    (e) C_6 (f) C_9

2. Reduce each of the following congruences as much as possible by cancellation.

    (a) $650 \equiv 350 \pmod{75}$

    (b) $320 \equiv 32 \pmod{12}$

    (c) $215 \equiv 5 \pmod{21}$

3. Determine which of the following relations on N, if any, is an equivalence relation. Justify your conclusion.

    (a) aRb if $a \mid b$ (b) aRb if $\frac{1}{2}(a + b)$ is an integer

    (c) aRb if $a + b$ is prime   (d) aRb is $ab = k^2$ for some integer k

4. Given that parallelism is an equivalence relation on the set of lines in the xy plane, what is the equivalence class of (a) the x axis? (b) the y axis? (Your answer should be a geometric description.)

5. Under the equivalence relation of Example 2 above, how many distinct equivalence classes are there?

6. In the xy plane, say that two points are equivalent if they are equidistant from the origin. What is the equivalence class of the point $(3, 4)$? (Describe it geometrically.)

7. Prove that if R is an equivalence relation on set S, then (a) no equivalence class is empty, (b) any two distinct equivalence classes are disjoint, and (c) the union of all the equivalence classes under R is S.

Returning to the subject of congruences, we see by way of the following theorem that for every natural number m, congruence (mod m) is an equivalence relation on Z.

Theorem 4.3   *Let m be a natural number. Then congruence (mod m) is an equivalence relation on Z.*

PROOF: Since $m \mid 0$, we have $m \mid (a - a)$ for all a in Z, so $a \equiv a \pmod{m}$, for all a in Z. Therefore, congruence (mod m) is reflexive. If $a \equiv b \pmod{m}$, then $m \mid (a - b)$, so $a - b = km$ for some k. This implies that $b - a = (-k)m$, so $m \mid (b - a)$; hence $b \equiv a \pmod{m}$. Therefore, congruence (mod m) is symmetrical. For all a, b, and c in Z, if $a \equiv b$ (mod m) and $b \equiv c$ (mod m), then $m \mid (a - b)$ and $m \mid (b - c)$. Therefore, $m \mid [(a - b) + (b - c)]$; that is, $m \mid (a - c)$. Therefore, $a \equiv c$ (mod m), so congruence (mod m) is transitive. Since congruence (mod m) is reflexive, symmetrical, and transitive, it is an equivalence relation on Z.   ■

Let us look at the equivalence classes under congruence (mod m). There are m such classes, namely, $[0]$, $[1]$, $[2]$, $\cdots$, $[m - 1]$. Now $[0] = \{0, \pm m, \pm 2m, \pm 3m, \cdots\}$. In other words, $[0] = \{km : k$ is in $Z\}$. More generally, if $0 \leq r \leq m - 1$, then $[r] = \{km + r : k$ is in $Z\}$. That is to say, $[r]$ consists of those integers which are r more than a multiple of m. Another way to put this would be to say that $[r]$ consists of those integers which leave a remainder r when divided by m.

For example, when $m = 2$, the two resulting equivalence classes should be familiar to you:

$$[0] = \{0, \pm 2, \pm 4, \pm 6, \cdots\} = \text{the set of all even integers}$$
$$[1] = \{\pm 1, \pm 3, \pm 5, \pm 7, \cdots\} = \text{the set of all odd integers}$$

Again, when $m = 3$, the three resulting equivalence classes are

$$[0] = \{\cdots , -6, -3, 0, 3, 6, \cdots\}$$
$$[1] = \{\cdots , -5, -2, 1, 4, 7, \cdots\}$$
$$[2] = \{\cdots , -4, -1, 2, 5, 8, \cdots\}$$

4.4   LINEAR CONGRUENCES

For all natural numbers m and b, we know by the division algorithm [Theorem 2.2] that there exist q and r such that $b = qm + r$ and $0 \leq r \leq m - 1$. This fact also holds for arbitrary integers b. We now introduce the concept of least positive residues.

Definition 4.5   ***Least Positive Residue* (mod m)**
If m is a natural number and b is an integer, let $b = qm + r$, where $0 \leq r \leq m - 1$. We say that r is the *least positive residue* of b (mod m).

For example, the least positive residue of 35   (mod 8) is 3, since $35 = 4(8) + 3$; the least positive residue of 35   (mod 7) is 0, since

$35 = 5(7) + 0$; and the least positive residue of -13 (mod 10) is 7, since $-13 = (-2)10 + 7$.

The set of least positive residues (mod m) is the set $\{0, 1, 2, \cdots, m - 1\}$. For example, the set of least positive residues (mod 4) is $\{0, 1, 2, 3\}$, and the set of least positive residues (mod 7) is $\{0, 1, 2, 3, 4, 5, 6\}$.

Note that $a \equiv b$ (mod m) if and only if a and b have the same least positive residue (mod m).

Congruences may be used to prove results regarding divisibility without the necessity of calculating large numbers. As a first example, we prove that $11 \mid (233^5 + 1)$. Since $233 > 11$, we begin by finding the least positive residue of 233 (mod 11). Since

$$233 = 21(11) + 2$$

we have

$$233 \equiv 2 \quad (\text{mod } 11)$$

Therefore, $233^5 \equiv 2^5 \equiv 32 \equiv 10$ (mod 11), so $233^5 + 1 \equiv 10 + 1 \equiv 11 \equiv 0$ (mod 11); that is, $11 \mid (233^5 + 1)$.

Note that whenever a number greater than or equal to the modulus is generated, it should be replaced by its least positive residue. (This rule does not apply to exponents, however.)

As a second example, we prove that $31 \mid (15^{10} - 1)$:

$$
\begin{aligned}
15^2 &= 225 \equiv 8 \quad (\text{mod } 31) \\
15^4 &\equiv (15^2)^2 \equiv 8^2 \equiv 64 \equiv 2 \quad (\text{mod } 31) \\
15^5 &\equiv 15(15^4) \equiv 15(2) \equiv 30 \quad (\text{mod } 31) \\
15^{10} &\equiv (15^5)^2 \equiv 30^2 \equiv 900 \equiv 1 \quad (\text{mod } 31)
\end{aligned}
$$

We have $15^{10} \equiv 1$ (mod 31), so $31 \mid (15^{10} - 1)$.

We shall often need to compute the least positive residue of a^m (mod n), where the integers a, m, and n are all greater than 1. One could obtain the desired result by successively computing the least positive residues (mod n) of a^2, a^3, a^4, $\cdots$, a^m, but this would require $m - 1$ multiplications. A more efficient procedure essentially makes use of the binary representation of the exponent m. Note that any even power of a given base can be computing by squaring, since $a^{2k} = (a^k)^2$. Also, any odd power of a given base can be computed by multiplying the base by a square, since $a^{2k+1} = a(a^{2k})$.

Suppose, for example, that we wish to compute the least positive residue of 7^{19} (mod 31). Now

$$19 = 2(9) + 1 = 18 + 1$$
$$18 = 2(9)$$
$$9 = 2(4) + 1 = 8 + 1$$
$$8 = 2(4)$$
$$4 = 2(2)$$
$$2 = 2(1)$$

Therefore, we need compute only $7^2, 7^4, 7^8, 7^9, 7^{18}$, and 7^{19} (mod 31):

$$7^2 = (7^1)^2 = 49 \equiv 18 \quad (\text{mod } 31)$$
$$7^4 = (7^2)^2 \equiv 18^2 \equiv 324 \equiv 14 \quad (\text{mod } 31)$$
$$7^8 = (7^4)^2 \equiv 14^2 \equiv 196 \equiv 10 \quad (\text{mod } 31)$$
$$7^9 = 7(7^8) \equiv 7(10) \equiv 70 \equiv 8 \quad (\text{mod } 31)$$
$$7^{18} = (7^9)^2 \equiv 8^2 \equiv 64 \equiv 2 \quad (\text{mod } 31)$$
$$7^{19} = 7(7^{18}) \equiv 7(2) \equiv 14 \quad (\text{mod } 31)$$

In general, the number of squaring operations needed to compute a^m in this manner is $[\log_2 m]$. For example, in order to compute the least positive residue of 7^{19} (mod 31), the number of squarings was $[(\log_2 19)] = 4$.

In elementary algebra, one learns how to solve linear equations, that is, equations of the type $ax = b$, where $a \neq 0$ and x is unknown. The corresponding problem in number theory involves linear congruences, which are defined below.

Definition 4.6   ***Linear Congruence***
If a, b, and m are integers such that $a \not\equiv 0$ (mod m) and x is unknown, then the congruence

$$ax \equiv b \quad (\text{mod } m)$$

is called a linear congruence (mod m).

Before we discuss how to solve a linear congruence, we must discuss what is meant by a solution. Suppose, for example, that we wish to solve the linear congruence

4.1 $3x \equiv 1 \quad (\text{mod } 5)$

If we consider a solution to be a number, then we are faced with a clumsy situation, since there are infinitely many numbers that are solutions, namely, $x = 2, 7, 12, 17$, etc., as well as $x = -3, -8, -13, -18$, etc.

These solutions all have the form $x = 5k + 2$, where k is an integer. Therefore, it makes sense to consider the solution of congruence 4.1 to be the congruence

4.2 $x \equiv 2 \pmod 5$

We also could have said that the solution of congruence 4.1 is the congruence

4.3 $x \equiv 7 \pmod 5$

but it is more convenient to use least positive residues. In general, the solution of a linear congruence

4.4 $ax \equiv b \pmod m$

if it exists, will have the form

4.5 $x \equiv t \pmod m$

where t is the least positive residue $\pmod m$; that is, $0 \le t \le m - 1$. Any solution not in this form can, and usually should, be reduced to this form. For example, the congruence

$$x \equiv 47 \pmod{20}$$

should be reduced to

$$x \equiv 7 \pmod{20}$$

Let us look at some additional examples. The congruence

4.6 $3x \equiv 7 \pmod{13}$

has the solution

4.7 $x \equiv 11 \pmod{13}$

since $3(11) = 33 = 2(13) + 7$. The congruence

4.8 $5x \equiv 2 \pmod{37}$

has the solution

4.9 $x \equiv 30 \pmod{37}$

since $5(30) = 150 = 4(37) + 2$. The congruence

4.10 $3x \equiv 2 \pmod 6$

has no solution.

We are about to begin the study of linear congruences, that is, polynomial congruences of degree 1. In section 4.6 we will see how to obtain all solutions, if any, of the congruence

4.11 $$f(x) \equiv 0 \quad (\mathrm{mod}\ p^n)$$

where $f(x)$ is a polynomial of degree at least 2, p is prime, and $n \geq 2$, given the solutions of the congruence

4.12 $$f(x) \equiv 0 \quad (\mathrm{mod}\ p)$$

In Chapter 7 we will study quadratic congruences, that is, polynomial congruences of degree 2.

Before we commence, some further comments are in order. Let $f(x)$ be a polynomial of degree at least 1 with integer coefficients. Any solution to the congruence

4.13 $$f(x) \equiv 0 \quad (\mathrm{mod}\ m)$$

will have the form

4.5 $$x \equiv t \quad (\mathrm{mod}\ m)$$

Therefore, it is always theoretically possible to obtain all solutions of congruence 4.11 by trial and error, that is, by the successive evaluation of $f(0), f(1), f(2), \cdots, f(m-1) \quad (\mathrm{mod}\ m)$. For example, suppose we wish to solve the congruence

4.14 $$x^3 + x + 2 \equiv 0 \quad (\mathrm{mod}\ 3)$$

Here $f(x) = x^3 + x + 2$. Now $f(0) = 2 \equiv 2 \quad (\mathrm{mod}\ 3)$, $f(1) = 4 \equiv 1 \quad (\mathrm{mod}\ 3)$, and $f(2) = 12 \equiv 0 \quad (\mathrm{mod}\ 3)$. Therefore, the unique solution of (14) is

4.15 $$x \equiv 2 \quad (\mathrm{mod}\ 3)$$

The trial-and-error method for solving polynomial congruences works well for small m but is impractical for large m.

We shall often have the need to solve a linear congruence. The two following theorems tells us when a linear congruence admits solutions and, if so, how many.

Theorem 4.4   *The congruence $ax \equiv b$ (mod m) has at least one solution if and only if $(a, m) \mid b$.*

PROOF: Let $d = (a, m)$. If $ax \equiv b$ (mod m), then $ax - b = km$ for some k, so $ax - km = b$. Since $d \mid a$ and $d \mid m$, we must have $d \mid b$. Conversely, suppose that $d \mid b$; that is, $b = jd$ for some j. We know by Theorem 2.4 that there exist u and v such that $d = au + mv$, so $b = j(au + mv) = a(ju) + m(jv)$. Letting $x = ju$, we have $ax + m(jv) = b$, so $ax + m(jv) \equiv b$ (mod m). This reduces to $ax \equiv b$ (mod m). ∎

For example, we stated earlier that the congruence

4.10 $3x \equiv 2$ (mod 6)

has no solution. This is true because $(3, 6) = 3$, and $3 \nmid 2$.

Theorem 4.5   *Let $d = (a, m)$. If $d \mid b$, then the congruence*

4.4 $ax \equiv b$ (mod m)

has exactly d solutions (mod m).

PROOF: By Theorem 4.4, we know that congruence 4.4 has no solution unless $d \mid b$. Suppose that $x \equiv y$ (mod m) and $x \equiv z$ (mod m) are solutions of congruence 4.4, where y and z are least positive residues (mod m). Then $ay \equiv az \equiv b$ (mod m). Now Theorem 4.1, part C$_8$, implies that $y \equiv z$ (mod m/d); that is, $y = z + k(m/d)$ for some k. By the division algorithm (Theorem 2.2), $k = qd + r$, with $0 \leq r \leq d - 1$, so $y = z + (qd + r)(m/q) = z + qm + r(m/d)$. Therefore,

4.16 $y \equiv z + r(m/d)$ (mod m)

Letting r assume all its possible values, we obtain at most d solutions of congruence 4.4 that are distinct, that is, incongruent (mod m). We now show that these d solutions are all distinct (mod m). If $z + i(m/d) \equiv z + j(m/d)$ (mod m), then $i(m/d) \equiv j(m/d)$ (mod m), so that $m \mid [i(m/d) - j(m/d)]$; that is, $m \mid (i - j)(m/d)$. This implies that $d \mid (i - j)$. However, this is impossible if $0 \leq i < j \leq d - 1$. ∎

For example, the congruence

4.17 $8x \equiv 12$ (mod 20)

has four solutions (mod 20), since $(8, 20) = 4$ and $4 \mid 12$. The congruence

4.18 $9x \equiv 15$ (mod 30)

has three solutions (mod 30), since $(9, 30) = 3$ and $3 \mid 15$. The congruence

4.19 $9x \equiv 15$ (mod 27)

has no solution, since $(9, 27) = 9$ but $9 \nmid 15$.

Suppose that $(a, m) = 1$, so that the congruence

4.4 $ax \equiv b$ (mod m)

has a unique solution. How do we actually find it without resorting to trial and error? The answer is found in the proof of Theorem 4.4. Using the Euclidean algorithm to compute (a, m), we find integers u and v such that $au + mv = 1$. This yields the congruence

4.20 $au \equiv 1$ (mod m)

Then multiplication by b produces the congruence

4.21 $a(ub) \equiv b$ (mod m)

so our solution of congruence 4.4 is

4.22 $x \equiv bu$ (mod m)

For example, suppose that we wish to solve the congruence

4.23 $7x \equiv 2$ (mod 37)

First, we compute $(7, 37)$. We have

$$37 = 5(7) + 2$$
$$7 = 3(2) + 1$$

Therefore, $(7, 37) = 1$, and congruence 4.23 has a unique solution (mod 37). Furthermore,

$$1 = 7 - 3(2)$$

and

$$2 = 37 - 5(7)$$

so $1 = 7 - 3[37 - 5(7)] = 7 - 3(37) + 15(7) = 7(16) + 37(-3)$. Here $u = 16$ and $b = 2$, so our solution of congruence 4.23 is given by

(24) $x \equiv 16(2) \equiv 32$ (mod 37)

CHECK: $7(32) = 224 = 6(37) + 2$, so $7(32) \equiv 2$ (mod 37).

As a second example, suppose that we wish to solve the congruence

4.25 $3x \equiv 4 \pmod{25}$

First, we compute $(3, 25)$. We have

$$25 = 8(3) + 1$$

Therefore, $(3, 25) = 1$, and congruence 4.25 has a unique solution (mod 25). Furthermore, $1 = 25 - 8(3) = 3(-8) + 25(1)$. Here $u = -8$ and $b = 4$, so our solution of congruence 4.25 is given by

4.26 $x \equiv 4(-8) \equiv -32 \pmod{25}$

Reducing to a least positive residue (mod 25), we obtain

4.27 $x \equiv 18 \pmod{25}$

CHECK: $3(18) = 54 = 2(25) + 4$, so $3(18) \equiv 4 \pmod{25}$.

Now suppose that $(a, m) = d, d \mid b$, and $d > 1$. We wish to find all d solutions of the congruence 4.4. By Theorem 4.1, part C_8 we know that

4.28 $(a/d)x \equiv b/d \pmod{m/d}$

Furthermore, by Theorem 2.3, part G_5 and Theorem 4.4, congruence 4.27 has a unique solution (mod m/d), namely

4.29 $x \equiv t \pmod{m/d}$

Let $x_k \equiv t + (k - 1)(m/d) \pmod{m}$, where $k = 1, 2, 3, \cdots, d$. Then each x_k is a solution of congruence 4.4, and all the x_k are distinct (mod m).
For example, let us solve the congruence

4.30 $8x \equiv 6 \pmod{10}$

Since $(8, 10) = 2$ and $2 \mid 6$, we expect two solutions (mod 10). Now congruence 4.30 reduces to the congruence

4.31 $4x \equiv 3 \pmod{5}$

whose unique solution is

4.32 $x \equiv 2 \pmod{5}$

Therefore, our desired solutions (mod 10) are

4.33 $x_1 \equiv 2 + 0(5) \equiv 2 \pmod{10}$

4.34 $x_2 \equiv 2 + 1(5) \equiv 7 \pmod{10}$

As a second example, let us solve the congruence

4.35 $9x \equiv 12 \pmod{21}$

Since $(9, 21) = 3$ and $3 \mid 12$, we expect three solutions (mod 21). Now congruence 4.35 reduces to the congruence

4.36 $3x \equiv 4 \pmod 7$

whose unique solution is

4.37 $x \equiv 6 \pmod 7$

Therefore, our desired solutions of congruence 4.35 are

4.38 $x_1 \equiv 6 + 0(7) \equiv 6 \pmod{21}$

4.39 $x_2 \equiv 6 + 1(7) \equiv 13 \pmod{21}$

4.40 $x_3 \equiv 6 + 2(7) \equiv 20 \pmod{21}$

To summarize, given a linear congruence

4.4 $ax \equiv b \pmod m$

where $(a, m) = d, d \mid b$, and $d > 1$, we first reduce congruence 4.4 to the congruence

4.28 $(a/d)x \equiv b/d \pmod{m/d}$

Then we obtain the unique solution of congruence 4.28, and finally, we use the solution of congruence 4.28 to obtain all d solutions of congruence 4.4.

SECTION 4.4 EXERCISES

8. Use congruences to verify each of the following divisibility statements.

   (a) $13 \mid (145^6 + 1)$ (b) $17 \mid (1855^4 + 1)$

   (c) $233 \mid (2^{29} - 1)$ (d) $223 \mid (2^{37} - 1)$

   (e) $431 \mid (2^{43} - 1)$ (f) $167 \mid (2^{83} - 1)$

9. Use trial and error to find all solutions, if any, of each of the following congruences.

   (a) $x^3 + x + 1 \equiv 0 \pmod 3$

   (b) $x^3 - 2x + 3 \equiv 0 \pmod 5$

   (c) $x^2 - 2 \equiv 0 \pmod 7$

   (d) $x^4 + 2x^3 + x - 2 \equiv 0 \pmod 7$

10. Find all solutions, if any, of each of the following linear congruences.

    (a) $3x \equiv 1 \pmod{23}$

    (b) $7x \equiv 1 \pmod{47}$

    (c) $5x \equiv 2 \pmod{210}$

    (d) $6x \equiv 7 \pmod{25}$

    (e) $8x \equiv 5 \pmod{31}$

    (f) $8x \equiv 12 \pmod{32}$

    (g) $12x \equiv 6 \pmod{21}$

    (h) $2x \equiv 2 \pmod{16}$

    (i) $20x \equiv 8 \pmod{24}$

11. If p is prime, prove that
    $(a + b)^p \equiv a^p + b^p \pmod{p}$.

12. If x is odd and $n \geq 1$, prove that
    $x^{2^n} \equiv 1 \pmod{2^{n+2}}$.

13. If m and n are natural numbers, prove that
    $m^{2n} \equiv 1 \pmod{m + 1}$.

14. If p is an odd prime and $2 \leq k \leq p - 1$,
    prove that $\binom{p+1}{k} \equiv 0 \pmod{p}$.

15. If p is an odd prime and $1 \leq k \leq p - 2$,
    prove that $\binom{p-1}{k} \equiv (-1)^k \pmod{p}$.

16. If $\prod_{i=1}^{r} a_i \equiv 2 \pmod{4}$, prove that (a) there
    exists j such that $1 \leq j \leq r$ and $a_j \equiv 2 \pmod{4}$ and (b) a_i is odd for all $i \neq j$.

SECTION 4.4 COMPUTER EXERCISES

17. Write a computer program that will perform modular exponentiation, that is, which given natural numbers a, m, and n will compute the least positive residue of $a^m \pmod{n}$.

18. Let p be an odd prime. Let $0 < a < p$. Let $f(x) = x^2 - a$. Write a computer program to find all solutions, if any, of the congruence $f(x) \equiv 0 \pmod{p}$ by trial and error, that is, by evaluating $f(1), f(2), f(3), \cdots , f(p - 1)$.

19. Using the program from the preceding exercise, for each prime p from 31 to 47, find the number of integers a such that $0 < a < p$ and the congruence $f(x) \equiv 0 \pmod{p}$ has solutions. Is there a pattern?

4.5   LINEAR DIOPHANTINE EQUATIONS AND THE CHINESE REMAINDER THEOREM

An equation in two or more unknown *integers* is called a *Diophantine equation*, in honor of Diophantus of Alexandria, who lived about 250 A.D. We shall now apply linear congruences to the solution of linear Diophantine equations in two unknowns.

Consider the equation

4.41 $ax + by = c$

where a, b, and c are given integers and x and y are unknown. Let $d = (a, b)$. Since $d \mid a$ and $d \mid b$, if equation 4.41 has a solution, then $d \mid c$. If

$d > 1$, then equation 4.41 may be simplified by dividing out d. Therefore, we assume henceforth that $d = 1$. We convert equation 4.41 to the congruence

4.42 $ax \equiv c \pmod{b}$

Since $d = 1$, $d \mid c$, so congruence 4.42 has a unique solution (mod b), namely

4.43 $x \equiv x_0 \pmod{b}$

with $0 \leq x_0 < b$. Converting congruence 4.42 to an equation, we get

4.44 $x = x_0 + bt$

where t is an arbitrary integer. Let $y_0 = (c - ax_0)/b$. Then $y = (c - ax)/b = [c - a(x_0 + bt)]/b = (c - ax_0)/b - at = y_0 - at$. Therefore, the complete solution of the equation 4.41 is given (in parametric form) by the equations:

4.45 $x = x_0 + bt$

4.46 $y = y_0 - at$

where the parameter t is an arbitrary integer.

For example, let us solve the Diophantine equation

4.47 $6x + 11y = 41$

We convert equation 4.47 to a congruence (mod 11), namely

4.48 $6x \equiv 41 \pmod{11}$

which reduces first to

4.44 $6x \equiv 8 \pmod{11}$

and then further to

4.50 $3x \equiv 4 \pmod{11}$

The unique solution of congruence 4.50 is given by

4.51 $x \equiv 5 \pmod{11}$

which we convert to the equation

4.52 $x = 5 + 11t$

Note that $x_0 = 5$. Now $y_0 = [41 - 6(5)]/11$, so we have

4.53 $y = 1 - 6t$

The complete solution of equation 4.47 is given by equations 4.52 and 4.53, where t denotes an arbitrary integer.

Next, we consider a "story" problem that leads to a linear Diophantine equation in two unknowns. Old McDonald (that farmer of reknown) sold chickens at \$5 each and geese at \$8 each. He collected a total of \$99. Assuming that he sold at least one bird of each kind, how many of each kind did he sell?

To solve this problem, let x be the number of chickens sold, and let y be the number of geese sold. Since the total amount of money collected is the sum of the value of the sold chickens and the value of the sold geese, we obtain the equation

4.54 $$5x + 8y = 99$$

where we further specify that $x \geq 1$ and $y \geq 1$. To solve equation 4.54, we first reduce it to a congruence (mod 8), namely

4.55 $$5x \equiv 99 \quad (\text{mod } 8)$$

which then reduces to

4.56 $$5x \equiv 3 \quad (\text{mod } 8)$$

The unique solution of congruence 4.56 is

4.57 $$x \equiv 7 \quad (\text{mod } 8)$$

so $x_0 = 7$, and we convert congruence 4.57 to the equation

4.58 $$x = 7 + 8t$$

Now $y_0 = [99 - 5(7)]/8 = 8$, so we obtain the equation

4.59 $$y = 8 - 5t$$

Since $x \geq 1$, equation 4.58 implies that $t \geq 0$; since $y \geq 1$, equation 4.59 implies that $t \leq 1$. Therefore, $t = 0$ or 1, so our problem has two solutions:

4.60 $$x = 7, \qquad y = 8$$

4.61 $$x = 15, \qquad y = 3$$

Suppose that we wish to solve a system of simultaneous linear congruences in one unknown, such as

4.62
$$x \equiv 3 \quad (\text{mod } 5)$$
$$x \equiv 2 \quad (\text{mod } 6)$$
$$x \equiv 4 \quad (\text{mod } 7)$$

Problems of this type were considered by the Chinese mathematician Sun Tzu in the first century A.D. The following theorem, known as the *Chinese remainder theorem*, guarantees the existence of a solution and tells us how to find it, provided that the moduli of the various given linear congruences are pairwise relatively prime.

Theorem 4.6   *(Chinese Remainder Theorem)*
Let m_1, m_2, m_3, $\cdots$, m_r (where $r \geq 2$) be natural numbers that are pairwise relatively prime and whose product is M. Then the system of r simultaneous linear congruences

4.63
$$
\begin{aligned}
x &\equiv a_1 \pmod{m_1} \\
x &\equiv a_2 \pmod{m_2} \\
x &\equiv a_3 \pmod{m_3} \\
&\vdots \\
x &\equiv a_r \pmod{m_r}
\end{aligned}
$$

has a unique solution (mod M).

PROOF: For each index i such that $1 \leq i \leq r$, let $M_i = M/m_i$. By hypothesis, $(m_i, m_j) = 1$ whenever $i \neq j$. Therefore, the result of Exercise 30 in Chapter 2 implies that $(M_i, m_i) = 1$ for all i. It follows from Theorem 4.5 that there exists x_i such that

4.64
$$M_i x_i \equiv 1 \pmod{m_i}$$

Now let

4.65
$$x = \sum_{i=1}^{r} M_i x_i a_i$$

Since $M_j \equiv 0 \pmod{m_i}$ for all $j \neq i$, this implies that

4.66
$$x \equiv M_i x_i a_i \equiv a_i \pmod{m_i}$$

for all i; that is, x is a solution of system 4.63.

If z is any other solution of system 4.63, then $z \equiv a_i \pmod{m_i}$ for all i, so $z - x \equiv 0 \pmod{m_i}$; that is, $m_i \mid (z - x)$ for all i. Therefore, $[m_1, m_2, m_3, \cdots, m_r] \mid (z - x)$. Since the m_i are pairwise relatively prime by hypothesis, it follows from the result of Exercise 43 in Chapter 2 that $[m_1, m_2, m_3, \cdots, m_r] = M$. Therefore, $M \mid (z - x)$; that is, $z \equiv x \pmod{M}$.   ∎

For example, let us solve the system of linear congruences

4.62
$$x \equiv 3 \quad (\text{mod } 5)$$
$$x \equiv 2 \quad (\text{mod } 6)$$
$$x \equiv 4 \quad (\text{mod } 7)$$

We are given $m_1 = 5$, $m_2 = 6$, and $m_3 = 7$, so $M = m_1 m_2 m_3 = 5 \cdot 6 \cdot 7 = 210$. Also, $M_1 = M/m_1 = 210/5 = 42$, $M_2 = M/m_2 = 210/6 = 35$, and $M_3 = M/m_3 = 210/7 = 30$. Now $M_1 x_1 \equiv 1 \quad (\text{mod } m_1)$ says that $42x_1 \equiv 1 \quad (\text{mod } 5)$ or $2x_1 \equiv 1 \quad (\text{mod } 5)$, so $x_1 \equiv 3 \quad (\text{mod } 5)$; $M_2 x_2 \equiv 1 \quad (\text{mod } m_2)$ says that $35x_2 \equiv 1 \quad (\text{mod } 6)$ or $5x_2 \equiv 1 \quad (\text{mod } 6)$, so $x_2 \equiv 5 \quad (\text{mod } 6)$; and $M_3 x_3 \equiv 1 \quad (\text{mod } m_3)$ says that $30x_3 \equiv 1 \quad (\text{mod } 7)$ or $2x_3 \equiv 1 \quad (\text{mod } 7)$, so $x_3 \equiv 4 \quad (\text{mod } 7)$. Finally, $x \equiv M_1 x_1 a_1 + M_2 x_2 a_2 + M_3 x_3 a_3 \quad (\text{mod } M)$, that is,

$$x \equiv (42)(3)(3) + (35)(5)(2) + (30)(4)(4)$$
$$\equiv 378 + 350 + 480 \equiv 1208 \quad (\text{mod } 210)$$

Reducing, we obtain

4.67
$$x \equiv 158 \quad (\text{mod } 210)$$

CHECK: $158 = 31(5) + 3 = 26(6) + 2 = 22(7) + 4$

SECTION 4.5 EXERCISES

20. Use congruences to find all positive integer solutions, if any, of each of the following equations.

(a) $2x + 5y = 47$ (b) $7x + 4y = 59$

(c) $8x + 5y = 46$ (d) $4x + 9y = 50$

(e) $13x + 10y = 499$ (f) $12x + 17y = 299$

21. Solve each of the following systems of simultaneous congruences.

(a) $x \equiv 1 \quad (\text{mod } 4)$ (b) $x \equiv 2 \quad (\text{mod } 5)$
     $x \equiv 2 \quad (\text{mod } 7)$ $x \equiv 5 \quad (\text{mod } 8)$

(c) $x \equiv 2 \quad (\text{mod } 3)$ (d) $x \equiv 2 \quad (\text{mod } 5)$
     $x \equiv 3 \quad (\text{mod } 4)$ $x \equiv 3 \quad (\text{mod } 7)$
                                               $x \equiv 1 \quad (\text{mod } 8)$

(e) $x \equiv 3 \quad (\text{mod } 4)$ (f) $x \equiv 1 \quad (\text{mod } 5)$
     $x \equiv 5 \quad (\text{mod } 7)$ $x \equiv 3 \quad (\text{mod } 6)$
     $x \equiv 2 \quad (\text{mod } 9)$ $x \equiv 2 \quad (\text{mod } 7)$

(g) $x \equiv 2 \quad (\text{mod } 4)$ (h) $x \equiv 4 \quad (\text{mod } 7)$
     $x \equiv 4 \quad (\text{mod } 5)$ $x \equiv 6 \quad (\text{mod } 11)$
     $x \equiv 3 \quad (\text{mod } 7)$ $x \equiv 9 \quad (\text{mod } 13)$

(i) $x \equiv 1 \quad (\text{mod } 9)$
     $x \equiv 2 \quad (\text{mod } 10)$
     $x \equiv 3 \quad (\text{mod } 11)$

22. A sack of gold coins is stolen by a gang of nine thieves. If each thief gets an equal share of the coins, then two coins remain. If one thief is caught before the coins are divided so that each of the others gets an equal share, then one coin remains. If two thieves are caught before the coins are divided so that each of the seven others gets an equal share, then no coins remain. Find the smallest possible number of coins in the sack.

23. Old McDonald collected $75 at the market by selling chickens and geese. He got $4 for each chicken and $7 for each goose. How many of each did he sell?

24. Prove that if $m_1, m_2, m_3, \cdots, m_r$ are odd natural numbers that are pairwise relatively prime and whose product is M, and if $x \equiv \frac{1}{2}(m_i - 1) \pmod{m_i}$ for each i, then $x \equiv \frac{1}{2}(M - 1) \pmod{M}$.

*25. Find an odd integer k such that $k(2^n) - 1$ is composite for all $n \geq 1$.

26. An alternate approach to solving the system 4.63 is as follows: Using the notation of Theorem 4.6, for each i such that $1 \leq i \leq r$, multiply the congruence $x \equiv a_i \pmod{m_i}$ by M_i to obtain $M_i x \equiv M_i a_i \pmod{M}$. Now add these r congruences to obtain a single congruence:

$$\left(\sum_{i=1}^{r} M_i \right) x \equiv \sum_{i=1}^{r} M_i a_i \pmod{M}$$

It is easily verified that $(\sum_{i=1}^{r} M_i, M) = 1$, so that the last congruence has a unique solution (mod M), which is the solution of system 4.63. Using this approach, solve each system of congruences from Exercise 21.

SECTION 4.5 COMPUTER EXERCISES

27. Write a computer program to solve systems of three simultaneous linear congruences with moduli that are pairwise relatively prime.

28. Write a computer program to solve the Diophantine equation $ax + by = c$, where a, b, and c are given and $(a, b) = 1$.

4.6 POLYNOMIAL CONGRUENCES (mod p^n)

Let $f(x)$ be a polynomial of degree $r \geq 1$ with integer coefficients. Let $m \geq 2$. There are various methods of determining all solutions, if any, of the congruence

4.13 $$f(x) \equiv 0 \pmod{m}$$

depending on the values of r and m. In Section 4.3 we saw how to solve congruence 4.13 by trial and error for arbitrary r and m. We also saw how to solve linear congruences (the case $r = 1$) by using the Euclidean algorithm. In Chapter 7 we will consider quadratic congruences (the case $r = 2$). In this section, where p denotes an arbitrary prime and n denotes any natural number, we will learn how to obtain the solutions of the congruence

4.68 $$f(x) \equiv 0 \pmod{p^{n+1}}$$

from the solutions of the congruence

4.69 $$f(x) \equiv 0 \pmod{p^n}$$

First of all, note that if congruence 4.68 has a solution, then by Theorem 4.1, part C_9, so does every congruence of the type

4.70 $$f(x) \equiv 0 \pmod{p^k}$$

where $1 \le k \le n$. In particular, suppose that congruence 4.69 has only the solutions $x \equiv x_1, x_2, x_3, \cdots, x_m \pmod{p^n}$. If y is a solution of congruence 4.68, then y is also a solution of congruence 4.69. Therefore, there exists i such that $1 \le i \le m$ and $y \equiv x_i \pmod{p^n}$. This implies that there exists t such that $0 \le t \le p - 1$ and $y = x_i + tp^n$. The following theorem tells us under what conditions such a t exists and how many such t's exist.

Theorem 4.7   *Let $f(x)$ be a polynomial with integer coefficients and degree $r \ge 1$. Let p be prime, $n \ge 1$. Let y be a solution of*

4.68 $$f(x) \equiv 0 \pmod{p^{n+1}}$$

Then $y = x_i + tp^n \pmod{p^{n+1}}$, where $0 \le x_i < p^n$, and x_i is a solution of

4.69 $$f(x) \equiv 0 \pmod{p^n}$$

such that $0 \le t \le p - 1$, and t satisfies the congruence

4.71 $$tf'(x_i) \equiv \frac{-f(x_i)}{p^n} \pmod{p}$$

Furthermore, if h is the number of solutions of congruence 4.71, then

$$h = \begin{cases} 1 & \text{if } p \nmid f'(x_i) \\ 0 & \text{if } p \mid f'(x_i) \text{ and } p^{n+1} \nmid f(x_i) \\ p & \text{if } p \mid f'(x_i) \text{ and } p^{n+1} \mid f(x_i) \end{cases}$$

PROOF: Let $x_1, x_2, x_3, \cdots, x_m$ be the only solutions of congruence 4.69. If $f(y) \equiv 0 \pmod{p^{n+1}}$, then also $f(y) \equiv 0 \pmod{p^n}$, so there exists i such that $1 \le i \le m$ and $y \equiv x_i \pmod{m}$. Therefore, $y = x_i + tp^n$ for some t such that $0 \le t \le p - 1$. Using a Taylor series expansion, we get

4.72 $$f(y) = f(x_i + tp^n) = f(x_i) + tp^n f'(x_i) + \frac{1}{2}t^2 p^{2n} f''(x_i) + \cdots$$

If we convert this equation to a congruence $\pmod{p^{n+1}}$ and reduce, since $2n \ge n + 1$, we get

4.73 $$f(y) \equiv f(x_i) + tp^n f'(x_i) \pmod{p^{n+1}}$$

Since $f(y) \equiv 0 \pmod{p^{n+1}}$ by hypothesis, we obtain

4.74
$$tp^n f'(x_i) \equiv -f(x_i) \quad (\bmod\ p^{n+1})$$

Since $f(x_i) \equiv 0 \quad (\bmod\ p^n)$, we have $p^n \mid f(x_i)$. Dividing congruence 4.74 by p^n and applying Theorem 4.1, part C_8, we have

4.75
$$tf'(x_i) \equiv \frac{-f(x_i)}{p^n} \quad (\bmod\ p)$$

Let b be the number of solutions of congruence 4.75. If $p \nmid f'(x_i)$, then Theorem 4.5 implies that $b = 1$. If $p \mid f'(x_i)$ and $p^{n+1} \nmid f(x_i)$, then Theorem 4.4 implies that $b = 0$. If $p \mid f'(x_i)$ and $p^{n+1} \mid f(x_i)$, then Theorem 4.5 implies that $b = p$, and the solutions of congruence 4.75 are $t = 0, 1, 2, \cdots, p - 1$. ∎

For example, suppose that we wish to solve the congruence

4.76
$$x^3 + 2x + 2 \equiv 0 \quad (\bmod\ 49)$$

According to Theorem 4.7, we must first solve the congruence

4.77
$$x^3 + 2x + 2 \equiv 0 \quad (\bmod\ 7)$$

By trial and error, the only solutions of congruence 4.77 are $x_1 \equiv 2 \quad (\bmod\ 7)$ and $x_2 \equiv 3 \quad (\bmod\ 7)$. In this problem $f(x) = x^3 + 2x + 2, f'(x) = 3x^2 + 2, p = 7$, and $n = 1$. If we let $i = 1$, then congruence 4.75 becomes

4.78
$$tf'(2) \equiv \frac{-f(2)}{7} \quad (\bmod\ 7)$$

That is,

4.79
$$14t \equiv -2 \quad (\bmod\ 7)$$

Here $7 \mid f'(2)$ but $7^2 \nmid f(2)$, so by Theorem 4.7, the congruence 4.79 has no solution, and we obtain no solution of congruence 4.76 corresponding to $x_1 = 2$.

If we let $i = 2$, then congruence 4.75 becomes

4.80
$$tf'(3) \equiv \frac{-f(3)}{7} \quad (\bmod\ 7)$$

That is,

4.81
$$29t \equiv -5 \quad (\bmod\ 7)$$

Here $7 \nmid f'(3)$, so congruence 4.81 has a unique solution, namely, $t \equiv 2 \quad (\bmod\ 7)$. Therefore, we obtain one corresponding solution of congruence 4.76, namely,

4.82
$$y \equiv 3 + 2(7^1) \equiv 17 \quad (\bmod\ 49)$$

As a second example, suppose we wish to solve the congruence

4.83 $$x^3 + 3x + 5 \equiv 0 \quad (\text{mod } 9)$$

According to Theorem 4.7, we must first solve the congruence

4.84 $$x^3 + 3x + 5 \equiv 0 \quad (\text{mod } 3)$$

By trial and error, $x_1 \equiv 1 \quad (\text{mod } 3)$ is the unique solution of congruence 4.84. In this problem, $f(x) = x^3 + 3x + 5, f'(x) = 3x^2 + 3, p = 3$, and $n = 1$. Now congruence 4.75 becomes

4.85 $$tf'(1) \equiv \frac{-f(1)}{3} \quad (\text{mod } 3)$$

That is,

4.86 $$6t \equiv -3 \quad (\text{mod } 3)$$

Here $3 \mid f'(1)$ and $3^2 \mid f(1)$, so congruence 4.86 has three solutions, namely, $t \equiv 0, 1, 2 \quad (\text{mod } 3)$. The corresponding solutions of congruence 4.83 are

$$y_1 \equiv 1 + 0(3) \equiv 1 \quad (\text{mod } 9)$$
$$y_2 \equiv 1 + 1(3) \equiv 4 \quad (\text{mod } 9)$$
$$y_3 \equiv 1 + 2(3) \equiv 7 \quad (\text{mod } 9)$$

As a third example, consider the congruence

4.87 $$x^3 - 5x + 1 \equiv 0 \quad (\text{mod } 27)$$

By repeated application of Theorem 4.7, if congruence 4.87 has a solution, then so does

4.88 $$x^3 - 5x + 1 \equiv 0 \quad (\text{mod } 3)$$

By inspection, the unique solution of congruence 4.88 is

4.89 $$x \equiv 1 \quad (\text{mod } 3)$$

We now seek solutions of the congruence

4.90 $$x^3 - 5x + 1 \equiv 0 \quad (\text{mod } 9)$$

According to Theorem 4.7, any solution of congruence 4.90 must have the form

4.91 $$x \equiv 1 + 3t \quad (\text{mod } 9)$$

where t satisfies the congruence

4.92
$$tf'(1) \equiv \frac{-f(1)}{3} \quad (\text{mod } 3)$$

Here $f(x) = x^3 - 5x + 1$ and $f'(x) = 3x^2 - 5$, so $f(1) = -3$ and $f'(1) = -2$. Therefore, congruence 4.92 becomes

4.93
$$-2t \equiv 1 \quad (\text{mod } 3)$$

or

4.94
$$t \equiv 1 \quad (\text{mod } 3)$$

Thus

$$x \equiv 4 \quad (\text{mod } 9)$$

is the unique solution of congruence 4.90.

According to Theorem 4.7, any solution of congruence 4.87 must have the form

4.95
$$x \equiv 4 + 9t \quad (\text{mod } 27)$$

where t satisfies the congruence

4.96
$$tf'(4) \equiv \frac{-f(4)}{9} \quad (\text{mod } 3)$$

That is,

4.97
$$43t \equiv -5 \quad (\text{mod } 3)$$

or

4.98
$$t \equiv 1 \quad (\text{mod } 3)$$

Therefore, the unique solution of congruence 4.87 is

4.99
$$x \equiv 13 \quad (\text{mod } 27)$$

SECTION 4.6 EXERCISES

Find all solutions, if any, of each of the following congruences.

29. $x^2 + 3x + 1 \equiv 0 \quad (\text{mod } 121)$

30. $x^3 - x + 4 \equiv 0 \quad (\text{mod } 25)$

31. $x^4 - 4x - 2 \equiv 0 \quad (\text{mod } 25)$

32. $x^3 - x^2 + 6 \equiv 0 \quad (\text{mod } 16)$

33. $x^3 + x + 2 \equiv 0 \quad (\text{mod } 8)$

34. $x^5 - 5x + 3 \equiv 0 \quad (\text{mod } 25)$

35. $x^3 + x + 1 \equiv 0 \quad (\text{mod } 81)$

36. $2x^3 - x + 1 \equiv 0 \quad (\text{mod } 169)$

37. $2x^3 - 3x^2 + 1 \equiv 0 \quad (\text{mod } 49)$

38. $2x^3 - 3x^2 - 6 \equiv 0 \quad (\text{mod } 49)$

SECTION 4.6 COMPUTER EXERCISES

39. Let p be an odd prime. Let a, b, c, and d be integers such that $(a, p) = 1$. Let $f(x) = ax^3 + bx^2 + cx + d$. Write a computer program that will find all solutions, if any, of the congruence $(f(x) \equiv 0 \pmod{p})$ by trial and error, that is, by successive evaluation of $f(0), f(1), f(2), \cdots, (p - 1)$.

40. With a, b, c, d, p and $f(x)$ as in Exercise 39, write a computer program that will find all solutions, if any, of the congruence $f(x) \equiv 0 \pmod{p^2}$.

41. Apply your program from Exercise 40 to find all solutions of the congruence $x^3 \equiv 1 \pmod{p}$ for each prime p such that $5 \leq p \leq 61$. Is there a pattern?

4.7   MODULAR ARITHMETIC AND FERMAT'S THEOREM

If the natural number $m > 1$, let us return to the set of least positive residues (mod m), namely, $\{0, 1, 2, \cdots, m - 1\}$, which we denote Z_m. Let us define addition (mod m) on Z_m as follows:

Definition 4.7   *Addition (mod m)*

If a and b belong to Z_m, then $a \oplus b$ is the least positive residue (mod m) of $a + b$. That is to say,

$$a \oplus b = \begin{cases} a + b & \text{if } 0 \leq a + b < m \\ a + b - m & \text{if } m \leq a + b < 2m \end{cases}$$

For example, in Z_7, we have

$$2 \oplus 4 = 6$$
$$3 \oplus 5 = 1 \quad \text{since } 3 + 5 = 8 \equiv 1 \pmod{7}$$
$$6 \oplus 1 = 0 \quad \text{since } 6 + 1 = 7 \equiv 0 \pmod{7}$$

Also, in Z_{10}, we have

$$2 \oplus 3 = 5$$
$$5 \oplus 7 = 2 \quad \text{since } 5 + 7 = 12 \equiv 2 \pmod{10}$$
$$8 \oplus 9 = 7 \quad \text{since } 8 + 9 = 17 \equiv 7 \pmod{10}$$

If m is small, it is convenient to describe addition (mod m) by means of an addition table. For example, if $m = 3$, we have

$\oplus$	0	1	2
0	0	1	2
1	1	2	0
2	2	0	1

In general, in such a table one finds $a \oplus b$ by looking for the entry that is in the same row as a and in the same column as b. Glancing at the table above, we see that $1 \oplus 2 = 0$ in Z_3.

One can verify that on Z_m, addition (mod m) has the following properties:

A$_1$: (Commutative property) $a \oplus b = b \oplus a$ for all a and b in Z_m.

A$_2$: (Associative property) $(a \oplus b) \oplus c = a \oplus (b \oplus c)$ for all a, b, and c in Z_m.

A$_3$: (Identity element) $a \oplus 0 = 0 \oplus a = a$ for all a in Z_m.

A$_4$: (Inverse element) For all a in Z_m, there is b in Z_m such that $a \oplus b = 0$.

For example, in Z_7, the additive inverse of 2 is 5, since $2 + 5 = 0$; likewise, the additive inverse of 6 is 1, since $6 + 1 = 0$.

If you are familiar with abstract algebra, you should realize that properties A$_1$ through A$_4$ imply that Z_m is an Abelian group of order m with respect to addition (mod m). In fact, if a is a nonzero element of Z_m, then a may be represented as a sum of a 1s. (For example, if $m \geq 4$, then $3 = 1 + 1 + 1$.) This implies that the additive group of Z_m is the cyclic group of order m.

Similarly, we define multiplication (mod m) on Z_m as follows:

Definition 4.8   ***Multiplication (mod m)***
If a and b belong to Z_m, then $a \odot b$ is the least positive residue (mod m) of ab.

For example, in Z_7, we have

$$4 \odot 6 = 3 \quad \text{since } 4(6) = 24 \equiv 3 \pmod 7$$
$$3 \odot 3 = 2 \quad \text{since } 3(3) = 9 \equiv 2 \pmod 7$$
$$2 \odot 4 = 1 \quad \text{since } 2(4) = 8 \equiv 1 \pmod 7$$

Also, in Z_{10}, we have

$$2 \odot 3 = 6$$
$$5 \odot 7 = 5 \quad \text{since } 5(7) = 35 \equiv 5 \quad (\text{mod } 10)$$
$$8 \odot 9 = 2 \quad \text{since } 8(9) = 72 \equiv 2 \quad (\text{mod } 10)$$

Again, if m is small, it is convenient to describe multiplication (mod m) by means of a multiplication table. For example, if $m = 3$, we have

$\odot$	0	1	2
0	0	0	0
1	0	1	2
2	0	2	1

One can verify that multiplication (mod m) on Z_m has the following properties:

M_1: (Commutative property) $a \odot b = b \odot a$ for all a and b in Z_m.

M_1: (Associative property) $(a \odot b) \odot c = a \odot (b \odot c)$ for all a, b, and c in Z_m.

M_1: (Identity element) $a \odot 1 = 1 \odot a = a$ for all a in Z_m.

M_1: (Distributive property) $a \odot (b \oplus c) = (a \odot b) \oplus (a \odot c)$ for all a, b, and c in Z_m.

Using the terminology of abstract algebra, one would say that under the operations of addition (mod m) and multiplication (mod m), Z_m is a *commutative ring with identity*.

If also $m = p$, a prime, then Z_p has an additional property:

M_5: (Multiplicative inverse)   For all a in Z_p such that $a \neq 0$, there exists b in Z_p such that $a \odot b = b \odot a = 1$.

In this case, the set of nonzero elements of Z_p forms an Abelian group under multiplication (mod p). This implies that Z_p is a special kind of ring, namely, a *field*.

For example, in Z_{11}, the multiplicative inverse of 3 is 4, since $3 \odot 4 = 1$; also, the multiplicative inverse of 5 is 9, since $5 \odot 9 = 1$.

Let us look at the multiplication table for multiplication (mod 7) in Z_7:

$\odot$	0	1	2	3	4	5	6
0	0	0	0	0	0	0	0
1	0	1	2	3	4	5	6
2	0	2	4	6	1	3	5
3	0	3	6	2	5	1	4
4	0	4	1	5	2	6	3
5	0	5	3	1	6	4	2
6	0	6	5	4	3	2	1

Suppose we wish to compute 5^{20} (mod 7). Using the table above, we see that

$$5^2 \equiv 4 \quad (\text{mod } 7)$$
$$5^4 \equiv (5^2)^2 \equiv 4^2 \equiv 2 \quad (\text{mod } 7)$$
$$5^5 \equiv 5 \cdot 5^4 \equiv 5 \cdot 2 \equiv 3 \quad (\text{mod } 7)$$
$$5^{10} \equiv (5^5)^2 \equiv 3^2 \equiv 2 \quad (\text{mod } 7)$$
$$5^{20} \equiv (5^{10})^2 \equiv 2^2 \equiv 4 \quad (\text{mod } 7)$$

There is a shortcut, however. If we notice that $5^6 \equiv 1$ (mod 7), then $5^{20} = 5^{6(3)+2} \equiv (5^6)^3 5^2 \equiv 1^3 5^2 \equiv 5^2 \equiv 4$ (mod 7). More generally, if $(a, 7) = 1$, then $a^6 \equiv 1$ (mod 7). This congruence is a special case of an important theorem that was discovered by the French mathematician Pierre de Fermat (1601–1665).

Theorem 4.8   *(Fermat's "Little" Theorem)*
If p is prime and $(a, p) = 1$, then $a^{p-1} \equiv 1$ (mod p).

PROOF: We will use induction on a to prove that if p is prime, then $a^p \equiv a$ (mod p) for all a. If also $(a, p) = 1$, then by Theorem 4.1, part C$_7$ a can be cancelled out of this congruence, leaving the desired result

$$a^{p-1} \equiv 1 \quad (\text{mod } p)$$

Since $0^p = 0$, it follows that $0^p \equiv 0$ (mod p). Now, assuming that $a^p \equiv a$ (mod p), we wish to show that $(a + 1)^p \equiv a + 1$ (mod p). However,

$$(a + 1)^p = a^p + 1 + \sum_{j=1}^{p-1} \binom{p}{j} a^{p-j}$$

by the binomial theorem (Theorem 1.13, part B$_3$). The result of Exercise 23 in Chapter 3 says that $\binom{p}{j} \equiv 0$ (mod p) if $1 \le j \le p - 1$. Therefore, $(a + 1)^p \equiv a^p + 1$ (mod p). Since $a^p \equiv a$ (mod p) by the induction hypothesis, we have $(a + 1)^p \equiv a + 1$ (mod p). ∎

Pierre de Fermat was born on August 20, 1601 in the village of Beaumont-de-Lomanges in southern France. His father was a prosperous leather merchant, and his mother came from a family with good legal connections. In his youth, Fermat received a good classical education: he learned Spanish, Italian, Latin, and Greek, in addition to French. In the late 1620s, he spent some time in Bordeaux, where he studied in depth the mathematical works of François Vieta (1540–1603).

The year 1631 was eventful for Fermat. He received a bachelor of civil law degree from the University of Orleans, married his mother's cousin, Louise de Long, and assumed the offices of Conseiller and Commissaire aux Enquetes in the provincial parliament of Toulouse. His legal and parliamentary career prospered as he rose through the ranks in a system based on seniority. In 1638 he became Conseiller aux Enquetes; in 1642 he became a member of the criminal court and the Grande Chambre; and in 1648, he was appointed King's Counselor. At home, he and his wife raised a family of two sons and three daughters.

All the while, Fermat remained devoted to mathematics, his great avocation. Before Newton and Leibniz were born, Fermat conceived some of the main ideas in differential and integral calculus, such as finding the maximum and minimum values of a polynomial function and finding the area of a region bounded by a curve. Independently from Descartes, he invented analytic geometry. Together with Pascal, he invented probability theory.

In addition, he made many significant contributions to number theory, such as Fermat's "little" theorem (Theorem 4.8), which was later proved by Euler; the method of "descent" for solving Diophantine equations; and the discovery that if p is prime and $p \equiv 1 \pmod 4$, then p is a sum of two squares. Fermat's conjecture, which we mentioned in Chapter 1 and will explore further in Chapter 11, states that the equation

$$x^n + y^n = z^n$$

has no solution in positive integers for any $n \geq 3$. To prove or disprove Fermat's conjecture remains a significant open problem.

Indeed, Fermat may be said to be responsible for the establishment of number theory as a mathematical discipline in its own right. In a letter to Frenicle de Bessy written in February of 1657, Fermat says:

> There is scarcely anyone who states purely arithmetic questions, scarcely anyone who understands them. Is this not because arithmetic has been treated up to this time geometrically rather than arithmetically? This is

certainly indicated by many works ancient and modern. Diophantus himself also indicates this. But he has freed himself from geometry a little more than the others have, in that he limits his analysis to rational numbers only; nevertheless the *Zetetica of Vieta*, in which the methods of Diophantus are extended to continuous magnitude and therefore to geometry, witness the insufficient separation of arithmetic from geometry.

Now arithmetic has a special domain of its own, the theory of numbers. This was touched upon but only to a slight degree by Euclid in his *Elements*, and by those who follow him it has not been sufficiently extended, unless perchance it lies hid in those books of Diophantus which the ravages of time have destroyed. Arithmeticians have now to develop or restore it.

Fermat communicated his mathematical discoveries via personal correspondences with other mathematicians such as Descartes, Mersenne, Frenicle, and Roberval. Although Fermat did not know English, he communicated by means of an intermediary with the prominent English mathematicians John Wallis and Lord William Brouncker. In 1657 Fermat challenged Wallis to prove that if D is not a square, then the equation

$$x^2 - Dy^2 = 1$$

has infinitely many integer solutions. (This equation, now known as Pell's equation, will be considered in some detail in Chapter 11.) Later that same year Brouncker provided the proof that Fermat requested. In a letter of congratulation to Brouncker, Wallis said that he had "preserved untarnished the fame which Englishmen have won in former times with Frenchmen and has shown that England's champions of wisdom are just as strong as those in war."

Fermat was the last amateur to make significant contributions to number theory. He died in Castres on January 12, 1665. His mathematical papers were published posthumously by his eldest son, Clement-Samuel.

SECTION 4.7 EXERCISES

42. (a) Construct the multiplication table for multiplication (mod 5) in Z_5.   (b) Find the multiplicative inverse of every nonzero element.

43. (a) Construct the multiplication table for multiplication (mod 11) in Z_{11}.   (b) Find the multiplicative inverse of every nonzero element.

44. Construct the multiplication table for multiplication (mod 10) in Z_{10}. Which elements have multiplicative inverses?

45. Prove that if p is prime and $(a, p) = 1$, then the congruence

$$ax \equiv b \quad (\text{mod } p)$$

has the solution

$$x \equiv a^{p-2}b \quad (\text{mod } p)$$

46. Prove that if p and q are primes and $q \mid (a^p - 1)$, then $q \mid (a - 1)$ or $p \mid (q - 1)$.

47. If p is prime and $a \geq 2$, prove that

$$\left(a - 1, \frac{a^p - 1}{a - 1}\right) = \begin{cases} p & \text{if } p \mid (a - 1) \\ 1 & \text{if } p \nmid (a - 1) \end{cases}$$

*48. We say that prime q is a *primitive* factor of $a^n - 1$ if $q \mid (a^n - 1)$ but $q \nmid (a^m - 1)$ for all m such that $0 < m < n$. Prove that if p is prime and $a \geq 2$, then $a^p - 1$ has a primitive factor.

49. Prove that if p is prime, $n \geq 1$, and $a^d \equiv 1 \quad (\text{mod } p^n)$, then $a^{pd} \equiv 1 \quad (\text{mod } p^{n+1})$.

SECTION 4.7 COMPUTER EXERCISES

50. Let $a \geq 2$, and let p be an odd prime. Write a computer program to find the least primitive factor of $a^p - 1$.

51. Suppose that n is a given odd integer. We wish to determine whether n is prime or composite. If $2^{n-1} \not\equiv 1 \quad (\text{mod } n)$, then by Theorem 4.8, n is certainly composite. For $n < 1000$, test the conjecture that if $2^{n-1} \equiv 1 \quad (\text{mod } n)$, then n is prime. That is, write a computer program that, for each odd composite value of n below 1000, verifies whether $2^{n-1} \equiv 1 \quad (\text{mod } n)$. (The exceptional values of n, if any, are called *pseudoprimes to the base 2*.)

4.8   WILSON'S THEOREM AND FERMAT NUMBERS

Quadratic congruences will be studied in detail in Chapter 7, but we are ready to obtain some preliminary results.

Theorem 4.9   Let p be an odd prime. Then $x^2 \equiv a^2 \quad (\text{mod } p)$ if and only if $x \equiv \pm a \quad (\text{mod } p)$.

PROOF: If $x \equiv \pm a \pmod{p}$, then Theorem 4.1, part C_4 implies that $x^2 \equiv a^2 \quad (\text{mod } p)$. Conversely, if $x^2 \equiv a^2 \quad (\text{mod } p)$, then $p \mid (x^2 - a^2)$; that is, $p \mid (x - a)(x + a)$. By Theorem 3.2, $p \mid (x - a)$ or $p \mid (x + a)$, so $x - a \equiv 0 \quad (\text{mod } p)$ or $x + a \equiv 0 \quad (\text{mod } p)$; that is, $x \equiv \pm a \quad (\text{mod } p)$. ∎

For example, if $x^2 \equiv 4 \pmod{11}$, then $x \equiv \pm 2 \pmod{11}$; also, if $x^2 \equiv 9 \pmod{17}$, then $x \equiv \pm 3 \pmod{17}$. Note that if m is composite and $x^2 \equiv a^2 \pmod{m}$, then it need not follow that $x \equiv \pm a \pmod{m}$. For example, $3^2 \equiv 1^2 \pmod{8}$, yet $3 \not\equiv \pm 1 \pmod{8}$.

Theorem 4.10   *If p is an odd prime, $p \mid (a^2 + b^2)$, and $(a, b) = 1$, then $p \equiv 1 \pmod{4}$.*

PROOF: By hypothesis, $p \mid (a^2 + b^2)$. If $p \mid a$, then $p \mid a^2$, so $p \mid b^2$ and $p \mid b$. However, then $(a, b) \geq p$, contrary to hypothesis. Therefore, $p \nmid a$. Similarly, $p \nmid b$. Since $p \mid (a^2 + b^2)$, we have $-a^2 \equiv b^2 \pmod{p}$. Therefore, $(-a^2)^{\frac{1}{2}(p-1)} \equiv (b^2)^{\frac{1}{2}(p-1)} \pmod{p}$; that is, $(-1)^{\frac{1}{2}(p-1)} a^{p-1} \equiv b^{p-1} \pmod{p}$. Since $p \nmid ab$, Fermat's "little" theorem (Theorem 4.8) implies that $a^{p-1} \equiv b^{p-1} \equiv 1 \pmod{p}$. Therefore, $(-1)^{\frac{1}{2}(p-1)} \equiv 1 \pmod{p}$. If $p \equiv 3 \pmod{4}$, then $(-1)^{\frac{1}{2}(p-1)} = -1$, so we would have $-1 \equiv 1 \pmod{p}$ and hence $p \mid 2$, an impossibility. Therefore, $p \equiv 1 \pmod{4}$.   ■

For example, $20^2 + 9^2 = 481 = 13 \cdot 37$. Note that each of the prime factors, 13 and 37, is congruent to 1. $\pmod{4}$.

The following theorem, which is named after the English mathematician John Wilson (1741–1793), was first proved by Lagrange.

Theorem 4.11   *(Wilson's Theorem)*
If p is prime, then $(p - 1)! \equiv -1 \pmod{p}$.

PROOF: One can easily verify that Wilson's theorem holds if $p = 2$ or 3. Now suppose that $p \geq 5$. Since $p - 1 \equiv -1 \pmod{p}$ and $(p - 1, p) = 1$, it suffices by virtue of Theorem 4.1, part C_7 to show that $(p - 2)! \equiv 1 \pmod{p}$. By Theorem 4.5, for each j such that $1 \leq j \leq p - 1$, there exists unique k such that $1 \leq k \leq p - 1$ and $jk \equiv 1 \pmod{p}$. If $k = j$, then $j^2 \equiv 1 \pmod{p}$. However, then Theorem 4.9 implies that $j \equiv \pm 1 \pmod{p}$, so $j = 1$ or $p - 1$. Therefore, if $2 \leq j \leq p - 2$, then there exists unique $k \neq j$ such that $2 \leq k \leq p - 2$ and $jk \equiv 1 \pmod{p}$. If we multiply together all $\frac{1}{2}(p - 3)$ congruences of this type, we obtain $(p - 2)! \equiv 1 \pmod{p}$.   ■

For example, if $p = 11$, we have

$$2 \cdot 6 \equiv 1 \quad (\text{mod } 11)$$
$$3 \cdot 4 \equiv 1 \quad (\text{mod } 11)$$
$$5 \cdot 9 \equiv 1 \quad (\text{mod } 11)$$
$$7 \cdot 8 \equiv 1 \quad (\text{mod } 11)$$

So

$$9! \equiv 2 \cdot 3 \cdot 4 \cdot 5 \cdot 6 \cdot 7 \cdot 8 \cdot 9 \equiv 1 \quad (\text{mod } 11)$$

Next, we apply Theorems 4.10 and 4.11 to solve a particular quadratic congruence.

Theorem 4.12   *If p is an odd prime, then the congruence*

$$x^2 \equiv -1 \quad (\text{mod } p)$$

has the solutions $x \equiv \pm[(p-1)/2]!$ (mod p) if $p \equiv 1$ (mod 4) and has no solution if $p \equiv 3$ (mod 4).

PROOF: If $x^2 \equiv -1$ (mod p), then $p \mid (x^2 + 1)$. Since $(x, 1) = 1$, Theorem 4.10 implies that $p \equiv 1$ (mod 4). By Theorem 4.9, it suffices to show that if $a = [(p-1)/2]!$, then $a^2 \equiv -1$ (mod p). Let $ab = (p - 1)!$ so that

$$b = \frac{(p-1)!}{a} = \prod_{j=1}^{\frac{1}{2}(p-1)} (p - j)$$

Now $b \equiv \prod_{j=1}^{\frac{1}{2}(p-1)} (-j)$ (mod p), so

$$b \equiv (-1)^{\frac{1}{2}(p-1)} \prod_{j=1}^{\frac{1}{2}(p-1)} j \equiv (-1)^{\frac{1}{2}(p-1)} \left(\frac{p-1}{2}\right)!$$
$$\equiv (-1)^{\frac{1}{2}(p-1)} a \quad (mod\ p)$$

Since $p \equiv 1$ (mod 4), $(-1)^{\frac{1}{2}(p-1)} = 1$, so $b \equiv a$ (mod p). Therefore, $a^2 \equiv ab \equiv (p - 1)!$ (mod p). Finally, Wilson's theorem (Theorem 4.11) implies that $a^2 \equiv -1$ (mod p).   ■

For example, let us solve the quadratic congruence

$$x^2 \equiv -1 \quad (\text{mod } 13)$$

Here $p = 13$, so $[(p-1)/2]! = 6! = 720$. Now $720 \equiv 5$ (mod 13), so the solutions are

$$x \equiv \pm 5 \quad (\text{mod } 13).$$

Note that while Theorem 4.12 gives an explicit formula for the solution of the quadratic congruence, the formula is impractical for large p because of the large amount of multiplication needed to compute $[(p-1)/2]! \pmod{p}$. In Chapter 7 we will present an efficient method for finding solutions of quadratic congruences when these solutions exist.

The result of Exercise 2 in Chapter 3 implies that if $2^m + 1$ is prime, then $m = 2^n$ for some $n \geq 0$. The numbers $2^{2^n} + 1$ were first discussed by Fermat and are now named after him.

Definition 4.9   ***Fermat Numbers***
If $n \geq 0$, let $f_n = 2^{2^n} + 1$. Then f_n is called the *n*th *Fermat number*. If f_n is prime, then we call it a *Fermat prime*.

Let us compute the first few Fermat numbers:

$$f_0 = 2^{2^0} + 1 = 2^1 + 1 = 2 + 1 = 3$$
$$f_1 = 2^{2^1} + 1 = 2^{2^1} + 1 = 2^2 + 1 = 4 + 1 = 5$$
$$f_2 = 2^{2^2} + 1 = 2^4 + 1 = 16 + 1 = 17$$
$$f_3 = 2^{2^3} + 1 = 2^8 + 1 = 256 + 1 = 257$$
$$f_4 = 2^{2^4} + 1 = 2^{16} + 1 = 65{,}536 + 1 = 65{,}537$$

Each of these Fermat numbers is prime. In a letter to Pascal dated August 29, 1654, Fermat stated that every Fermat number is prime. This is the only statement ever made by Fermat that is known to be false. In 1732 Euler showed that f_5 is composite. Indeed, $f_5 = 2^{2^5} + 1 = 2^{32} + 1 = 4{,}294{,}967{,}297 = 641 \cdot 6{,}700{,}417$. It is known that f_n is composite for all n such that $5 \leq n \leq 19$ and for many other values of n. It is not known whether f_n is prime for any $n \geq 20$.

There is an interesting connection between Fermat primes and a classical problem in plane geometry. The problem is to find all odd integers n such that a regular n-sided polygon can be inscribed in a circle using just a straight edge and a compass. (A *regular polygon* has sides that all have the same length as well as angles that all have the same measure.) Amazingly, if n is odd and $n \geq 3$, then a regular n-sided polygon can be inscribed in a circle if and only if n is a product of one or more distinct Fermat primes. The proof of this assertion involves some deep results of abstract algebra and is therefore omitted. (See Gauss, Article 365).

Next, we show that distinct Fermat numbers are relatively prime.

Theorem 4.13   *If $m \neq n$, then $(f_m, f_n) = 1$.*

PROOF: Without loss of generality, assume that $m > n \geq 0$, so $m = n + k$ for some $k \geq 1$. Suppose that p is prime and $p \mid (f_m, f_n)$. Since f_m and f_n are odd, p is odd. Since $p \mid f_n$, we have

$$2^{2^n} \equiv -1 \pmod{p}$$

Therefore,

$$(2^{2^n})^{2^k} \equiv (-1)^{2^k} \pmod{p}$$

That is,

$$2^{2^{n+k}} \equiv 1 \pmod{p}$$

or

$$2^{2^m} \equiv 1 \pmod{p}$$

However, since $p \mid f_m$, we have

$$2^{2^m} \equiv -1 \pmod{p}$$

Therefore, $-1 \equiv 1 \pmod{p}$, which implies that $p = 2$, an impossibility. Therefore, (f_m, f_n) has no prime divisor, so $(f_m, f_n) = 1$.   ∎

We can now prove a theorem that complements Theorem 3.11.

Theorem 4.14   *There exist infinitely many primes of the form $4k + 1$.*

PROOF: If $n \geq 1$, let $f_n = 2^{2^n} + 1 = (2^{2^{n-1}})^2 + 1$. Let p_n be a prime factor of f_n. Since f_n is odd, p_n must be odd. Theorem 4.10 implies that $p_n \equiv 1 \pmod 4$. Theorem 4.13 implies that $p_n \nmid f_m$ if $m \neq n$. Therefore, $p_n \neq p_m$ if $m \neq n$. Thus the prime factors p_n of the Fermat numbers f_n with $n \geq 1$ yield an infinite sequence of primes of the form $4k + 1$.   ∎

SECTION 4.8 EXERCISES

52. Prove that if n is composite and $n > 4$, then $(m - 1)! \equiv 0 \pmod n$. (This is the converse of Wilson's theorem.)

53. Obtain an alternate proof of Theorem 4.14 as follows: Suppose that $q_1, q_2, q_3, \cdots, q_n$ are the only primes of the form $4k + 1$. Let Q be the product of these primes, and look at $4Q^2 + 1$.

54. Prove that if p is prime, $p \equiv 3 \pmod 4$, and $p \mid (a^2 + b^2)$, then $p^2 \mid (a^2 + b^2)$.

55. Let p be prime. Prove that $p \equiv 1 \pmod 4$ if and only if there exist two consecutive squares whose sum is divisible by p.

56. Prove that if $2^j + 3 = 7^k$, then $j = 2$ and $k = 1$.

57. Let f_n be the nth Fermat number. Prove that if $m > n$, then $f_m \equiv 2 \pmod{f_n}$.

*58. Let p be an odd prime. Say that p has the inverse property if for every j such that $1 < j < \frac{1}{2}p$ there exists k such that $\frac{1}{2}p < k < p - 1$ and $jk \equiv 1 \pmod p$. Prove that 5, 7, and 13 are the only primes that have the inverse property.

SECTION 4.8 COMPUTER EXERCISES

Consider the congruence

$$x^2 \equiv -1 \pmod p$$

where p is a prime such that $p \equiv 1 \pmod 4$.

59. Write a computer program to solve the congruence above using the formula of Theorem 4.12, namely,

$$x \equiv \pm[(p-1)/2]! \pmod p.$$

60. Write a computer program to solve the congruence above as follows: Using trial and error, find an integer $q \geq 2$ such that $q^{\frac{1}{2}(p-1)} \equiv -1 \pmod p$. Then the solutions are $x \equiv \pm q^{\frac{1}{4}(p-1)} \pmod p$.

61. Solve the congruence above with $p = 241$ using the computer programs from Exercises 59 and 60. Which is faster?

4.9   PYTHAGOREAN EQUATION

If a right triangle has legs of lengths x and y and hypotenuse of length z, then x, y, and z satisfy the Pythagorean equation

4.100
$$x^2 + y^2 = z^2$$

This statement is one of the most important theorems of plane geometry, and it was introduced to the Western world by Pythagoras (589–490 B.C.). In number theory, we are interested in positive integer solution of this equation, which we call *Pythagorean triples*. The Pythagorean triple with minimal z is $x = 4$, $y = 3$, and $z = 5$.

Suppose that x, y, and z are a Pythagorean triple such that $(x, y, z) = d > 1$. Then there exist integers r, s, and t such that $x = dr$, $y = ds$, $z = dt$, and $(r, s, t) = 1$. Now $(dr)^2 + (ds)^2 = (dt)^2$, so $r^2 + s^2 = t^2$. This implies that all solutions of the Pythagorean equation (4.100) may be obtained from just those solutions which we call *primitive*, that is, those such that $(x, y, z) = 1$. For example, the minimal solution is primitive, since $(4, 3, 5) = 1$, but the solution $x = 8$, $y = 6$, and $z = 10$ is not primitive, since $(8, 6, 10) = 2 > 1$. If $D = (x, y)$ and the Pythagorean equation (4.100) holds, then also $D \mid z$, so $D \mid (x, y, z)$; that is, $D \mid d$. However, in any case, $d \mid D$. Therefore, $D = d$; that is, $(x, y) = (x, y, z)$. We therefore note that a solution of the Pythagorean equation (4.100) is primitive if and only if $(x, y) = 1$. Similarly $(x, z) = (y, z) = 1$.

Suppose that x, y, and z are a primitive Pythagorean triple. Therefore, x and y are not both even. If x and y are both odd, then $x^2 \equiv y^2 \equiv 1 \pmod 4$, so $z^2 \equiv 2 \pmod 4$, an impossibility. Therefore, one of x and y, say x, is even, while y is odd. This implies that z^2 is odd, and hence z is odd. If we write $x^2 = z^2 - y^2$ and let $x = 2w$, we get $4w^2 = z^2 - y^2 = (z + y)(z - y)$. Since y and z are both odd, $\frac{1}{2}(z + y)$ and $\frac{1}{2}(z - y)$ are integers. Therefore, $w^2 = \left(\frac{z+y}{2}\right)\left(\frac{z-y}{2}\right)$. Let $t = \left(\frac{z+y}{2}, \frac{z-y}{2}\right)$. Therefore, $t \mid \frac{z+y}{2}$ and $t \mid \frac{z-y}{2}$, so $t \mid z$ and $t \mid y$. Thus $t \mid (y, z)$. However, $(y, z) = (x, y, z) = 1$, so $t = 1$. Now the result of Exercise 21 in Chapter 3 implies that $\frac{1}{2}(z + y) = a^2$ and $\frac{1}{2}(z - y) = b^2$, with $(a, b) = 1$. Adding these equations, we get $z = a^2 + b^2$; subtracting, we get $y = a^2 - b^2$. Since $y < z$, we must have $b > 0$. Since $w^2 = a^2 b^2$, we have $w = ab$, so $x = 2ab$. Since $y > 0$, we must have $a > b$. Since z is odd, we must have $a \not\equiv b \pmod 2$.

To recapitulate, if x, y, and z are natural numbers such that $(x, y) = 1$ and $x^2 + y^2 = z^2$, then $x = 2ab, y = a^2 - b^2$, and $z = a^2 + b^2$, where a and b are natural numbers such that $a > b$, $(a, b) = 1$, and $a \not\equiv b \pmod 2$. Clearly, the converse also holds. We therefore conclude the following:

Theorem 4.15   *Let x and y be natural numbers such that $(x, y) = 1$. Then $x^2 + y^2 = z^2$ if and only if $x = 2ab, y = a^2 - b^2$, and $z = a^2 + b^2$, where a and b are integers such that $a > b > 0$, $(a, b) = 1$, and $a \not\equiv b \pmod 2$.*

We can easily generate infinitely many primitive solutions of the Pythagorean equation (4.100) by letting $a = 2n$ and $b = 1$ or $a = 2n + 1$ and $b = 2$ and letting n range through all natural numbers. We list all solutions of the Pythagorean equation (4.100) such that $a \leq 7$ in Table 4.1.

Next, we consider the related equation

4.101 $$x^4 + y^4 = z^2$$

Let $d = (x, y)$, so $d \mid x, d \mid y, d^4 \mid z^2$, and $d^2 \mid z$. This implies that $x = dr$ and $y = ds$ for integers r and s such that $r^4 + s^4 = (z/d^2)^2$. Therefore, it suffices to consider only solutions of equation 4.101 that are primitive, that is, solutions such that $(x, y) = 1$.

Table 4.1. Solutions to the Pythagorean Equation (4.100) such that $a \leq 7$

a	b	x	y	z
2	1	4	3	5
3	2	12	5	13
4	1	8	15	17
4	3	24	7	25
5	2	20	21	29
5	4	40	9	41
6	1	12	35	37
6	5	60	11	61
7	2	28	45	53
7	4	56	33	65
7	6	84	13	85

If equation 4.101 has any solution, then it must have a solution such that z is minimal. Also, Theorem 4.15 implies that $x^2 = 2ab$, $y^2 = a^2 - b^2$, and $z = a^2 + b^2$, where a and b are integers such that $a > b > 0$, $(a, b) = 1$, and $a \not\equiv b \pmod 2$.

Now $b^2 + y^2 = a^2$ and $(b, y) = (a, b) = 1$, so Theorem 4.15 implies that a is odd; hence b is even. Again, by Theorem 4.15 we have $b = 2uv$, $y = u^2 - v^2$, and $a = u^2 + v^2$ for integers u and v such that $u > v > 0$, $(u, v) = 1$, and $u \not\equiv v \pmod 2$. However, $(\frac{1}{2}x)^2 = \frac{1}{2}ab$ and $(a, \frac{1}{2}b) = (a, b) = 1$, so by the result of Exercise 21 in Chapter 3, we have $a = s^2$ and $\frac{1}{2}b = t^2$ for natural numbers s and t such that $(s, t) = 1$. Now $t^2 = uv$ and $(u, v) = 1$, so $u = m^2$ and $v = n^2$ for integers m and n such that $(m, n) = 1$. We therefore have $m^4 + n^4 = u^2 + v^2 = a = s^2$; that is, $m^4 + n^4 = s^2$. However, $s \leq a < z$, which contradicts the minimality of z. We conclude the following:

Theorem 4.16   *The equation $x^4 + y^4 = z^2$ has no solution in natural numbers.*

We might note that the method used to prove Theorem 4.16, namely, using a given solution to generate a smaller solution, is known as *Fermat's method of descent*. As an immediate consequence of Theorem 4.16, we have the following:

Theorem 4.17   *The equation $x^4 + y^4 = z^4$ has no solution in natural numbers.*

SECTION 4.9 EXERCISES

62. Find all Pythagorean triples such that one member is 17.

63. Find all Pythagorean triples, not necessarily primitive, such that one member is 20.

64. If x, y, and z are a Pythagorean triple, prove that (a) $3 \mid xyz$ and (b) $5 \mid xyz$.

65. Extend Table 4.1 so that it includes all $a \leq 12$.

66. Prove that if $n \geq 2$, then the equation $x^{2^n} + y^{2^n} = z^{2^n}$ has no solution in natural numbers.

67. Prove that if $x^2 + y^2 = 2z^2$, then there exist a and b such that $x = (a + b)^2 - 2b^2$, $y = |(a - b)^2 - 2b^2|$, and $z = a^2 + b^2$.

68. Prove that the equation $x^2 + y^2 = z^2 + 1$ has infinitely many solutions in natural numbers.

69. Prove that the equation $x^4 + 4y^4 = z^2$ has no solution in natural numbers.

70. Find all solutions in natural numbers of the equation $3^x + 4^y = 5^z$.

71. Prove that the equation $x^4 + y^2 = z^2$ has infinitely many primitive solutions in natural numbers. Find three solutions with x even and three solutions with x odd.

*72. The *inradius* of a triangle is the radius of the inscribed circle. Prove that if x, y, and z are a Pythagorean triple, then the inradius of the corresponding triangle is an integer.

73. Prove that for every odd prime p there are exactly two right triangles with integers sides and inradius p.

SECTION 4.9 COMPUTER EXERCISES

74. Write a computer program to find all primitive solutions of the equation $x^2 + y^2 = z^2$ such that $z \leq 10,000$.

75. Modify the computer program of Exercise 74 so as to identify all solutions of the equations $x^2 + y^2 = z^2$ such that $z \leq 10,000$ and $|x - y| = 1$.

CHAPTER FIVE

ARITHMETIC FUNCTIONS

5.1 INTRODUCTION

An *arithmetic function* (also known as a *number theoretic function*) is a function whose domain is the set of natural numbers. In this chapter we study four such functions: $\tau(n)$, the *number-of-divisors* function; $\sigma(n)$, the *sum-of-divisors* function; $\mu(n)$, the *Moebius* function; and $\phi(n)$, *Euler's* function. We conclude with Euler's theorem, which generalizes Fermat's theorem. In order to develop the properties of these functions, we introduce a concept known as the *Dirichlet product*, or *Dirichlet convolution of two functions*.

5.2 SIGMA FUNCTION, TAU FUNCTION, DIRICHLET PRODUCT

Definition 5.1   ***Summation Function***

Let $f(n)$ and $g(n)$ be functions defined on N. We say that $g(n)$ is the *summation function* of $f(n)$ if $g(n)$ is the summation of all $f(d)$ as d runs through the divisors of n. If so, we write

$$g(n) = \sum_{d \mid n} f(d)$$

For example, let $n = 6$. Since the divisors of 6 are 1, 2, 3, and 6, we have

$$g(6) = \sum_{d \mid 6} f(d) = f(1) + f(2) + f(3) + f(6)$$

Our first specific example of a summation function is the sum-of-divisors function, denoted $\sigma(n)$.

Definition 5.2 **Sum-of-Divisors Function**

Let $\sigma(n)$ denote the *sum of the divisors* of n. We use the notation

$$\sigma(n) = \sum_{d \mid n} d$$

For example,

$$\sigma(6) = \sum_{d \mid 6} d = 1 + 2 + 3 + 6 = 12$$

Also

$$\sigma(4) = \sum_{d \mid 4} d = 1 + 2 + 4 = 7$$

Note that if $M(n) = n$, that is, M is the identity function with respect to multiplication, then

$$\sigma(n) = \sum_{d \mid n} d = \sum_{d \mid n} M(d)$$

that is, $\sigma(n)$ is the summation function of $M(n)$.

Note that if p is prime and $k \geq 1$, then the divisors of p^k are 1, p, $p^2, \cdots, p^k$, so

$$\sigma(p^k) = 1 + p + p^2 + \cdots + p^k = \frac{p^{k+1} - 1}{p - 1}$$

Another summation function is the number-of-divisors function $\tau(n)$.

Definition 5.3 **Number-of-Divisors Function**

Let $\tau(n)$ denote the *number of divisors* of n. We use the notation

$$\tau(n) = \sum_{d \mid n} 1$$

This notation means that to compute $\tau(n)$, start from zero and add a 1 for every divisor of n.

For example,

$$\tau(6) = \sum_{d \mid 6} 1 = 1 + 1 + 1 + 1 = 4$$

Also

$$\tau(4) = \sum_{d|4} 1 = 1 + 1 + 1 = 3$$

Note that if $u(n)$ is the constant function: $u(n) = 1$ for all n, then

$$\tau(n) = \sum_{d|n} 1 = \sum_{d|n} u(d)$$

that is, $\tau(n)$ is the summation function of $u(n)$. Note that if p is prime and $k \geq 1$, then $\tau(p^k) = k + 1$.

Suppose that we wish to evaluate $\tau(n)$ and $\sigma(n)$ for composite integers n that have two or more prime factors. For example, let $n = 12$. First, we list the divisors of 12, namely, 1, 2, 3, 4, 6, and 12. Then, by direct evaluation, $\tau(12) = 6$ and $\sigma(12) = 1 + 2 + 3 + 4 + 6 + 12 = 28$. Clearly, direct evaluation of $\tau(n)$ and $\sigma(n)$ would be cumbersome if n has a large number of divisors, for example, if $n = 360$.

Note that $\tau(4)\tau(3) = 3 \cdot 2 = 6 = \tau(12)$; also, $\sigma(4)\sigma(3) = 7 \cdot 4 = 28 = \sigma(12)$. Generalizing, it appears that $\tau(mn) = \tau(m)\tau(n)$ and $\sigma(mn) = \sigma(m)\sigma(n)$. We shall see that this is always the case, provided that $(m, n) = 1$. An arithmetic function f such that $f(mn) = f(m)f(n)$ whenever $(m, n) = 1$ is called *multiplicative*. We shall see that if $f(n)$ is multiplicative, and if $g(n)$ is the summation function of $f(n)$, then g is also multiplicative. This will provide us with a more efficient alternative to the direct evaluation of $g(n)$.

Definition 5.4   ***Multiplicative Function***
If $f(mn) = f(m)f(n)$ whenever $(m, n) = 1$, then we say f is a *multiplicative function*.

We shall see that the functions $\tau(n)$ and $\sigma(n)$, as well as the functions $\mu(n)$ and $\phi(n)$, which will be defined later on, are all multiplicative functions. The evaluation of a multiplicative function is greatly simplified by applying the following theorem.

Theorem 5.1   *If f is a multiplicative function and $n = \prod_{i=1}^{r} p_i^{e_i}$, then*

$$f(n) = \prod_{i=1}^{r} f(p_i^{e_i})$$

PROOF: Exercise (induction on r)

REMARK: Theorem 5.1 says that a multiplicative function is completely determined by what it does to powers of primes. For example, $\sigma(20) = \sigma(2^2 5^1) = \sigma(2^2)\sigma(5^1) = 7 \cdot 6 = 42$. Some functions have the even stronger property of being totally multiplicative.

Definition 5.5

Totally Multiplicative Function
If $f(mn) = f(m)f(n)$ for all m and n, then we say f is a *totally multiplicative function*.

For example, let $M(n) = n$. $M(n)$ is the identity function with respect to ordinary multiplication. Now $M(mn) = mn = M(m)\,M(n)$ for all m and n, so $M(n)$ is a totally multiplicative function. Other such functions include

The unit function: $u(n) = 1$ for all n
The zero function: $z(n) = 0$ for all n

$$\text{The function} \qquad I(n) = \begin{cases} 1 & \text{if } n = 1 \\ 0 & \text{if } n > 1 \end{cases}$$

Clearly, a function that is totally multiplicative is multiplicative. The converse does not hold, since each of $\tau(n)$ and $\sigma(n)$ is multiplicative but not totally multiplicative. The following theorem establishes another property of multiplicative functions.

Theorem 5.2

If f is multiplicative and $f(n) \neq z(n)$, then $f(1) = 1$.

PROOF: By hypothesis, there exists n such that $f(n) \neq 0$. Since $(n, 1) = 1$ and f is multiplicative by hypothesis, we have $f(n) = f(n \cdot 1) = f(n)f(1)$. Therefore, $f(n)[1 - f(1)] = 0$. Since $f(n) \neq 0$, we have $1 - f(1) = 0$, so $f(1) = 1$.   ∎

In order to prove that $\tau(n)$ and $\sigma(n)$ are multiplicative functions, we next introduce the concept of the *Dirichlet product*, which is also known as the *Dirichlet convolution of two functions*.

Definition 5.6   **Dirichlet Product**

Let f and g be arithmetic functions. We define their *Dirichlet product* as

$$(f \cdot g)(n) = \sum_{d|n} f(d)g\left(\frac{n}{d}\right)$$

That is, $(f \cdot g)(n)$ is the sum of all products $f(d)g\left(\frac{n}{d}\right)$ as d ranges through the divisors of n.

For example,

$$(\tau \cdot \sigma)(4) = \sum_{d|4} \tau(d)\sigma\left(\frac{4}{d}\right)$$
$$= \tau(1)\sigma(4) + \tau(2)\sigma(2) + \tau(4)\sigma(1)$$
$$= 1(7) + 2(3) + 3(1)$$
$$= 7 + 6 + 3 = 16$$

Alternatively, we write

$$(f \cdot g)(n) = \sum_{ab=n} f(a)g(b)$$

That is, $(f \cdot g)(n)$ is the sum of all products $f(a)g(b)$ as a and b range over all integers whose product is n. We next develop some properties of the Dirichlet product.

Theorem 5.3   *Properties of the Dirichlet Product*

DP_1: $f \cdot g = g \cdot f$
DP_2: $(f \cdot g) \cdot h = f \cdot (g \cdot h)$
DP_3: $f \cdot I = f$

DP_4: $f \cdot u = \sum_{d|n} f(d)$

PROOF OF DP_1: $(f \cdot g)(n) = \sum_{d|n} f(d)g\left(\frac{n}{d}\right)$

Let $c = n/d$, so $d = n/c$. As d ranges through the divisors of n, so does c. Therefore,

$$(f \cdot g)(n) = \sum_{c|n} f\left(\frac{n}{c}\right)g(c) = \sum_{c|n} g(c)f\left(\frac{n}{c}\right) = (g \cdot f)(n) \qquad \blacksquare$$

PROOF OF DP$_2$:

$$[(f \cdot g) \cdot h](n) = \sum_{dc=n} (f \cdot g)(d)h(c) = \sum_{dc=n} \left[\sum_{ab=d} f(a)g(b) \right] h(c)$$

$$= \sum_{abc=n} f(a)g(b)h(c)$$

Also

$$[f \cdot (g \cdot h)](n) = \sum_{ad=n} f(a)(g \cdot h)(d) = \sum_{ad=n} f(a) \left[\sum_{bc=d} g(b)h(c) \right]$$

$$= \sum_{abc=n} f(a)g(b)h(c) \qquad \blacksquare$$

PROOF OF DP$_3$:

$$(f \cdot I)(n) = \sum_{d|n} f(d)I\left(\frac{n}{d}\right)$$

If $d < n$, then $n/d > 1$, so $I(n/d) = 0$. Therefore,

$$(f \cdot I)(n) = f(n)I(1) = f(n)1 = f(n) \qquad \blacksquare$$

PROOF OF DP$_4$:

$$(f \cdot u)(n) = \sum_{d|n} f(d)u\left(\frac{n}{d}\right) = \sum_{d|n} f(d)1 = \sum_{d|n} f(d) \qquad \blacksquare$$

Some remarks are in order. If we consider the Dirichlet product as a binary operation on the set of all arithmetic functions, then DP$_1$ says that the Dirichlet product is commutative, DP$_2$ says that the Dirichlet product is associative, and DP$_3$ says that the function $I(n)$ is an identity element with respect to the Dirichlet product. Furthermore, DP$_4$ says that the summation function of $f(n)$ can be represented as the Dirichlet product of $f(n)$ and $u(n)$. In proving that certain functions are multiplicative, the following theorem is quite useful.

Theorem 5.4   *If f and g are both multiplicative, then so is f · g.*

PROOF: Let $h = f \cdot g$. Then $h(mn) = (f \cdot g)(mn) = \sum_{d|mn} f(d)g\left(\frac{mn}{d}\right)$. We must show that if $(m, n) = 1$, then $h(mn) = h(m)h(n)$. If $d \mid mn$ and $(m, n) = 1$, then $d = ab$, where $a \mid m$ and $b \mid n$. Therefore,

$$h(mn) = \sum_{a|m, b|n} f(ab)g\left(\frac{mn}{ab}\right)$$

$$= \sum_{a|m} \sum_{b|n} f(ab)g\left(\frac{mn}{ab}\right)$$

[We can compute $h(mn)$ by summing first over divisors of m.] Since f and g are multiplicative, we have

$$h(mn) = \sum_{a|m}\sum_{b|n} f(a)f(b)g\left(\frac{m}{a}\right)g\left(\frac{n}{b}\right)$$

When summing over divisors of n, terms involving divisors of m are constant. Therefore,

$$h(mn) = \sum_{a|m} f(a)g\left(\frac{m}{a}\right)\left[\sum_{b|n} f(b)g\left(\frac{n}{b}\right)\right] = \sum_{a|m} f(a)g\left(\frac{m}{a}\right)h(n)$$

When summing over divisors of m, $h(n)$ is constant. Therefore,

$$h(mn) = h(n)\sum_{a|m} f(a)g\left(\frac{m}{a}\right) = h(n)h(m) = h(m)h(n) \qquad \blacksquare$$

For example,

$$
\begin{aligned}
h(6) &= (f \cdot g)(6) \\
&= f(1)g(6) + f(2)g(3) + f(3)g(2) + f(6)g(1) \\
&= 1g(6) + f(2)g(3) + f(3)g(2) + f(6)1 \\
&= g(2)g(3) + f(2)g(3) + f(3)g(2) + f(2)f(3) \\
&= [g(2) + f(2)]g(3) + [g(2) + f(2)]f(3) \\
&= [g(2) + f(2)][g(3) + f(3)] \\
&= [1g(2) + f(2)1][1g(3) + f(3)1] \\
&= [f(1)g(2) + f(2)g(1)][f(1)g(3) + f(3)g(1)] \\
&= h(2)h(3)
\end{aligned}
$$

As an immediate consequence, we obtain the following:

Theorem 5.5   *If f is multiplicative and $g(n) = \sum_{d|n}f(d)$, then g is multiplicative.*

PROOF: Since $u(n)$ is totally multiplicative, hence multiplicative, the conclusion follows from hypothesis, Theorem 5.3, part DP_4, and Theorem 5.4.

$\blacksquare$

Now we are ready to prove that the functions $\tau(n)$ and $\sigma(n)$ are multiplicative.

Theorem 5.6 $\tau(n)$ *is multiplicative.*

PROOF: $$\tau(n) = \sum_{d|n}1 = \sum_{d|n}u(d)u\left(\frac{n}{d}\right) = (u \cdot u)(n)$$

Since $u(n)$ is multiplicative, the conclusion follows from Theorem 5.4.

∎

We can now use the canonical factorization of n to compute $\tau(n)$.

Theorem 5.7   *If $n = \prod_{i=1}^{r} p_i^{e_i}$, where the p_i are distinct primes, then $\tau(n) = \prod_{i=1}^{r}(e_i + 1)$.*

PROOF: This follows from Theorems 5.6 and 5.1, since $\tau(p^e) = e + 1$ if p is prime.

∎

For example, $\tau(360) = \tau(2^3 3^2 5^1) = 4 \cdot 3 \cdot 2 = 24$; also, $\tau(112) = \tau(2^4 7^1) = 5 \cdot 2 = 10$.

Theorem 5.8 $\sigma(n)$ *is multiplicative.*

PROOF: $$\sigma(n) = \sum_{d \mid n} d = \sum_{d \mid n} M(d)$$

Since $M(n)$ is multiplicative, the conclusions follows from Theorem 5.5.

∎

We can now use the canonical factorization of n to compute $\sigma(n)$.

Theorem 5.9   *If $n = \prod_{i=1}^{r} p_i^{e_i}$, then*

$$\sigma(n) = \prod_{i=1}^{r} \frac{p_i^{e_i+1} - 1}{p_i - 1}$$

PROOF: This follows from Theorems 5.8 and 5.1, since $\sigma(p^e) = \frac{p^{e+1} - 1}{p - 1}$.

∎

For example,

$$\begin{aligned}
\sigma(360) &= \sigma(2^3 3^2 5^1) \\
&= \left(\frac{2^4 - 1}{2 - 1}\right)\left(\frac{3^3 - 1}{3 - 1}\right)\left(\frac{5^2 - 1}{5 - 1}\right) \\
&= 15 \cdot 13 \cdot 6 = 1170
\end{aligned}$$

Also

$$\sigma(784) = \sigma(2^4 7^2)$$
$$= \left(\frac{2^5 - 1}{2 - 1}\right)\left(\frac{7^3 - 1}{7 - 1}\right)$$
$$= 31 \cdot 57 = 1767$$

There is a connection between the σ function and one of the oldest unsolved problems in number theory. In Book IX, Proposition 36, of his *Elements*, which was written about 300 B.C., Euclid said that a number is "perfect" if it is equal to the sum of its proper divisors. We can redefine perfect numbers in terms of σ.

Definition 5.7 ***Perfect Number***
n is perfect if $\sigma(n) = 2n$.

The smallest perfect number is $6 = 1 + 2 + 3$; the next smallest is $28 = 1 + 2 + 4 + 7 + 14$. The following theorem characterizes even perfect numbers.

Theorem 5.10 *(Euclid–Euler)*
Let n be even. Then n is perfect if and only if $n = 2^{p-1}(2^p - 1)$, where p and $2^p - 1$ are prime.

PROOF: If $n = 2^{p-1}(2^p - 1)$, where the odd factor is prime, then Theorem 5.9 implies that $\sigma(n) = \sigma(2^{p-1})\sigma(2^p - 1) = (2^p - 1)(2^p - 1 + 1) = 2^p(2^p - 1) = 2n$. Conversely, suppose that $n = 2^{k-1}m$, where $k \geq 2$, m is odd, and $\sigma(n) = 2n$. Now Theorem 5.9 implies that $\sigma(2^{k-1})\sigma(m) = 2^k m$, so $(2^k - 1)\sigma(m) = 2^k m$. Since $(2^k - 1, 2^k) = 1$, Euclid's lemma (Theorem 2.5) implies that $(2^k - 1) \mid m$; that is, $m = (2^k - 1)d$ for some $d \geq 1$. Since $(2^k - 1)\sigma(m) = (2^k - 1)m + m$, dividing by $2^k - 1$, we get $\sigma(m) = m + d$. If $d > 1$, then $\sigma(m) \geq m + d + 1$. Therefore, $d = 1$, so $m = 2^k - 1$. Since $\sigma(m) = m + 1$, m is prime, so k is prime. ∎

For example, since $2^5 - 1 = 31$ is prime, it follows that $2^{5-1}(2^5 - 1) = 16 \cdot 31 = 496$ is perfect. Likewise, since $2^7 - 1 = 127$ is prime, it follows that $2^{7-1}(2^7 - 1) = 64 \cdot 127 = 8128$ is perfect.

REMARKS: If $2^k - 1$ is prime, then k is prime. This is the result of Exercise 1 in Chapter 3. The converse of this statement is false, however. For example, 11 is prime, yet $2^{11} - 1 = 2047 = 23 \cdot 89$ is composite. In fact, after studying the contents of Chapter 7, we will be able to show that p is a prime such that $p \equiv 3 \pmod 4$, $p \geq 11$, and $q = 2p + 1$ is prime, then q is a nontrivial factor of $2^p - 1$. The first few primes that satisfy these conditions are 11, 23, 83, 131, and 179.

Let us take a closer look at those special primes which occur as factors of even perfect numbers.

Definition 5.8   ***Mersenne Prime***
If p and q are primes such that $q = 2^p - 1$, then we say that q is a *Mersenne prime*.

Marin Mersenne (1588–1648) was a French monk who was the first to compile a list of primes of the form $2^p - 1$. At present, 32 Mersenne primes are known to exist. These correspond to the following values of p: 2, 3, 5, 7, 13, 17, 19, 31, 61, 89, 107, 127, 521, 607, 1279, 2203, 2281, 3217, 4253, 4423, 9689, 9941, 11213, 19937, 21701, 23209, 44497, 86243, 110503, 132049, 216091, and 756839. All but the first 12 of these results were obtained using computers. According to Theorem 5.10, there is a one-to-one correspondence between Mersenne primes and even perfect numbers. It has been conjectured that there exist infinitely many Mersenne primes.

Euclid claimed that all perfect numbers are even. Now, 23 centuries later, it is still not known whether an odd number can be perfect, but many necessary conditions for the existence of an odd perfect number have been established. For example, an odd perfect number must have at least eight distinct prime factors. (See Guy.) We now develop some results leading to Euler's theorem regarding the canonical factorization of an odd perfect number.

Theorem 5.11   *If the prime $p \equiv 1 \pmod 4$, then $\sigma(p^e) \equiv e + 1 \pmod 4$.*

PROOF:
$$\sigma(p^e) = \sum_{j=0}^{e} p^j \equiv \sum_{j=0}^{e} 1^j \equiv \sum_{j=0}^{e} 1 \equiv e + 1 \pmod 4$$

∎

Theorem 5.12   *If the prime $p \equiv -1 \pmod{4}$, then*

$$\sigma(p^e) \equiv \begin{cases} 1 \pmod{4} & \text{if } 2 \mid e \\ 0 \pmod{4} & \text{if } 2 \nmid e \end{cases}$$

PROOF: $\sigma(p^e) = \dfrac{p^{e+1}-1}{p-1} \equiv \dfrac{(-1)^{e+1}-1}{-2} \pmod{4}$

The conclusion follows easily from this.   ∎

Theorem 5.13   *(Euler)*
If n is odd and perfect, then

$$n = p^a \prod_{i=1}^{r} q_i^{2b_i}$$

where p and the q_i are odd primes and $p \equiv a \equiv 1 \pmod{4}$.

PROOF: By hypothesis, $\sigma(n) = 2n \equiv 2 \pmod{4}$. If $n = \prod_{i=1}^{k} p_i^{e_i}$, then Theorem 5.9 implies that $\sigma(n) = \prod_{i=1}^{k}\sigma(p_i^{e_i})$. The result of Exercise 16 in Chapter 4 implies that there exists unique j such that $\sigma(p_j^{e_j}) \equiv 2 \pmod{4}$, while $\sigma(p_i^{e_i})$ is odd for all $i \neq j$. Theorem 5.12 implies that $p_j \equiv -1 \pmod{4}$, so $p_j \equiv 1 \pmod{4}$. Theorem 5.11 implies that $e_j + 1 \equiv 2 \pmod{4}$, so $e_j \equiv 1 \pmod{4}$. Now suppose that $i \neq j$. If $p_i \equiv 1 \pmod{4}$, then Theorem 5.11 implies that e_i is even. If $p_i \equiv -1 \pmod{4}$, then Theorem 5.12 implies that e_i is even. The conclusion now follows if we let $r = k - 1$, $p_j = p$, $e_j = a$, $p_i = q_i$, and $e_i = 2b_i$ for all $i \neq j$.   ∎

REMARKS: We regret that, at present, no examples of odd perfect numbers have yet been discovered. Such a number, if it exists, would have more than 100 decimal digits. Theorem 5.13 has been useful in eliminating certain candidates from consideration.

SECTION 5.2 EXERCISES

1. Evaluate $\tau(n)$ and $\sigma(n)$ for each n such that $40 \leq n \leq 50$.

2. Prove that $\prod_{d \mid n} d = n^{1/2\tau(n)}$.

3. Prove that if n is a natural number such that $\tau(n) = q$, where q is prime, then $n = p^{q-1}$ for some prime p.

4. Determine the canonical factorization (in terms of unknown primes) of all n such that $\tau(n) = 12$. Find the least such number.

5. If $\omega(n)$ denotes the number of distinct prime factors of n, prove hat $\tau(n) \geq 2^{\omega(n)}$.

6. Prove that $\tau(n)$ is odd if and only if n is a square.

7. Prove that if $n \equiv 7 \pmod 8$, then $\sigma(n) \equiv 0 \pmod 8$.

*8. Prove that if $n \equiv 23 \pmod{24}$, then $\sigma(n) \equiv 0 \pmod{24}$.

9. If $\sigma(n)$ is prime, prove that $n = q^k$, where q is prime and $k \geq 1$.

10. Prove that if $\sigma(p^k) = n$, where p is prime, then $p \mid (n - 1)$.

11. Prove that if n is odd, then $\tau(n) \equiv \sigma(n) \pmod 2$.

12. Prove that if p is prime and $n \geq 2$, then $\sigma(p^{n^2 - 1})$ is composite.

13. Find all even perfect numbers of the form $m^3 + 1$.

14. Prove that each of the following functions is totally multiplicative.
    (a) $u(n)$ (b) $z(n)$ (c) $I(n)$

15. Prove that the ordinary (not Dirichlet) product of two multiplicative functions is multiplicative.

16. Prove Theorem 5.1.

17. Prove that if $mn > 1$, then $\sigma(mn) > m\sigma(n)$.

18. A natural number n is said to be *abundant* if $\sigma(n) > 2n$. Prove that every multiple of an abundant number is also abundant.

19. Prove that if n is perfect, then $\prod_{p \mid n} \frac{p}{p-1} > 2$.

20. Prove that $(M \cdot M)(n) = n\tau(n)$.

21. Prove that if m is a perfect number, then the equation $\sigma(n) = 2(n + m)$ has infinitely many solutions.

22. Prove that if p and q are distinct primes such that $\sigma(p^2) = \sigma(q^4)$, then $p = 5$ and $q = 2$.

23. Prove that there are no primes p and q such that $\sigma(p^2) = \sigma(q^6)$.

24. Prove that if p is prime and $\sigma(p^m) = 2^n$, then $m = 1$ and n is prime.

25. Find the least exponent k such that $k \geq 12$ and the equation $\sigma(n) = 2^k$ has no solution.

SECTION 5.2 COMPUTER EXERCISES

26. Suppose that p is prime such that $2^p - 1$ is composite. Write a computer program to find the least prime factor q of $2^p - 1$ using the facts that (i) $q \equiv 1 \pmod{2p}$, (ii) $2^p \equiv 1 \pmod q$, and (iii) $q^2 \leq 2^p - 1$.

27. Let the odd primes be numbered $p_1 = 3$, $p_2 = 5, p_3 = 7$, etc. Write a computer program that, given an integer i such that

$1 \leq i \leq 20$, computes the least value of r such that

$$\prod_{j=1}^{i+r-1} \frac{p_i}{p_i - 1} > 2$$

(This value of r is the minimum number of distinct prime factors of an odd perfect number whose least prime factor is p_i.)

5.3   DIRICHLET INVERSE, MOEBIUS FUNCTION, EULER'S FUNCTION, AND EULER'S THEOREM

We now return to the study of Dirichlet product. Let f, g, and h be arithmetic functions such that $h = f \cdot g$. We saw in Theorem 5.4 that if f and g are both multiplicative, then so is h. We now prove the more difficult result that if g and h are both multiplicative, then so is f.

Theorem 5.14   *Let $h = f \cdot g$. If g and h are multiplicative, then so is f.*

PROOF: We will show that if g is multiplicative but f isn't, then neither is h. Let mn be the smallest number such that $(m, n) = 1$ yet $f(mn) \neq f(m)f(n)$. Since g and h are multiplicative by hypothesis, assuming $g(n) \neq z(n)$ and $h(n) \neq z(n)$, we have $g(1) = h(1) = 1$ by Theorem 5.2. Since $h(1) = f(1)g(1)$, this implies that $f(1) = 1$. Therefore, $mn > 1$. Now

$$h(mn) = \sum_{d|mn} f(d)g\left(\frac{mn}{d}\right) = f(mn)g(1) + \sum_{d|mn,\, d \neq mn} f(d)g\left(\frac{mn}{d}\right)$$

As in the proof of Theorem 5.4, we have $d = ab$, where $a \mid m$ and $b \mid n$. Therefore,

$$h(mn) = f(mn) + \sum_{a|m, b|n, ab \neq mn} f(ab)g\left(\frac{mn}{ab}\right)$$
$$= f(mn) + \sum_{a|m, b|n, ab \neq mn} f(a)f(b)g\left(\frac{m}{a}\right)g\left(\frac{n}{b}\right)$$
$$= f(mn) - f(m)f(n) + \sum_{a|m}\sum_{b|n} f(a)g\left(\frac{m}{a}\right)f(b)g\left(\frac{n}{b}\right)$$

As in the proof of Theorem 5.4, the double sum simplifies to a product of two sums, so we have

$$h(mn) = f(mn) - f(m)f(n) + \left[\sum_{a|m} f(a)g\left(\frac{m}{a}\right)\right]\left[\sum_{b|n} f(b)g\left(\frac{n}{b}\right)\right]$$
$$= f(mn) - f(m)f(n) + h(m)h(n)$$

Since $f(mn) \neq f(m)f(n)$, it follows that $h(mn) \neq h(m)h(n)$, so h is not multiplicative.   ∎

Recall that the function

$$I(n) = \begin{cases} 1 & \text{if } n = 1 \\ 0 & \text{if } n > 1 \end{cases}$$

is the identity function with respect to the Dirichlet product. We next introduce the concept of the Dirichlet inverse of an arithmetic function. We determine sufficient conditions for the Dirichlet inverse of a function f to exist and show how to compute the inverse.

Definition 5.9   *Dirichlet Inverse*

Given an arithmetic function f, if there exists a function g such that $f \cdot g = I$, then we say g is the *Dirichlet inverse* of f, and we write $g = f^{-1}$.

The following theorem guarantees the existence of the Dirichlet inverse for most functions of interest, especially multiplicative functions, and is constructive.

Theorem 5.15   *If f is an arithmetic function such that $f(1) \neq 0$, then f^{-1} exists.*

PROOF: Let $g(1) = 1/f(1)$. If $n > 1$, let

$$g(n) = - \sum_{d|n, d \neq n} f\left(\frac{n}{d}\right) g(d)$$

Then $(f \cdot g)(1) = f(1)g(1) = 1$, whereas if $n > 1$, then

$$(f \cdot g)(n) = (g \cdot f)(n)$$
$$= \sum_{d|n} g(d) f\left(\frac{n}{d}\right)$$
$$= g(n) + \sum_{d|n, d \neq n} g(d) f\left(\frac{n}{d}\right) = 0$$

Therefore, $(f \cdot g)(n) = I(n)$.   ∎

Next, we see that every nontrivial multiplicative function has a nontrivial multiplicative inverse.

Theorem 5.16   *If f is multiplicative and $f(n) \neq z(n)$, then f^{-1} exists, f^{-1} is multiplicative, and $f^{-1}(n) \neq z(n)$.*

PROOF: Since $f(n)$ is multiplicative and $f(n) \neq z(n)$ by hypothesis, Theorem 5.2 implies that $f(1) = 1$. Therefore, Theorem 5.15 implies that f^{-1} exists. Now $f \cdot f^{-1} = I$. Since f and I are multiplicative, Theorem 5.14 implies that f^{-1} is multiplicative. Finally, if $f^{-1}(n) = z(n)$, then $I = f \cdot f^{-1} = f \cdot z = z$, an impossibility.   ∎

Let us now construct the Dirichlet inverse of the unit function $u(n) = 1$. We obtain a function that is known as the *Moebius function* and is denoted $\mu(n)$.

Definition 5.10   **Moebius Function**

Let $\mu(n) = u^{-1}(n)$. Since μ is multiplicative, $\mu(1) = 1$, $\mu(p) = -\mu(1)u(p) = -1$, and $\mu(p^2) = -[\mu(1)u(p^2) + \mu(p)u(p)] = -(1 - 1) = 0$. If $k \geq 3$, then

$$\mu(p^k) = -\sum_{d|p^k, d \neq p^k} \mu(d)u\left(\frac{p^k}{d}\right)$$

$$= -\sum_{j=0}^{k-1} \mu(p^j)u(p^{k-j})$$

$$= -[\mu(1)u(p^k) + \mu(p)u(p^{k-1}) + \sum_{j=2}^{k-1} \mu(p^j)u(p^{k-j})]$$

$$= -(1 - 1 + 0) = 0$$

(The last summation is 0 by induction on k.) We have seen that

$$\mu(p^k) = \begin{cases} -1 & \text{if } k = 1 \\ 0 & \text{if } k > 1 \end{cases}$$

Since μ is multiplicative, we can now compute $\mu(n)$ for any n.

Theorem 5.17   *If $n = \prod_{i=1}^{r} p_i^{e_i}$, then*

$$\mu(n) = \begin{cases} (-1)^r & \text{if all } e_i = 1 \\ 0 & \text{otherwise} \end{cases}$$

PROOF: Theorem 5.15 implies that μ is multiplicative. Theorem 5.1 implies that $\mu(n) = \prod_{i=1}^{r} \mu(p_i^{e_i})$. If all $e_i = 1$, then $\mu(n) = \prod_{i=1}^{r}(-1) = (-1)^r$. If any $e_i \geq 2$, then $\mu(p_i^{e_i}) = 0$, so $\mu(n) = 0$.   ∎

For example,

$$\mu(42) = \mu(2 \cdot 3 \cdot 7) = (-1)^3 = -1$$
$$\mu(44) = \mu(2^2 \cdot 11) = 0$$
$$\mu(46) = \mu(2 \cdot 23) = (-1)^2 = 1$$
$$\mu(47) = -1$$

If $g(n)$ is the summation function of $f(n)$, then the Moebius inversion formula, which follows below, allows us to express $f(n)$ in terms of a summation involving $g(n)$.

Theorem 5.18   *(Moebius Inversion Formula)*

$$g(n) = \sum_{d|n} f(d) \qquad \text{if and only if} \qquad f(n) = \sum_{d|n} g(d)\mu\left(\frac{n}{d}\right)$$

PROOF: By Theorem 5.3, part DP_4 and Definition 5.6, it suffices to show that $g = f \cdot u$ if and only if $f = g \cdot \mu$. If $g = f \cdot u$, then $g \cdot \mu = (f \cdot u) \cdot \mu = f \cdot (u \cdot \mu) = f \cdot I = f$, by Theorem 5.3, part DP_2, and DP_3. Similarly, if $f = g \cdot \mu$, then $f \cdot u = (g \cdot \mu) \cdot u = g \cdot (\mu \cdot u) = g \cdot I = g$.   ■

For example, since $\sigma(n) = \sum_{d|n} d$, the Moebius inversion formula implies that $n = \sum_{d|n} \sigma(d)\mu(n/d)$. Similarly, since $u(n) = 1 = \sum_{d|n} I(d)$, the Moebius inversion formula implies that $I(n) = \sum_{d|n} u(d)\mu(n/d) = \sum_{d|n} \mu(n/d)$. We can simplify this result by noting that $I = u \cdot \mu = \mu \cdot u$, so $I(n) = \sum_{d|n} \mu(d)$.

We next define and develop the properties of Euler's totient function. Euler's totient occurs in many places, including Euler's theorem, which generalizes Fermat's theorem.

Definition 5.11   ***Euler's Totient Function***

Let $\phi(n)$ denote the number of integers k such that $1 \le k < n$ and $(k, n) = 1$.

For example, $\phi(9) = 6$, since $(k, 9) = 1$ for $k = 1, 2, 4, 5, 7,$ and 8; Also, $\phi(10) = 4$, since $(k, 10) = 1$ for $k = 1, 3, 7,$ and 9. Note that $\phi(p) = p - 1$ and $\phi(p^k) = p^k - p^{k-1} = p^{k-1}(p - 1)$.

Theorem 5.19 $$\sum_{d|n} \phi(d) = n$$

PROOF: If $d \mid n$, let $S(d) = \{k : 1 \le k \le n$ and $(k, n) = d\}$. Let $f(d)$ be the number of elements of the set $S(d)$. Let $1 \le k \le n$. If k belongs to both $S(d_1)$ and $S(d_2)$, where d_1 and d_2 are distinct divisors of n, then $d_1 = (k, n) = d_2$, and impossibility. Therefore, the sets $S(d)$ are pairwise disjoint. Furthermore, there exists d such that $d \mid n$ and $(k, n) = d$; that is, k belongs to $S(d)$. Therefore, the union of all sets $S(d)$ such that $d \mid n$ is the set $\{1, 2, 3, \cdots, n\}$. This implies that $\sum_{d|n} f(d) = n$.

Theorem 2.3, parts G_5 and G_6 imply that $(k, n) = d$ if and only if $(k/d, n/d) = 1$. By Definition 5.11, the number of such k is $\phi(n/d)$; that is, $f(d) = \phi(n/d)$. Now $M(n) = n = \sum_{d|n} \phi(n/d) = \sum_{d|n} u(d)\phi(n/d) = (u \cdot \phi)(n)$. Theorem 5.3, part DP_1 implies that $M(n) = (\phi \cdot u)(n)$. Therefore, Theorem 5.3, part DP_4 implies that $n = M(n) = \sum_{d|n} \phi(d)$.   ■

For example, the divisors of 12 are 1, 2, 3, 4, 6, and 12. Thus,

$$\sum_{d|12} \phi(d) = \phi(1) + \phi(2) + \phi(3) + \phi(4) + \phi(6) + \phi(12)$$
$$= 1 + 1 + 2 + 2 + 2 + 4 = 12$$

As a by-product of the proof of Theorem 5.19, we can now show that ϕ is a multiplicative function.

Theorem 5.20 ϕ *is multiplicative.*

PROOF: Let $M(n) = n$ and $u(n) = 1$. We saw in the proof of Theorem 5.19 that $M(n) = (\phi \cdot u)(n)$. Since $M(n)$ and $u(n)$ are multiplicative, it follows from Theorem 5.14 that $\phi(n)$ is multiplicative.   ∎

In the next theorem we give a formula that allows one to compute $\phi(n)$ from the canonical factorization of n.

Theorem 5.21   *If $n = \prod_{i=1}^{r} p_i^{e_i}$, where the p_i are distinct primes, then*

$$\phi(n) = \prod_{i=1}^{r} p_i^{e_i - 1}(p_i - 1)$$

PROOF: This follows from Theorems 5.20 and 5.1, since $\phi(p^e) = p^{e-1}(p - 1)$ if p is prime.   ∎

For example, $\phi(675) = \phi(3^3 5^2) = \phi(3^3)\phi(5^2) = (3^2 \cdot 2)(5 \cdot 4) = 360$; also, $\phi(735) = \phi(3 \cdot 5 \cdot 7^2) = \phi(3)\phi(5)\phi(7^2) = 2 \cdot 4 \cdot (7 \cdot 6) = 336$.

The following two theorems, whose proofs are left as exercises, provide alternate ways of computing $\phi(n)$.

Theorem 5.22

$$\phi(n) = \sum_{d|n} d\mu\left(\frac{n}{d}\right)$$

PROOF: Exercise

Theorem 5.23

$$\phi(n) = n\prod_{p|n}\left(1 - \frac{1}{p}\right)$$

PROOF: Exercise

Euler's theorem (Theorem 5.26 below) is a generalization of Fermat's theorem (Theorem 4.8). We proceed to prove Euler's theorem by means of the following two preliminary theorems.

Theorem 5.24   *If p is prime, $(a, p) = 1$, and $r \geq 1$, then $a^{\phi(p^r)} \equiv 1 \pmod{p^r}$.*

PROOF: (Induction on r)   Let $(a, p) = 1$. Then Fermat's theorem implies that $a^{p-1} \equiv 1 \pmod{p}$. However, $p - 1 = \phi(p)$, so $a^{\phi(p)} \equiv 1 \pmod{p}$; that is, Theorem 5.24 holds for $r = 1$. Now we must show that if $a^{\phi(p^r)} \equiv 1 \pmod{p^r}$, then $a^{\phi(p^{r+1})} \equiv 1 \pmod{p^{r+1}}$. To simplify our task, let $b = a^{\phi(p^r)} = a^{p^{r-1}(p-1)}$, so $b^p = (a^{p^{r-1}(p-1)})^p = a^{p^r(p-1)} = a^{\phi(p^{r+1})}$. In terms of b, we must show that if $b \equiv 1 \pmod{p^r}$, then $b^p \equiv 1 \pmod{p^{r+1}}$; that is, if $p^r \mid (b - 1)$, then $p^{r+1} \mid (b^p - 1)$. Since $(b - 1) \mid (b^p - 1)$, by Theorem 2.1, part D_6 it suffices to show that $p \mid \left(\frac{b^p - 1}{b - 1}\right)$. Now $\frac{b^p - 1}{b - 1} = \sum_{j=0}^{p-1} b^j \equiv \sum_{j=0}^{p-1} 1^j \equiv p \pmod{p^r}$, so Theorem 4.1, part C_9 implies that $\frac{b^p - 1}{b - 1} \equiv p \equiv 0 \pmod{p}$.   ∎

For example, let $p = 7$, $r = 2$, and $(a, 7) = 1$. Then $a^{\phi(7^2)} \equiv 1 \pmod{7^2}$; that is, $a^{42} \equiv 1 \pmod{49}$. Also, if $p = 5$, $r = 3$, and $(a, 5) = 1$, then $a^{\phi(5^3)} \equiv 1 \pmod{5^3}$; that is, $a^{100} \equiv 1 \pmod{125}$.

Theorem 5.25   *If $(m, n) = 1$, $a^{\phi(m)} \equiv 1 \pmod{m}$, $a^{\phi(n)} \equiv 1 \pmod{n}$, and $(a, mn) = 1$, then $a^{\phi(mn)} \equiv 1 \pmod{mn}$.*

PROOF: Since $(m, n) = 1$ by hypothesis, Theorem 5.20 implies that $\phi(mn) = \phi(m)\phi(n)$. Therefore, $a^{\phi(mn)} = a^{\phi(m)\theta(n)} = (a^{\phi(m)})^{\phi(n)}$. Since $a^{\phi(m)} \equiv 1 \pmod{m}$ by hypothesis, we have $a^{\phi(mn)} \equiv 1^{\phi(n)} \equiv 1 \pmod{m}$. Similarly, $a^{\phi(mn)} \equiv 1 \pmod{n}$. Now Theorem 2.9, part L_6 and the result of Exercise 45 in Chapter 2 imply that $a^{\phi(mn)} \equiv 1 \pmod{mn}$.   ∎

For example, suppose that $(a, 72) = 1$. Since $72 = 8 \cdot 9$, this implies that $(a, 8) = (a, 2) = 1$ and $(a, 9) = (a, 3) = 1$. Theorem 5.24 implies that $a^{\phi(8)} \equiv 1 \pmod{8}$ and $a^{\phi(9)} \equiv 1 \pmod{9}$. Since $(8, 9) = 1$, Theorem 5.25 implies that $a^{\phi(72)} \equiv 1 \pmod{72}$; that is, $a^{24} \equiv 1 \pmod{72}$.

Theorem 5.26   *(Euler's Theorem)*
If $m \geq 2$ and $(a, m) = 1$, then $a^{\phi(m)} \equiv 1 \pmod{m}$.

PROOF: Let $m = \prod_{i=1}^{r} p_i^{e_i}$, where the p_i are distinct primes, and use induction on r. By Theorem 5.24, Euler's theorem holds for $r = 1$. If $m = \prod_{i=1}^{r+1} p_i^{e_i}$, let $j = \prod_{i=1}^{r} p_i^{e_i}$ and $k = p_r^{e_r+1}$, so $m = jk$. Since $(a, m) = 1$ by hypothesis, that is $(a, jk) = 1$, it follows that $(a, j) = (a, k) = 1$. By induction hypothesis, $a^{\phi(j)} \equiv 1 \pmod{j}$ and $a^{\phi(k)} \equiv 1 \pmod{k}$. Since $(j, k) = 1$, Theorem 5.25 implies that $a^{jk} \equiv 1 \pmod{jk}$; that is, $a^{\phi(m)} \equiv 1 \pmod{m}$. ∎

For example, since $(3, 10) = 1$ and $\phi(10) = 4$, it follows from Euler's Theorem that $3^4 \equiv 1 \pmod{10}$. Similarly, since $(5, 18) = 1$ and $\phi(18) = 6$, it follows that $5^6 \equiv 1 \pmod{18}$.

Let $(a, m) = 1$, where either (i) $m = 4n$ for some $n \geq 2$ or (ii) $pq \mid m$, where p and q are distinct odd primes. In the next chapter we shall see that under these circumstances a stronger result than Euler's theorem holds, namely, $a^{\frac{1}{2}\phi(m)} \equiv 1 \pmod{m}$. For example, since $15 = 3 \cdot 5$ and $(7, 15) = 1$, we have $7^{\frac{1}{2}\phi(15)} \equiv 1 \pmod{15}$; that is, $7^4 \equiv\ = 1 \pmod{15}$. Also, since $12 = 4 \cdot 3$ and $(5, 12) = 1$, we have $5^{\frac{1}{2}\phi(12)} \equiv 1 \pmod{12}$; that is, $5^2 \equiv 1 \pmod{12}$.

Leonhard Euler

Leonhard Euler was born in Basel, Switzerland, on April 15, 1707. His father, a Protestant minister, was fond of mathematics and had attended the lectures of Jakob Bernoulli. Euler entered the University of Basel in 1720. At that time, Johann Bernoulli was on the faculty. Bernoulli did not give private lessons to Euler but told him what books to study on his own.

In 1723 Euler received a master's degree in philosophy. He preferred mathematics, however. In 1727 Euler applied for a vacant position at the University of Basel but was turned down. He then accepted a position at the newly organized St. Petersburg Academy in Russia, where he had been recommended by the Bernoulli family.

In St. Petersburg, Euler was active not only in mathematics, but also in many practical matters: mapping the Russian territory, shipbuilding and navigation, the testing of pumps and scales, and so on. He wrote on mechanics, hydromechanics, and lunar and planetary motion. His reputation grew, and he became a member of the scientific academies of the major European nations. In 1738 he lost the sight in his right eye.

Due to a change of regime in Russia, Euler's position became less secure. In 1741 he accepted an offer from Frederick II, King of Prussia, to come to the Berlin Academy. In Berlin, in addition to his ongoing research program, Euler had many administrative duties. He supervised the botanical garden and the observatory, hired personnel, and managed the publication of maps and calendars. He was even responsible for the plumbing in the king's summer palace. Euler did not get along that well with King Frederick. When the presidency of the Berlin Academy became vacant, Frederick refused to appoint Euler to the position.

In 1766 Euler accepted an invitation from Queen Katherine of Russia to return to St. Petersburg, where he spent the remaining years of his life. In 1771 Euler became completely blind. This did not curtail his scientific activities. Euler died on September 18, 1783.

Euler's contributions to number theory include the discovery of what is now known as Euler's function, the discovery (but not the proof) of the law of quadratic reciprocity (which we shall encounter in Chapter 7), the use of continued fractions to solve Pell's equation, and the proof of Fermat's conjecture for the case $n = 3$.

Euler also was active in algebra, analysis, differential equations, the calculus of variations, and geometry. In addition, Euler introduced some important features of modern mathematical notation, including the use of the letter e to denote the base of the natural logarithms, the use of the letter i to denote the square root of -1, and the use of $f(x)$ to denote functional notation.

We summarize our results on arithmetic functions in Table 5.1.

Table 5.1. Summary of Results on Arithmetic Functions

Function	Definition	$F(p^e)$	Dirichlet Form	Name
$I(n)$	$\begin{array}{ll}1 & \text{if } n = 1 \\ 0 & \text{if } n > 1\end{array}$	0	—	Dirichlet identity
$M(n)$	n	p^e	—	Multiplicative identity
$u(n)$	1	1	—	Unit function (constant)
$\tau(n)$	$\sum_{d \mid n} 1$	$e + 1$	$(u \cdot u)(n)$	Number-of-divisors function
$\sigma(n)$	$\sum_{d \mid n} d$	$\frac{p^{e+1}-1}{p-1}$	$(M \cdot u)(n)$	Sum-of-divisors function
$\mu(n)$	$\begin{array}{ll}1 & \text{if } n = 1 \\ 0 & \text{if } p^2 \mid n \\ (-1)^r & \text{if } n = \prod_{i=1}^{r} p_i\end{array}$	$\begin{array}{ll}-1 & \text{if } e = 1 \\ 0 & \text{if } e > 1\end{array}$	$u^{-1}(n)$	Moebius function
$\phi(n)$	$\sum_{1 \le k \le n, (k,n)=1} 1$	$p^{e-1}(p - 1)$	$(M \cdot \mu)(n)$	Euler's totient function

Note: $I, M,$ and u are totally multiplicative; $\tau, \sigma, \mu,$ and ϕ are multiplicative.

SECTION 5.3 EXERCISES

28. Prove that $\sum_{d|n} |\mu(d)| = 2^{\omega(n)}$.

29. Prove that $\sum_{d|n} \mu(d)\tau(d) = (-1)^{\omega(n)}$.

30. Prove that $\sum_{d|n} \mu(d)\tau(n/d) = 1$.

31. Prove that $\sum_{d|n} \sigma(d)\mu(n/d) = n$.

32. Find a formula for $\sum_{d|n} \mu(d)\sigma(d)$ in terms of the canonical factorization of n.

33. Let $h(n) = \sigma^{-1}(n)$. Find (a) $h(p)$, (b) $h(p^2)$, and (c) $h(p^k)$, where $k \geq 3$.

34. Let $M(n) = n$ and $g(n) = M^{-1}(n)$. Find (a) $g(p)$, (b) $g(p^2)$, and (c) $g(p^k)$, where $k \geq 3$. Prove that $g(n) = \mu(n)M(n)$.

35. Prove that if f and g are functions such that f^{-1} and g^{-1} exist, then $(f \cdot g)^{-1} = f^{-1} \cdot g^{-1}$.

36. Evaluate $\mu(n)$ and $\phi(n)$ for each of the following values of n.

    (a) 42   (b) 79

    (c) 1000   (d) 945

    (e) 91   (f) 144

37. Prove that the number of reduced fractions a/b such that $1 \leq a < b \leq n$ is $\sum_{k=1}^{n} \phi(k)$.

38. Prove that if n is odd, then $2^{\omega(n)} \mid \phi(n)$.

39. Solve the equation $\phi(x) = n$ for each of the following values of n.

    (a) 1   (b) 2

    (c) 4   (d) 6

    (e) 8   (f) 10

    (g) 12

Hint: if $p \mid x$, then $(p - 1) \mid \phi(x)$.

40. Prove that if p is prime and $p \equiv 1 \pmod 3$, then the equation $\phi(x) = 2p$ has no solution.

41. Prove that $\phi(n) \mid n$ if and only if $n = 1, 2^a$, or $2^a 3^b$, where a and b are positive.

*42. Prove that if $\phi(x) = n$, then $x \leq 2(3^{\log_2 n})$.

43. If $m \geq 2$, find a formula for the sum of all integers k such that $0 < k < m$ and $(k, m) = 1$.

44. Prove that $I(n) = [1/n]$.

45. Prove that $\tau(n) = \sum_{k=1}^{n} \left[\frac{k}{n}\left[\frac{n}{k}\right]\right]$.

46. Prove that $\phi(n) = \sum_{k=1}^{n} \left[\frac{1}{(k,n)}\right]$.

47. Prove that $\frac{\phi(n)}{n} = \sum_{d|n} \frac{\mu(d)}{d}$.

48. Find a formula for $\sum_{d|n} \mu(d)\mu(n/d)$.

49. Prove that $n \mid \phi(2^n - 1)$ for all $n \geq 1$.

50. Prove that the equation $\phi(x) = 2^{32}$ has no odd solutions.

51. Prove Theorem 5.22.

52. Prove Theorem 5.23.

53. If you are familiar with abstract algebra, obtain an alternate proof of Euler's theorem as follows: For $m \geq 2$, let $Z_m = \{0, 1, 2, 3, \cdots, m - 1\}$. (a) Prove that multiplication (mod m) is an associative operation on Z_m. (b) Let S_m be the set of elements of Z_m that are relatively prime to m. Prove that S_m is a group of order $\phi(m)$ with respect to multiplication (mod m). (c) If t is any element of S_m, consider the cyclic subgroup generated by t, and quote an appropriate theorem of group theory.

54. Prove that if $m \mid n$, then $\phi(m) \mid \phi(n)$.

55. Prove that if $\phi(n) = \phi(n + 1)$, then $\mu(n) \neq 0$.

56. Let m and n be integers such that $x = m$ is the unique solution of the equation $\phi(x) = n$. Prove that $(2 \cdot 3 \cdot 7 \cdot 43)^2 \mid m$. (It is not known whether such an m exists.)

SECTION 5.3 COMPUTER EXERCISES

57. Write a computer program to compute $\phi(n)$ using the formula of Exercise 46.

58. Write a computer program that, given n, finds all solutions, if any, of the equation $\phi(x) = 2^n$.

59. Write a computer program to find all n such that $1 \leq n \leq 300$ and $\phi(n) = \phi(n + 1)$. Is there a pattern?

CHAPTER SIX

PRIMITIVE ROOTS AND INDICES

6.1 INTRODUCTION

Suppose that g and m are integers such that $0 < g < m$ and $(g, m) = 1$. Euler's theorem implies that $g^{\phi(m)} \equiv 1 \pmod{m}$. In many cases there is a smaller positive exponent k such that $g^k \equiv 1 \pmod{m}$. For example, if $g = 4$ and $m = 7$, then $(4, 7) = 1$ and $\phi(7) = 6$, but $4^3 \equiv 1 \pmod 7$. If h is the least positive exponent such that $g^h \equiv 1 \pmod{m}$, then we say g *has order* h (mod m) or g *belongs to* h (mod m). For example, 4 has order 3 (mod 7); also, 5 has order 6 (mod 7). By Euler's theorem, h exists and $h \leq \phi(m)$. In fact, we shall see that $h \mid \phi(m)$. If $h = \phi(m)$, then we call g a *primitive root* (mod m). We shall see below that 3 and 5 are primitive roots (mod 7).

If g is a primitive root (mod m) and $(t, m) = 1$, then t can be represented as a power of g (mod m). For example, $4 \equiv 5^2 \pmod 7$; also, $3 \equiv 5^5 \pmod 7$. Not all natural numbers have primitive roots. For example, there are no primitive roots (mod 12) or (mod 15). We will determine for a given m how many primitive roots exist, if any, and how to generate all primitive roots from a single primitive root. Finally, we develop the properties of indices, which are related to primitive roots and which enable us to solve certain polynomial congruences.

6.2 PRIMITIVE ROOTS: DEFINITION AND PROPERTIES

If $(a, m) = 1$, let S be the set of positive exponents k such that $a^k \equiv 1 \pmod{m}$. Euler's theorem implies that $\phi(m)$ belongs to S, so S is nonempty. We now investigate the properties of the least element of S.

Definition 6.1   *Order (mod m)*

If $(a, m) = 1$, let h be the least natural number such that $a^h \equiv 1 \pmod{m}$. We say that *a has order h* (mod m) or that *a belongs to h* (mod m).

For example,

$$2^1 \equiv 2 \not\equiv 1 \pmod{7}$$
$$2^2 \equiv 4 \not\equiv 1 \pmod{7}$$
$$2^3 \equiv 8 \equiv 1 \pmod{7}$$

Therefore, 2 has order 3 (mod 7). Also,

$$5^1 \equiv 5 \not\equiv 1 \pmod{12}$$
$$5^2 \equiv 25 \equiv 1 \pmod{12}.$$

Therefore, 5 has order 2   (mod 12).

Note that if a has order h (mod m), then $a^h \equiv 1$ (mod m) *and* $a^k \not\equiv 1$ (mod m) for all k such that $0 < k < h$. Furthermore, by Euler's theorem, $h \leq \phi(m)$. The two following theorems enable us to say more about h, namely, that $h \mid \phi(m)$.

Theorem 6.1   *If $(a, m) = 1$, a has order h   (mod m), and $a^k \equiv 1$ (mod m), then $h \mid k$.*

PROOF: Let $k = qh + r$, where $0 \leq r < h$. Now $a^k = a^{qh+r} = (a^h)^q a^r$, so $(a^h)^q a^r \equiv a^k$ (mod m). By hypothesis, $a^k \equiv a^h \equiv 1$ (mod m), so $a^r \equiv 1$ (mod m). By definition of h, we must have $r = 0$, so $h \mid k$.   ■

Theorem 6.2   *If $(a, m) = 1$ and a has order h   (mod m), then $h \mid \phi(m)$.*

PROOF: This follows from Euler's theorem (Theorem 5.26) and Theorem 6.1.   ■

For example, we have seen that 2 has order 3   (mod 7). Now $\phi(7) = 6$, and $3 \mid 6$. Also, 5 has order 2   (mod 12). Now $\phi(12) = 4$, and $2 \mid 4$. Theorem 6.2 is useful in computing the order of an integer a (mod m), where $(a, m) = 1$. For example, if $1 \leq t \leq 4$ and we wish to compute the order of t (mod 5), Theorem 6.2 tells us that the only possible candidates are the divisors of $\phi(5) = 4$, namely, 1, 2, and 4. Let us compute the order (mod 5) of each of 1, 2, 3, and 4, making use of the multiplication table for multiplication (mod 5):

*	0	1	2	3	4
0	0	0	0	0	0
1	0	1	2	3	4
2	0	2	4	1	3
3	0	3	1	4	2
4	0	4	3	2	1

We see that $1^1 \equiv 1$ (mod 5), so 1 has order 1   (mod 5); also,

$$2^1 \equiv 2 \not\equiv 1 \quad (\text{mod } 5)$$
$$2^2 \equiv 4 \not\equiv 1 \quad (\text{mod } 5)$$
$$2^4 \equiv (2^2)^2 \equiv 4^2 \equiv 1 \quad (\text{mod } 5)$$

so 2 has order 4   (mod 5). In addition,

$$3^1 \equiv 3 \not\equiv 1 \quad (\text{mod } 5)$$
$$3^2 \equiv 4 \not\equiv 1 \quad (\text{mod } 5)$$
$$3^4 \equiv (3^2)^2 \equiv 4^2 \equiv 1 \quad (\text{mod } 5)$$

so 3 has order 4   (mod 5). Also,

$$4^1 \equiv 4 \not\equiv 1 \quad (\text{mod } 5)$$
$$4^2 \equiv 1 \quad (\text{mod } 5)$$

so 4 has order 2   (mod 5).

We now introduce the concept of a primitive root   (mod m).

Definition 6.2   ***Primitive Root (*mod *m)***
If $(g, m) = 1$ and g has order $\phi(m)$ (mod m), then we say g is a primitive root (mod m).

For example, the preceding computations show that 2 and 3 are primitive roots (mod 5). Similarly, one can show that 3 and 5 are primitive roots (mod 7).

There are moduli for which primitive roots fail to exist. For example, let $m = 8$. If $(t, 8) = 1$, then $t \equiv 1, 3, 5,$ or 7   (mod 8). However, $1^1 \equiv 3^2 \equiv 5^2 \equiv 7^2 \equiv 1$ (mod 8). Now $\phi(8) = 4$, but no integer has order 4   (mod 8), so there are no primitive roots (mod 8). The reader can verify that there are no primitive roots (mod 12).

In the work ahead we will see that (1) if p is an odd prime and $k \geq 1$, then primitive roots exist (mod p^k) and (mod $2p^k$); (2) if g is a primitive root (mod m), then there are precisely $\phi[\phi(m)]$ primitive roots (mod m), all of which can be obtained from g; and (3) if $m \neq 2, 4, p^k$ or $2p^k$, then m has no primitive roots.

If $(a, m) = 1$ and the order of a (mod m) is known, the next theorem tells us how to compute the order of any power of a.

Theorem 6.3   *If $(a, m) = 1$ and a has order h (mod m), then a^k has order $h/(h, k)$ (mod m).*

PROOF: Let a^k belong to j (mod m). Let $d = (h, k)$, so $h = bd$ and $k = cd$ for some b and c such that $(b, c) = 1$. Now $(a^k)^b = a^{(cd)b} = (a^{bd})^c = (a^h)^c \equiv 1^c \equiv 1$ (mod m), by definition of h. Therefore, by definition of j and Theorem 6.1, we have $j \mid b$. On the other hand, $a^{kj} = (a^k)^j \equiv 1$ (mod m) by definition of j, so Theorem 6.1 implies that $h \mid kj$; that is, $bd \mid cdj$, so $b \mid cj$. Since $(b, c) = 1$, Euclid's lemma (Theorem 2.5) implies that $b \mid j$. Therefore, $j = b = h/d = h/(h, k)$.   ∎

For example, the reader can verify that 2 has order 8   (mod 17). Since $4 = 2^2$, 4 has order $8/(8, 2) = 8/2 = 4$ (mod 17); since $8 = 2^3$, 8 has order $8/(8, 3) = 8/1 = 8$ (mod 17); and since $16 = 2^4$, 16 has order $8/(8, 4) = 8/4 = 2$ (mod 17).

In elementary algebra, one learns that if $a > 1$, then $a^j = a^k$ if and only if $j = k$. The following theorem states a corresponding property regarding congruences.

Theorem 6.4   *If a has order h (mod m), then $a^j \equiv a^k$ (mod m) if and only if $j \equiv k$ (mod h).*

PROOF: By hypothesis, $(a, m) = 1$. First, assume that $a^j \equiv a^k$ (mod m). Without loss of generality, let $j \geq k$, so $j - k \geq 0$. Canceling a^k from each side of the last congruence, we get $a^{j-k} \equiv 1$ (mod m). Now Theorem 6.1 implies that $h \mid (j - k)$, since a has order h (mod m) by hypothesis. Therefore, $j \equiv k$ (mod h). Conversely, suppose that $j \equiv k$ (mod h), so $j = k + th$ for some $t \geq 0$. Now $a^j = a^{k+th} = a^k(a^h)^t$, so $a^j \equiv a^k(a^h)^t$ (mod m). However, $a^h \equiv 1$ (mod m) by hypothesis, so we conclude that $a^j \equiv a^k$ (mod m).   ∎

For example, since $2^{26} \equiv 2^2$ (mod 17) and 2 belongs to 8 (mod 17), it follows that $26 \equiv 2$ (mod 8). Also, since $3^{15} \equiv 3^3$ (mod 7), and 3 belongs to 6 (mod 7), it follows that $15 \equiv 3$ (mod 6). In the opposite direction, since $10 \equiv 2$ (mod 8), it follows that $2^{10} \equiv 2^2 \equiv 4$ (mod 17). Also, since $13 \equiv 1$ (mod 6), it follows that $3^{13} \equiv 3^1 \equiv 3$ (mod 7).

Note that Theorem 6.4 provides a convenient way to evaluate the least positive residue of a^j (mod m) if $(a, m) = 1$. First, find the order h of a (mod m). Then divide j by h to obtain a remainder k. Finally, compute a^k (mod m).

For example, suppose we wish to compute 3^{100} (mod 7). Now 3 has order 6 (mod 7). Since $100 = 16(6) + 4$, it follows that $3^{100} \equiv 3^4 \equiv 4$ (mod 7). Or suppose we wish to compute 2^{100} (mod 17). Since 2 has order 8 (mod 17) and $100 = 12(8) + 4$, it follows that $2^{100} \equiv 2^4 \equiv 16$ (mod 17).

Theorem 6.4 also has the following consequence.

Theorem 6.5   *If $(a, m) = 1$ and a has order h (mod m), then $1, a, a^2, a^3, \cdots, a^{h-1}$ are distinct (mod m).*

PROOF: We will show that if $a^j \equiv a^k$ (mod m), where $0 \leq k \leq j \leq h - 1$, then $j = k$. Since $a^j \equiv a^k$ (mod m), Theorem 6.4 implies that $j \equiv k$ (mod h), so $j - k = th$ for some $t \geq 0$. Also, $0 \leq j - k \leq h - 1$, so $0 \leq th \leq h - 1$. This implies that $t = 0$, so $j = k$.   ∎

For example, since $(2, 7) = 1$ and 2 has order 3 (mod 7), it follows that $1, 2$, and 2^2 are distinct (mod 7). Also, since $(5, 7) = 1$ and 5 has order 6 (mod 7), that is, 5 is a primitive root (mod 7), it follows that $1, 5, 5^2, 5^3, 5^4$, and 5^5 are all distinct (mod 7). Note that

$$1 \equiv 5^0 \quad (\text{mod } 7)$$
$$2 \equiv 5^4 \quad (\text{mod } 7)$$
$$3 \equiv 5^5 \quad (\text{mod } 7)$$
$$4 \equiv 5^2 \quad (\text{mod } 7)$$
$$5 \equiv 5^1 \quad (\text{mod } 7)$$
$$6 \equiv 5^3 \quad (\text{mod } 7)$$

This phenomenon may be generalized as follows:

Theorem 6.6   *If g is a primitive root (mod m) and $(a, m) = 1$, then there exists k such that $0 \leq k \leq \phi(m) - 1$ and $a \equiv g^k$ (mod m).*

PROOF: Let $S = \{j : 1 \leq j < m$ and $(j, m) = 1\}$. Note that S has $\phi(m)$ elements. If $(a, m) = 1$, let r be the least positive residue of a (mod m), so r belongs to S. Let T be the set of least positive residues (mod m) of the set $\{1, g, g^2, g^3, \cdots, g^{\phi(m)-1}\}$. Since g has order $\phi(m)$ by hypothesis, Theorem 6.5 implies that the $\phi(m)$ elements of T are distinct (mod m). Now T is a subset of S, and T has the same number of elements as S. It follows that $T = S$, so $r \equiv g^k$ (mod m) for some k such that $0 \leq k \leq \phi(m) - 1$. Since $a \equiv r$ (mod m), we have $a \equiv g^k$ (mod m).   ∎

For example, if $m = 9$, then $S = \{1, 2, 4, 5, 7, 8\}$. Now 2 is a primitive root (mod 9), and we have

$$1 \equiv 2^0 \quad (\text{mod } 9)$$
$$2 \equiv 2^1 \quad (\text{mod } 9)$$
$$4 \equiv 2^2 \quad (\text{mod } 9)$$
$$5 \equiv 2^5 \quad (\text{mod } 9)$$
$$7 \equiv 2^4 \quad (\text{mod } 9)$$
$$8 \equiv 2^3 \quad (\text{mod } 9)$$

Given g, a primitive root (mod m), how many others are there, and how do we find them? The following theorem answers this question.

Theorem 6.7   *If m has at least one primitive root, namely, g, then m has exactly $\phi(\phi(m))$ primitive roots, namely, all g^k such that $0 < k < \phi(m)$ and $(k, \phi(m)) = 1$.*

PROOF: Suppose $0 < a < m$ and $(a, m) = 1$. Since g is a primitive root (mod m), we know by Theorem 6.6 that $a \equiv g^k$ (mod m) for some k such that $0 \leq k < \phi(m)$. Now Theorem 6.3 implies that g^k has order

$\phi(m)/(k, \phi(m))$ (mod m). Therefore, g^k is a primitive root (mod m) if and only if $\phi(m)/(k, \phi(m)) = \phi(m)$, that is, if and only if $(k, \phi(m)) = 1$. The number of such k is $\phi(\phi(m))$.   ■

For example, 2 is a primitive root (mod 11). Therefore, all primitive roots (mod 11), which are $\phi(\phi(11)) = \phi(10) = 4$ in number, may be obtained from the least positive residues (mod 11) of 2^k, where $0 < k < 10$ and $(k, 10) = 1$, namely, $k = 1, 3, 7,$ and 9. We therefore obtain the primitive roots

$$2^1 \equiv 2 \quad (\text{mod } 11)$$
$$2^3 \equiv 8 \quad (\text{mod } 11)$$
$$2^7 \equiv 7 \quad (\text{mod } 11)$$
$$2^9 \equiv 6 \quad (\text{mod } 11)$$

Thus we see that the primitive roots (mod 11) are 2, 6, 7, and 8.

SECTION 6.2 EXERCISES

1. For each integer t such $1 \le t \le 12$, find the order of t (mod 13). Which of these t are primitive roots (mod 13)?

2. For each integer t such that $0 < t < 20$ and $(t, 20) = 1$, find the order of t (mod 20).

3. Verify that 2 is a primitive root (mod 19). Find all primitive roots (mod 19).

4. For each of the following values of n, assuming that primitive roots (mod n) exist, determine how many exist.

   (a) $n = 49$ (b) $n = 50$

   (c) $n = 53$ (d) $n = 54$

   (e) $n = 58$

5. Assuming that every prime has primitive roots, for each of the following values of k, find all primes that have exactly k primitive roots.

   (a) $k = 1$ (b) $k = 2$

   (c) $k = 4$ (d) $k = 6$

   (e) $k = 8$ (f) $k = 10$

   (g) $k = 12$

6. Prove that if p is an odd prime and t is arbitrary, then t^2 is *not* a primitive root (mod p).

7. Prove that if p is an odd prime and g is a primitive root (mod p), then $g^{\frac{1}{2}(p-1)} \equiv -1$ (mod p).

8. If p is a prime that has 2^k primitive roots for some $k \ge 1$, what must be true about p?

9. Let n be the number of primitive roots (mod p), where p is an odd prime and $p \ge 7$. Prove that if $n \equiv 2$ (mod 4), then $p \equiv 3$ (mod 4).

10. Prove that if t^k is a primitive root (mod p) for some $k > 1$, where p is an odd prime, then t itself is a primitive root (mod p).

11. If g and b are both primitive roots (mod p), where p is an odd prime, can gb also be a primitive root (mod p)? Explain.

12. Let g be a primitive root (mod p), where p is an odd prime. Prove that $p - g$ is also a primitive root (mod p) if and only if $p \equiv 1$ (mod 4).

13. Let 2 be a primitive root (mod p), where p is an odd prime. Find necessary and sufficient conditions on p so that 8 is also a primitive root (mod p).

14. If x is a real number such that $0 < x < 1$, then x has a decimal expansion

$$x = \frac{\sum_{k=1}^{\infty} a_k}{10^k} = .a_1 a_2 a_3 \cdots$$

that is essentially unique. If x is rational, then the decimal expansion of x is periodic; that is, there exist $t \geq 1$ such that $a_{t+k} = a_k$ for all $k \geq 1$. The least such t is called the *period* of the decimal expansion. For example,

$$\frac{1}{7} = 0.142857142857142857\ldots$$

has period $t = 6$. Prove that if p is prime and $(p, 10) = 1$, then the period of the decimal expansion of $1/p$ is the order of 10 (mod p).

*15. Prove that if p is an odd prime, $p \nmid xy$, and each of x and y has order 2^d (mod p), where $d \geq 1$ and $2^d \mid (p - 1)$, then xy has order 2^c (mod p), where $0 \leq c \leq d - 1$.

SECTION 6.2 COMPUTER EXERCISE

16. Write a computer program that, when given an integer t such that $1 \leq t \leq 256$, will find the order of t (mod 257).

17. Write a computer program that, given an odd prime p, (a) finds the least primitive root (mod p) by trial and error and (b) finds all primitive roots (mod p).

18. Write a computer program to find the order of 2 (mod p) for all primes p such that $p \equiv \pm 1$ (mod 8) and $p < 1000$. What seems to be true about the order of 2 (mod p) for these primes?

6.3   PRIMITIVE ROOTS: EXISTENCE

Let p be an odd prime. A key result regarding primitive roots, which is given in Theorem 6.12 below, is that p has primitive roots, in fact, $\phi(p - 1)$ of them. In order to obtain this result, we must first develop some properties of polynomial congruences (mod p). It is well known that in the field of real numbers or in the field of complex numbers, a polynomial of degree n can have at most n distinct roots. The following theorem says that this property also holds for polynomials in Z_p, that is, for polynomials (mod p).

Theorem 6.8   *Let p be an odd prime, $n \geq 1$, and $f(x) = \sum_{i=0}^{n} a_i x^i$, where the a_i are integers and $(a_n, p) = 1$. Then the congruence*

(*) $f(x) \equiv 0$ (mod p)

has at most n distinct solutions (mod p).

PROOF: (Induction on n)   If $n = 1$, then $f(x) = a_1x + a_0$ and $(a_1, p) = 1$ by hypothesis. Therefore, Theorem 4.5 implies that the congruence (*) has exactly one solution (mod p). Now assume by induction hypothesis that Theorem 6.8 holds for all polynomials of degree k, where $1 \le k \le n - 1$. Suppose that the congruence (*) has $n + 1$ distinct solutions (mod p), namely, $x \equiv r_1, r_2, r_3, \cdots, r_n, r_{n+1}$ (mod p). Let

$$g(x) = f(x) - a_n \prod_{i=1}^{n} (x - r_i)$$

Therefore,

$$g(r_j) = f(r_j) - a_n \prod_{i=1}^{n} (r_j - r_i)$$

so $g(r_j) \equiv 0$ (mod p) for all j such that $1 \le j \le n$. That is to say, $g(x)$ has n distinct roots (mod p). Let m be the degree of $g(x)$. Clearly, $m < n$, so $0 \le m \le n - 1$. By induction hypothesis, it is impossible that $1 \le m \le n - 1$. Therefore, $m = 0$, so $g(x)$ is a constant function. Since $g(r_j) \equiv 0$ (mod p), we must have $g(x) \equiv 0$ (mod p) for all x. This implies that

$$f(x) \equiv a_n \prod_{i=1}^{n} (x - r_i)   \pmod{p}$$

However, then

$$f(r_{n+1}) \equiv a_n \prod_{i=1}^{n} (r_{n+1} - r_i)   \pmod{p}$$

Since $f(r_{n+1}) \equiv 0$ (mod p) by hypothesis, we have

$$a_n \prod_{i=1}^{n} (r_{n+1} - r_i) \equiv 0   \pmod{p}$$

Since $(a_n, p) = 1$ by hypothesis, we have

$$\prod_{i=1}^{n} (r_{n+1} - r_i) \equiv 0   \pmod{p}$$

Since p is prime, we must have $r_{n+1} - r_i \equiv 0$ (mod p) for some i such that $1 \le i \le n$. This contradicts the assumption that all the r_i are distinct (mod p).   ∎

For example, the congruence $x^3 + x \equiv 0 \pmod{17}$ has the three solutions: $x \equiv 0, 4$, and $13 \pmod{17}$.

REMARK: If $f(x)$ is a polynomial of degree n and m is an integer that has no primitive roots, then the congruence $f(x) \equiv 0 \pmod{m}$ may have as many as $2n$ distinct roots $\pmod{m}$. For example, the congruence

$$x^2 - 1 \equiv 0 \pmod{8}$$

has four roots, namely, $x \equiv 1, 3, 5$, and $7 \pmod{8}$.

Next, we apply Theorem 6.8 to determine the exact number of roots $\pmod{p}$ of two specific polynomials.

Theorem 6.9   *If p is prime, then the congruence*

(2) $$x^{p-1} - 1 \equiv 0 \pmod{p}$$

has exactly $p - 1$ solutions.

PROOF: Fermat's theorem (Theorem 4.8) implies that the congruence (2) has at least $p - 1$ solutions $\pmod{p}$, namely, $x \equiv 1, 2, 3, \ldots, p - 1 \pmod{p}$. Theorem 6.8 implies that the congruence (2) has at most $p - 1$ solutions $\pmod{p}$. Therefore, the congruence (2) has exactly $p - 1$ solutions $\pmod{p}$.   ■

For example, the congruence $x^4 - 1 \equiv 0 \pmod{5}$ has four solutions, namely, $x \equiv 1, 2, 3$, and $4 \pmod{5}$.

The next theorem states yet a stronger result.

Theorem 6.10   *If p is prime and $d \mid (p - 1)$, then the congruence*

(3) $$x^d - 1 \equiv 0 \pmod{p}$$

has exactly d solutions.

PROOF: If $d = p - 1$, then the conclusion follows from Theorem 6.9. Otherwise, by hypothesis, we have $p - 1 = bd$, with $0 < b < p$. Let

$$f(x) = x^d - 1$$

and

$$g(x) = \frac{x^{p-1} - 1}{x^d - 1} = x^{(b-1)d} + x^{(b-2)d} + \cdots + x^{2d} + x^d + 1$$

Suppose that the congruence (3) has j solutions, while the congruence

(4) $g(x) \equiv 0 \pmod{p}$

has k solutions. We wish to show that $j = d$. Note that the degree of $g(x)$ is $(b - 1)d = p - 1 - d$. Theorem 6.8 implies that $j \le d$ and $k \le p - 1 - d$. If t is a solution of the congruence (3), then $t^d \equiv 1 \pmod{p}$, so $g(t) \equiv b \not\equiv 0 \pmod{p}$. Therefore, the congruences (3) and (4) have no common solution. Since $x^{p-1} - 1 = f(x)g(x)$ and p is prime, $x^{p-1} - 1 \equiv 0 \pmod{p}$ if and only if $f(x) \equiv 0 \pmod{p}$ or $g(x) \equiv 0 \pmod{p}$. Therefore, the number of solutions of the congruence (2), namely, $p - 1$, is the sum of the number of solutions of each of the congruences (3) and (4); that is, $p - 1 = j + k$. Since $k \le p - 1 - d$, it follows that $j \ge d$. Therefore, $j = d$.   ∎

For example, the congruence $x^3 - 1 \equiv 0 \pmod{7}$ has three solutions, namely, $x \equiv 1, 2,$ and $4 \pmod{7}$. Also, the congruence $x^4 - 1 \equiv 0 \pmod{13}$ has four solutions, namely, $x \equiv 1, 5, 8,$ and $12 \pmod{13}$.

The fact that every prime has primitive roots follows from the following more general theorem, whose proof is an elaborate counting argument.

Theorem 6.11   *If p is prime and $d \mid (p - 1)$, then the number of integers k such that $1 \le k \le p - 1$ and k has order $d \pmod{p}$ is $\phi(d)$.*

PROOF: If $d \mid (p - 1)$, where p is prime, let $f(d)$ be the number of integers k such that $1 \le k \le p - 1$ and k has order $d \pmod{p}$. If no such k exists, the $f(d) = 0 < \phi(d)$. If such a k does exist, then k is a solution of the congruence

(3) $x^d - 1 \equiv 0 \pmod{p}$

In fact, for each j such that $0 \le j \le d - 1$, k^j is a solution of the congruence (3). Theorem 6.10 implies that the congruence (3) has exactly d solutions. Therefore, every solution of the congruence (3) must have the form k^j with $0 \le j \le d - 1$. If r has order $d \pmod{p}$, then r is a solution of the congruence (3), so $r \equiv k^j \pmod{p}$ for some j such that $0 \le j \le d - 1$. Since $(k, p) = 1$ by hypothesis, Theorem 6.3 implies that k^j has order d if and only if $(j, d) = 1$. The number of such j is $\phi(d)$. Therefore, in this case, $f(d) = \phi(d)$. Whether or not any k has order $d \pmod{p}$, we have $f(d) \le \phi(d)$. According to Theorem 6.2, if $1 \le t \le p - 1$ and t has order $d \pmod{p}$, then $d \mid (p - 1)$.

Therefore, $\sum_{d|(p-1)} f(d) = p - 1$. However, Theorem 5.19 implies that $\sum_{d(p-1)} \phi(d) = p - 1$. Therefore, $\sum_{d|(p-1)} f(d) = \sum_{d|(p-1)} \phi(d)$. Since for each d we have $0 \leq f(d) \leq \phi(d)$, it follows that $f(d) = \phi(d)$ for all d such that $d \mid (p - 1)$. ∎

For example, if $p = 7$ and $d \mid (p - 1)$, that is, $d \mid 6$, then $d = 1, 2, 3$, or 6. Among the integers k such that $1 \leq k \leq 6$, we have

$$\phi(1) = 1 \quad \text{which has order 1 (mod 7), namely, 1}$$
$$\phi(2) = 1 \quad \text{which has order 2 (mod 7), namely, 6}$$
$$\phi(3) = 2 \quad \text{which have order 3 (mod 7), namely, 2 and 4}$$
$$\phi(6) = 2 \quad \text{which have order 6 (mod 7), namely, 3 and 5}$$

Also, if $p = 11$ and $d \mid (p - 1)$, that is, $d \mid 10$, then $d = 1, 2, 5$, or 10. Among the integers k such that $1 \leq k \leq 10$, we have

$$\phi(1) = 1 \quad \text{which has order 1 (mod 11), namely, 1}$$
$$\phi(2) = 1 \quad \text{which has order 2 (mod 11), namely, 10}$$
$$\phi(5) = 4 \quad \text{which have order 5 (mod 11), namely, 3, 4, 5 and 9}$$
$$\phi(10) = 4 \quad \text{which have order 10 (mod 11), namely, 2, 6, 7, and 8}$$

Applying Theorem 6.11 to the case $d = p - 1$, we obtain the following:

Theorem 6.12   *If p is prime, then p has $\phi(p - 1)$ primitive roots.*

PROOF: This follows from Definition 6.2 and Theorem 6.11, taking $d = p - 1$. ∎

For example, the prime 7 has $\phi(6) = 2$ primitive roots, namely, 3 and 5. Also, the prime 11 has $\phi(10) = 4$ primitive roots, namely, 2, 6, 7, and 8.

Recall from Theorem 6.7 that if we have one primitive root g (mod m), then we can obtain all the primitive roots (mod m) by raising g to appropriate exponents. For example, 3 is a primitive root (mod 31). Now $\phi(31) = 30$, so all primitive roots (mod 31) are of the form 3^j (mod 31), where $1 \leq j \leq 30$ and $(j, 30) = 1$. These j are 1, 7, 11, 13, 17, 19, 23, and 29. The corresponding primitive roots (mod 31) are $3^1 \equiv 3$ (mod 31), $3^7 \equiv 2187 \equiv 17$ (mod 31), $3^{11} \equiv 13$ (mod 31), $3^{13} \equiv 24$ (mod 31), $3^{17} \equiv 22$ (mod 31), $3^{19} \equiv 12$ (mod 31), $3^{23} \equiv 11$ (mod 31), and $3^{29} \equiv 21$ (mod 31).

Let g be the least positive primitive root (mod p). We have seen how all primitive roots (mod p) can be obtained from g. Table 2 in Appendix C lists the least primitive root (mod p) for all primes below 1000. Unfortunately, if p is a larger prime, there is no convenient formula to compute g. We can, however, determine g by the following "brute force" algorithm. Set $k = 2$. Compute b, the order of k (mod p). If $b = p - 1$, then $k = g$ and we are done. If $b < p - 1$, then increment k by 1 and repeat the process. We know by Theorem 6.12 that p has primitive roots, so this algorithm is guaranteed to terminate. The algorithm can be refined by eliminating from consideration numbers that are powers, such as 4, 8, and 9, as well as integers that are "quadratic residues" (mod p). The latter are defined in Chapter 7.

For example, let us compute g, the least primitive root (mod 41). Since $2^{20} \equiv 3^8 \equiv 5^{20} \equiv 1$ (mod 41), we know that $g \geq 6$. Since $6^d \not\equiv 1$ (mod 41) for $0 < d < 40$, it follows that $g = 6$.

Next, we consider primitive roots (mod p^n), where p is an odd prime and $n \geq 2$. We shall see that such primitive roots always exist, and we will learn how to find them.

Usually, if g is a primitive root (mod p), then g is also a primitive root (mod p^2). For example, 2 is a primitive root (mod 3) and 2 is also a primitive root (mod 9); also, 3 is a primitive root (mod 7) and 3 is a primitive root (mod 49). However, exceptions do exist. For example, 14 is a primitive root (mod 29), but 14 is not a primitive root (mod 29^2); also, 18 is a primitive root (mod 37), but 18 is not a primitive root (mod 37^2). The following theorem specifies how to find a primitive root (mod p^2), given a primitive root (mod p).

Theorem 6.13   *Let g be a primitive root* (mod p), *where p is an odd prime and $0 < g < p$. Then there exists b such that $0 < b < p$, $gb \equiv 1$ (mod p), and b is a primitive root* (mod p). *Moreover, either g or b is a primitive root* (mod p^2).

PROOF: By hypothesis, $(g, p) = 1$. Therefore, Theorem 4.5 implies that there exists b such that $gb \equiv 1$ (mod p). Without loss of generality, we may choose b so that $0 < b < p$. By Fermat's theorem, $b \equiv g^{p-1}b \equiv g^{p-2}(gb) \equiv g^{p-2}(1) \equiv g^{p-2}$ (mod p). Since $(p - 2, p - 1) = 1$, Theorem 6.3 implies that b has order $p - 1$ (mod p); that is, b is also a primitive root (mod p).

Let g have order j (mod p^2). Then Theorem 6.2 implies that $j \mid \phi(p^2)$; that is, $j \mid p(p - 1)$. Now $g^j \equiv 1$ (mod p^2), so Theorem 4.1, part C_9 implies that $g^j \equiv 1$ (mod p). Since g is a primitive root (mod p) by hypothesis, Theorem 6.1 implies that $(p - 1) \mid j$. Therefore, $j/(p - 1)$ is an integer and $j/(p - 1)$ divides p, so $j/(p - 1) = 1$

or p; that is $j = p - 1$ or $p(p - 1)$. Similarly, h has order $p - 1$ or $p(p - 1)$ $\pmod{p^2}$.

Now $1 < g < p$ and $1 < h < p$, so $1 < gh < p^2$. However, $gh = 1 + kp$, so $0 < k < p$. Furthermore, $(gh)^{p-1} = (1 + kp)^{p-1} = 1 + (p - 1)kp + \sum_{i=2}^{p-1} \binom{p-1}{i}(kp)^i$, so $(gh)^{p-1} \equiv 1 - kp \pmod{p^2}$. If neither g nor h is a primitive root $\pmod{p^2}$, then both g and h have order $p - 1$ $\pmod{p^2}$. In this case, $(gh)^{p-1} \equiv g^{p-1}h^{p-1} \equiv 1 \pmod{p^2}$. However, this implies that $p \mid k$, an impossibility. Therefore, at least one of g and h is a primitive root $\pmod{p^2}$. ∎

For example, suppose we wish to find a primitive root $\pmod{25}$. Now 2 is a primitive root $\pmod 5$, and $2^{5-1} \equiv 16 \not\equiv 1 \pmod{25}$, so 2 is also a primitive root $\pmod{25}$.

As a second example, suppose we wish to find a primitive root $\pmod{29^2}$, given that 14 is a primitive root $\pmod{29}$. Now $14^{29-1} \equiv 1 \pmod{29^2}$, so 14 is not a primitive root $\pmod{29^2}$. In this case, we compute $14^{-1} \pmod{29}$, that is, solve the congruence $14h \equiv 1 \pmod{29}$, where $0 < h < 29$. The solution, namely, $h = 27$, is a primitive root $\pmod{29}$ and $\pmod{29^2}$.

Generalizing, Theorem 6.13 leads to the following algorithm for computing a primitive root $\pmod{p^2}$. Let g be a primitive root $\pmod p$. Compute $g^{p-1} \pmod{p^2}$. If $g^{p-1} \not\equiv 1 \pmod{p^2}$, then g is a primitive root $\pmod{p^2}$. If $g^{p-1} \equiv 1 \pmod{p^2}$, then the solution of the congruence $gh \equiv 1 \pmod p$ is a primitive root $\pmod{p^2}$.

Note that Theorem 6.13 guarantees that the least primitive root $\pmod{p^2}$ is less than p.

There are other ways to find a primitive root $\pmod{p^2}$, given a primitive root $\pmod p$. We state the following relevant theorem:

Theorem 6.14   *Let g be a primitive root $\pmod p$, where p is an odd prime. Then each of the following is a primitive root $\pmod{p^2}$: (i) $g^p - p$, (ii) $g^p - gp$, and (iii) at least one of g or $g + p$.*

PROOF: Omitted, but see Robbins (1975), Rosen, and Trost.

For example, if we apply Theorem 6.14 to obtain a primitive root $\pmod{29^2}$ using the fact that 14 is a primitive root $\pmod{29}$, we get the following:

Using (i): $14^{29} - 29 \equiv 826 \pmod{29^2}$
Using (ii): $14^{29} - 14(29) \equiv 449 \pmod{29^2}$
Using (iii): $14^{28} \equiv 1 \pmod{29^2}$

Therefore, 14 is not a primitive root $\pmod{29^2}$. Thus $14 + 29 = 43$ is a primitive root $\pmod{29^2}$.

Fortunately, once a primitive root $\pmod{p^2}$ has been found, no further computation is needed to find a primitive root $\pmod{p^n}$ because of the following theorem:

Theorem 6.15 *If p is an odd prime and g is a primitive root $\pmod{p^2}$, then g is a primitive root $\pmod{p^n}$ for all $n \geq 2$.*

PROOF: (Induction on n) By hypothesis, g is a primitive root $\pmod{p^2}$. Now we must show that if g is a primitive root $\pmod{p^n}$, where $n \geq 2$, then g is also a primitive root $\pmod{p^{n+1}}$. Let g have order j $\pmod{p^{n+1}}$. Then Theorem 6.2 implies that $j \mid \phi(p^{n+1})$. Now $g^j \equiv 1 \pmod{p^{n+1}}$, so Theorem 4.1, part C9 implies that $g^j \equiv 1 \pmod{p^n}$. Since g is a primitive root $\pmod{p^n}$ by induction hypothesis, Theorem 6.1 implies that $\phi(p^n) \mid j$. Since $\phi(p^{n+1}) = p\phi(p^n)$, we must have $j = \phi(p^n)$ or $j = \phi(p^{n+1})$. Since g is a primitive root $\pmod{p^n}$, we know that $g^{\phi(p^{n-1})} \not\equiv 1 \pmod{p^n}$. However, Euler's theorem (Theorem 5.26) implies that $g^{\phi(p^{n-1})} \equiv 1 \pmod{p^{n-1}}$, since $(g, p^n) = (g, p^{n-1}) = 1$. Therefore, $g^{\phi(p^{n-1})} = 1 + kp^{n-1}$, where $p \nmid k$. Now $g^{\phi(p^n)} = (g^{\phi(p^{n-1})})^p = (1 + kp^{n-1})^p = 1 + kp^n + \sum_{i=2}^{p} \binom{p}{i}(kp^{n-1})^i$. Therefore, $g^{\phi(p^n)} \equiv 1 + kp^n \pmod{p^{n+1}}$. Since $p \nmid k$, $g^{\phi(p^n)} \not\equiv 1 \pmod{p^{n+1}}$. Therefore, $j \neq \phi(p^n)$. Hence, $j = \phi(p^{n+1})$, so g is a primitive root $\pmod{p^{n+1}}$. ∎

For example, since 2 is a primitive root $\pmod 9$, 2 is a primitive root $\pmod{3^n}$ for all $n \geq 2$. Also, since 35 is a primitive root $\pmod{37^2}$, 35 is a primitive root $\pmod{37^n}$ for all $n \geq 2$.

The next theorem shows that numbers that are twice an odd prime or twice a power of an odd prime also have primitive roots.

Theorem 6.16 *Let g be a primitive root $\pmod{p^n}$, where p is an odd prime and $n \geq 1$. Then (i) if g is odd, then g is also a primitive root $\pmod{2p^n}$, and (ii) if g is even, then $g + p^n$ is a primitive root $\pmod{2p^n}$.*

PROOF: Exercise.

For example, 2 and 5 are primitive roots (mod 9), so $2 + 9 = 11$ and 5 are primitive roots (mod 18). Also, 2, 6, 7, and 11 are primitive roots (mod 13), so $2 + 13 = 15$, $6 + 13 = 19$, 7, and 11 are primitive roots (mod 26).

So far we have seen that if p is an odd prime and $n \geq 1$, then primitive roots exist (mod p^n) and (mod $2p^n$). It is easily verified that 1 is a primitive root (mod 2) and that 3 is a primitive root (mod 4). It remains to be shown, via the following four theorems, that no other natural numbers have primitive roots.

Theorem 6.17   *If $m \geq n \geq 3$ and $(m, n) = 1$, then mn has no primitive roots.*

PROOF: Let $d = [\phi(m), \phi(n)]$ and $t = [\phi(m), \phi(n)]$. Since $m \geq 3$ by hypothesis, either $m = 2^k$ for some $k \geq 2$ or $p \mid m$, where p is an odd prime. In either case, $2 \mid \phi(m)$. Similarly, $2 \mid \phi(n)$. Therefore, $2 \mid d$, so $2 \leq d$. Now $t = \phi(m)\phi(n)/d$ by Theorem 2.9, part L$_7$, so $t \leq \frac{1}{2}\phi(m)\phi(n)$. However, $(m, n) = 1$ by hypothesis, so Theorem 5.20 implies that $\phi(m)\phi(n) = \phi(mn)$. Therefore, $t \leq \frac{1}{2}\phi(mn)$. Now suppose that $(a, mn) = 1$, so $(a, m) = (a, n) = 1$. Euler's theorem (Theorem 5.26) implies that $a^{\phi(m)} \equiv 1 \pmod{m}$. Therefore, $a^t \equiv 1 \pmod{m}$. Similarly, $a^t \equiv 1 \pmod{n}$. By the result of Exercise 45 in Chapter 2, $a^t \equiv 1 \pmod{mn}$. Since $t \leq \frac{1}{2}\phi(mn)$, it follows that a is not a primitive root (mod mn).   ∎

This leads to the following:

Theorem 6.18   *Let p and q be odd primes. If $4p \mid n$ or $pq \mid n$, then n has no primitive roots.*

PROOF: If $4p \mid n$, then $n = 2^j m$ with $j \geq 2$, m odd, and $m \geq 3$; if $pq \mid n$, then $n = p^j m$ with $j \geq 1$, $(p, m) = 1$, $p \geq 3$, and $m \geq 3$. In either case, Theorem 6.17 implies that n has no primitive roots.   ∎

For example, we saw earlier that $12 = 4 \cdot 3$ has no primitive roots. Also, $35 = 5 \cdot 7$ has no primitive roots.

The only integers not yet considered are $n = 2^k$, where $k \geq 3$. These are disposed of as follows:

Theorem 6.19   *If x is odd and $n \geq 3$, then $x^{2^{n-2}} \equiv 1$ (mod 2^n).*

PROOF: (Induction on n)   If x is odd, then $x \equiv 1, 3, 5,$ or 7 (mod 8). In each case, $x^2 \equiv 1$ (mod 8). Therefore, Theorem 6.19 holds for $n = 3$. Now assume that $n \geq 3$, x is odd, and $x^{2^{n-2}} \equiv 1$ (mod 2^n); that is, $2^n \mid (x^{2^{n-2}} - 1)$. Since x is odd, $2 \mid (x^{2^{n-2}} + 1)$. Therefore, Theorem 2.1, part D_6 implies that $2^{n+1} \mid (x^{2^{n-1}} - 1)$; that is, $x^{2^{n-1}} \equiv 1$ (mod 2^{n+1}). ∎

For example, $3^2 \equiv 1$ (mod 8), $3^4 \equiv 1$ (mod 16), $3^8 \equiv 1$ (mod 32), etc. As a consequence of Theorem 6.19, we obtain the following:

Theorem 6.20   *If $n = 2^k$, where $k \geq 3$, then n has no primitive roots.*

PROOF: If x is odd, let x have order t (mod 2^k), where $k \geq 3$. Theorem 6.19 implies that $t \leq 2^{k-2}$. However, $\phi(2^k) = 2^{k-1}$. Therefore, $n = 2^k$ has no primitive roots. ∎

For example, 8, 16, 32, 64, etc. have no primitive roots. Combining our results, we obtain the following:

Theorem 6.21   *Let m be a natural number. Then m has a primitive root if and only if $m = 2, 4, p^n,$ or $2p^n$, where p is an odd prime and $n \geq 1$.*

PROOF: This follows from Theorems 6.12, 6.13, 6.15, 6.17, 6.18, and 6.20. ∎

SECTION 6.3 EXERCISES

19. Find all roots of each of the following congruences.

    (a) $x^3 \equiv 1$ (mod 13)

    (b) $x^4 \equiv 1$ (mod 17)

    (c) $x^6 \equiv 1$ (mod 19)

20. For each d such that $d \mid 18$, find all t such that $0 < t < 19$ and t has order d (mod 19).

21. Given that 3 is the least primitive root (mod 43), find all primitive roots (mod 43).

22. Find all primitive roots (mod m), where (a) $m = 27$, (b) $m = 49$, and (c) $m = 50$.

23. Prove Theorem 6.16.

24. Find a formula for the number of primitive roots (mod p^n), where p is an odd prime and $n \geq 2$, in terms of p and n.

25. Prove that if g is a primitive root (mod p^n), where p is an odd prime and $n \geq 2$, then g is also a primitive root (mod p). (*Hint:* Use the results of Exercise 49 in Chapter 4.)

26. Prove that if g is a primitive root (mod m), where $m \geq 3$, then $g^{\frac{1}{2}\phi(m)} \equiv -1$ (mod m). Note that this generalizes the result of Exercise 7.

27. Let m be an integer such that $m \geq 5$ and m has primitive roots. Let G be the product of all integers g such that $0 < g < m$ and g is a primitive root (mod m). Prove that $G \equiv 1$ (mod m).

28. Let the congruence $x^2 \equiv 1$ (mod m) have r distinct solutions (mod m). Prove that if m has no primitive roots, then $4 \mid r$. (*Hint:* The Chinese remainder theorem is helpful.)

*29. Prove the following generalization of Wilson's Theorem: If $m \geq 2$, let P be the product of all integers k such that $0 < k < m$ and $(k, m) = 1$. If m has a primitive root, then $P \equiv -1$ (mod m); otherwise, $P \equiv 1$ (mod m).

30. If p is an odd prime, let S_p be the sum of all integers g such that $0 < g < p$ and g is a primitive root (mod p). Prove that $S_p \equiv \mu(p - 1)$ (mod p).

31. Let S_p be defined as in Exercise 30. Prove that if $p \equiv 1$ (mod 4), then $S_p = \frac{1}{2}p\phi(p - 1)$.

32. If p is an odd prime, let $p - 1 = \prod_{i=1}^{r} q_i^{e_i}$, where q_i are distinct odd primes. Prove that (a) for each i there exists h_i such that $0 < h_i < p$ and h_i has order $q_i^{e_i}$ (mod p) and (b) if $h = \prod_{i=1}^{r} h_i$, then h has order $p - 1$ (mod p); that is, h is a primitive root (mod p).

SECTION 6.3 COMPUTER EXERCISE

33. Write a computer program that for a given prime, p, finds all primitive roots (mod p), if any, that are not also primitive roots (mod p^2).

6.4   INDICES

We conclude this chapter by considering *indices*, which are related to primitive roots. We shall see that indices bear some resemblance to logarithms and are useful in solving certain polynomial congruences.

Definition 6.3   *Index*

Let g be a primitive root (mod p), where p is an odd prime. Let $(t, p) = 1$. The *index of t relative to g* is that exponent k such that $0 \leq k \leq p - 2$ and $t \equiv g^k$ (mod p). We write $\text{ind}_g t = k$, or, simply, $\text{ind } t = k$.

For example, recall that 5 is a primitive root (mod 7), and

$$1 \equiv 5^0 \pmod 7$$
$$2 \equiv 5^4 \pmod 7$$
$$3 \equiv 5^5 \pmod 7$$
$$4 \equiv 5^2 \pmod 7$$
$$5 \equiv 5^1 \pmod 7$$
$$6 \equiv 5^3 \pmod 7$$

Therefore, we have

$$\text{ind}_5 1 = 0$$
$$\text{ind}_5 2 = 4$$
$$\text{ind}_5 3 = 5$$
$$\text{ind}_5 4 = 2$$
$$\text{ind}_5 5 = 1$$
$$\text{ind}_5 6 = 3$$

The following theorem states some properties of indices, which recall similar properties of logarithms.

Theorem 6.23 *(Properties of Indices)*
Let p be an odd prime, and let g be a primitive root (mod *p*). *Then*

I_1: $a \equiv b \pmod p$ if and only if $\text{ind}_g a = \text{ind}_g b$.
I_2: $\text{ind}_g (g^k) \equiv k \pmod{p - 1}$
I_3: $\text{ind}_g 1 = 0$ and $\text{ind}_g g = 1$
I_4: $\text{ind}_g (ab) \equiv \text{ind}_g a + \text{ind}_g b \pmod{p - 1}$
I_5: $\text{ind}_g (a^b) \equiv b(\text{ind}_g a) \pmod{p - 1}$.

PROOF: Let $\text{ind}_g a = j$ and $\text{ind}_g b = k$. Then $0 \leq j, k \leq p - 2$, $a \equiv g^j \pmod p$, and $b \equiv g^k \pmod p$. ∎

PROOF OF I_1: If $a \equiv b \pmod p$, then $g^j \equiv g^k \pmod p$. By Theorem 6.4 and Definition 6.2, we have $j \equiv k \pmod{p - 1}$. Since $|j - k| \leq p - 2$, we must have $j = k$; that is, $\text{ind}_g a = \text{ind}_g b$. Conversely, if $\text{ind}_g a = \text{ind}_g b$, then $j = k$, so $g^j = g^k$. Hence, $g^j \equiv g^k \pmod p$; that is, $a \equiv b \pmod p$. ∎

PROOF OF I_2: Let $r = \text{ind}_g (g^k)$. Then Definition 6.3 implies that $g^r \equiv g^k \pmod p$. Now Theorem 6.4 and Definition 6.2 imply that $r \equiv k \pmod{p - 1}$. ∎

PROOF OF I_3:   This follows from the fact that $g^0 = 1$ and $g^1 = g$.   ∎

PROOF OF I_4:   Since $a \equiv g^j$ (mod p) and $b \equiv g^k$ (mod p), Theorem 4.1, part C_4 implies that $ab \equiv g^j g^k \equiv g^{j+k}$ (mod p). Therefore, $\text{ind}_g ab = \text{ind}_g (g^{j+k})$, but part I_2 implies that $\text{ind}_g (g^{j+k}) \equiv j + k$ (mod $p - 1$), so $\text{ind}_g ab \equiv j + k \equiv \text{ind}_g a + \text{ind}_g b$ (mod $p - 1$).   ∎

PROOF OF I_5:   (Induction on b)

Part I_5 is true when $b = 1$, since $\text{ind}_g (a^1) = \text{ind}_g a = 1 (\text{ind}_g a)$. Now part I_4 implies that $\text{ind}_g (a^{b+1}) = \text{ind}_g (a^b a) \equiv \text{ind}_g (a^b) + \text{ind}_g a$ (mod $p - 1$). However, by induction hypothesis, we have $\text{ind}_g (a^b) \equiv b(\text{ind}_g a)$ (mod $p - 1$). Therefore,
$\text{ind}_g (a^{b+1}) \equiv b(\text{ind}_g a) + \text{ind}_g a \equiv (b + 1) \text{ind}_g a$ (mod $p - 1$).   ∎

As an example of part I_4, again with $p = 7$ and $g = 5$, note that ind 2 + ind 3 = 5 + 4 = 9, while ind 6 = 3. Now $9 \equiv 3$ (mod 6); that is, ind 2 + ind 3 $\equiv$ ind 6   (mod 6).

Also ind 3 + ind 4 = 5 + 2 = 7 $\equiv$ 1   (mod 6) and ind 12 = ind 5 = 1, so ind 3 + ind 4 $\equiv$ ind 12   (mod 6).

As an example of part I_5, note that ind (2^3) = ind 8 = ind 1 = 0, while   3(ind 2) =   3(4) =   12 $\equiv$ 0   (mod 6),   so   ind (2^3) $\equiv$ 3(ind 2)   (mod 6).   Also,   ind (3^4) =   ind 81 =   ind 4 =   2,   while 4(ind 3) =   4(5) =   20. Now   20 $\equiv$ 2   (mod 6),   so   4(ind 3) $\equiv$ ind (3^4) (mod 6).

The numerical data from page 124 is conveniently presented in a table of indices as follows:

Table 6.2

a	$\text{ind}_5 a$
1	0
2	4
3	5
4	2
5	1
6	3

Indices are useful in solving congruences of the type

$$ax^n \equiv b \quad (\text{mod } p)$$

where p is prime and $(a, p) = 1$.

Example 6.1: Solve

$$3x^4 \equiv 4 \quad (\text{mod } 7)$$

Solution: Taking indices with respect to $g = 5$, we have

$$\text{ind } (3x^4) \equiv \text{ind } 4 \quad (\text{mod } 6)$$
$$\text{ind } 3 + 4(\text{ind } x) \equiv \text{ind } 4 \quad (\text{mod } 6)$$
$$5 + 4(\text{ind } x) \equiv 2 \quad (\text{mod } 6)$$
$$4(\text{ind } x) \equiv -3 \quad (\text{mod } 6)$$

Since $(4, 6) = 2$ and $2 \nmid 3$, the last linear congruence has no solution, according to Theorem 4.4. Therefore, the original exponential congruence has no solution.

Example 6.2: Solve

$$3x^5 \equiv 5 \quad (\text{mod } 7)$$

Solution:

$$\text{ind } (3x^5) \equiv \text{ind } 5 \quad (\text{mod } 6)$$
$$\text{ind } 3 + 5(\text{ind } x) \equiv \text{ind } 5 \quad (\text{mod } 6)$$
$$5 + 5(\text{ind } x) \equiv 1 \quad (\text{mod } 6)$$
$$5(\text{ind } x) \equiv -4 \quad (\text{mod } 6)$$
$$5(\text{ind } x) \equiv 2 \quad (\text{mod } 6)$$
$$\text{ind } x \equiv 4 \quad (\text{mod } 6)$$
$$x \equiv 2 \quad (\text{mod } 7) \qquad (\text{from Table 6.2})$$

Example 6.3: Solve

$$2x^3 \equiv 5 \quad (\text{mod } 7)$$

Solution:

$$\text{ind } (2x^3) \equiv \text{ind } 5 \quad (\text{mod } 6)$$
$$\text{ind } 2 + 3(\text{ind } x) \equiv \text{ind } 5 \quad (\text{mod } 6)$$
$$4 + 3(\text{ind } x) \equiv 1 \quad (\text{mod } 6)$$
$$3(\text{ind } x) \equiv -3 \quad (\text{mod } 6)$$
$$3(\text{ind } x) \equiv 3 \quad (\text{mod } 6)$$
$$\text{ind } x \equiv 1 \quad (\text{mod } 2)$$
$$\text{ind } x \equiv 1, 3, 5 \quad (\text{mod } 6)$$
$$x \equiv 3, 5, 6 \quad (\text{mod } 7) \qquad (\text{from Table 6.2})$$

REMARKS: Note that the use of indices converts an exponential congruence $(\text{mod } p)$, where x is the unknown, into a linear congruence $(\text{mod } p - 1)$, where $\text{ind } x$ is the unknown. Given a value of $\text{ind } x$ $(\text{mod } p - 1)$, we obtain a corresponding value of x $(\text{mod } p)$ from the table of indices, in this case, Table 6.2. In general, to solve a congruence of the type $ax^n \equiv b$ $(\text{mod } p)$ using indices, one needs a table of indices with respect to a primitive root $(\text{mod } p)$.

SECTION 6.4 EXERCISES

34. Using 2 as a primitive root (mod 11), construct a table of indices to be used in the following exercises.

35. Solve the congruence $3x^3 \equiv 5$ (mod 11).

36. Solve the congruence $6x^4 \equiv 7$ (mod 11).

37. Solve the congruence $5x^2 \equiv 6$ (mod 11).

38. Using 3 as a primitive root (mod 17), construct a table of indices to be used in the following exercises.

39. Solve the congruence $2x^3 \equiv 7$ (mod 17).

40. Solve the congruence $7x^4 \equiv 6$ (mod 17).

41. Solve the congruence $3x^5 \equiv 11$ (mod 17).

42. Solve the congruence $x^4 \equiv 2$ (mod 17).

43. Prove that if p is prime and $(a, p) = (n, p - 1) = 1$, then the congruence $ax^n \equiv b$ (mod p) always has exactly one solution.

SECTION 6.4 COMPUTER EXERCISE

44. If p is a prime such that 2 is a primitive root (mod p), write a computer program to construct a corresponding table of indices.

CHAPTER SEVEN

QUADRATIC CONGRUENCES

7.1 INTRODUCTION

In Chapter 4, among other items, we studied linear congruences, that is, congruences of the type

7.1
$$ax \equiv b \pmod{m}$$

where a, b, and m are given and x is unknown. This chapter is devoted to the study of *quadratic* congruences. We begin by considering the case where the modulus is an odd prime p. That is, we consider congruences of the type

7.2
$$Ay^2 + By + C \equiv 0 \pmod{p}$$

where A, B, C, and p are given and y is unknown.

Our problem may be simplified by some algebraic manipulation. Let the discriminant of the quadratic polynomial that appears in congruence 7.2 be $D = B^2 - 4AC$. Multiply both sides of congruence 7.2 by $4A$ and then add $B^2 - 4AC$. This yields

7.3
$$4A^2y^2 + 4ABy + B^2 \equiv B^2 - 4AC \pmod{p}$$

That is,

7.4
$$(2Ay + B)^2 \equiv D \pmod{p}$$

Now let $x = 2Ay + B$, and let a be the least positive residue of D (mod p). This yields the pure quadratic congruence

7.5
$$x^2 \equiv a \quad (\bmod\ p)$$

where $0 \leq a < p$. If $a = 0$, then congruence 7.5 has the unique solution

7.6
$$x \equiv 0 \quad (\bmod\ p)$$

We consider this case to be degenerate, and we specify henceforth that $0 < a < p$ in congruence 7.5.

For example, suppose that we wish to solve the congruence

7.7
$$3y^2 + 2y + 3 \equiv 0 \quad (\bmod\ 11)$$

Here $A = 3$, $B = 2$, and $C = 3$, so $D = B^2 - 4AC = -32$ and $a = -32 + 3(11) = 1$. Letting $x = 6y + 2$, we get

7.8
$$x^2 \equiv 1 \quad (\bmod\ 11)$$

Now Theorem 4.9 implies that $x \equiv \pm 1$ (mod 11); that is,

7.9
$$x \equiv 1,\ 10 \quad (\bmod\ 11).$$

We have, therefore, $6y + 2 \equiv 1, 10$ (mod 11), from which it follows that

7.10
$$y \equiv 5,\ 9 \quad (\bmod\ 11)$$

Similarly, let us try to solve the congruence

7.11
$$2y^2 + y + 3 \equiv 0 \quad (\bmod\ 5)$$

Here $A = 2$, $B = 1$, and $C = 3$, so $D = B^2 - 4AC = -23$ and $a = -23 + 5(5) = 2$. Letting $x = 4y + 1$, we have

7.12
$$x^2 \equiv 2 \quad (\bmod\ 5)$$

However, $0^2 \equiv 0$ (mod 5), $1^2 \equiv 4^2 \equiv 1$ (mod 5), and $2^2 \equiv 3^2 \equiv 4$ (mod 5). Therefore, congruence 7.12 has no solution, and neither does congruence 7.11.

7.2 QUADRATIC RESIDUES AND THE LEGENDRE SYMBOL

Given an odd prime p and an integer a such that $0 < a < p$, we will be concerned with (1) determining whether the congruence

7.5
$$x^2 \equiv a \quad (\bmod\ p)$$

has any solutions and (2) finding the solutions when they exist.

According to Theorem 6.8, congruence 7.5 has at most two distinct solutions. If $x \equiv b \pmod{p}$ is a solution, then so is $x \equiv -b \pmod{p}$. Therefore, congruence 7.5 either has two solutions

7.13
$$x \equiv \pm b \pmod{p}$$

or has no solution.

We now introduce the concept of quadratic residues and nonresidues, which facilitate the study of quadratic congruences.

Definition 7.1   *Quadratic Residue* (mod *p*)

If Congruence 7.5 has solutions, then we say that *a* is a *quadratic residue* (mod *p*).

For example, 3 is a quadratic residue (mod 11), because $5^2 \equiv 3 \pmod{11}$. Also, 2 is a quadratic residue (mod 17), because $6^2 \equiv 2 \pmod{17}$.

Definition 7.2   *Quadratic Nonresidue* (mod *p*)

If Congruence 7.5 has no solution, then we say that *a* is a *quadratic nonresidue* (mod *p*).

For example, 2 is a quadratic nonresidue (mod 5). The reader can verify, by trial and error, that 3 is a quadratic nonresidue (mod 7).

The following theorem tells us that there are equally many quadratic residues and quadratic nonresidues (mod *p*).

Theorem 7.1   *If p is an odd prime, then there are $\frac{1}{2}(p-1)$ quadratic residues (mod p) and also $\frac{1}{2}(p-1)$ quadratic nonresidues (mod p).*

PROOF: Let n be the number of quadratic residues (mod p). It suffices to show that $n = \frac{1}{2}(p-1)$, since then the number of quadratic nonresidues (mod p) is $p - 1 - n = p - 1 - \frac{1}{2}(p-1) = \frac{1}{2}(p-1)$. If $(x, p) = 1$, then either $x \equiv i \pmod{p}$ for some i such that $1 \le i \le \frac{1}{2}(p-1)$ or $x \equiv j \pmod{p}$ for some j such that $\frac{1}{2}(p+1) \le j \le p-1$. In the latter case, we have $j = p - i$, where $1 \le i \le \frac{1}{2}(p-1)$. Therefore, $x \equiv i$ or $p - i \pmod{p}$, where $1 \le i \le \frac{1}{2}(p-1)$; that is, $x \equiv \pm i \pmod{p}$, where $1 \le i \le \frac{1}{2}(p-1)$. However, then $x^2 \equiv i^2 \pmod{p}$. Therefore, if a is a quadratic residue (mod p), then $a \equiv i^2 \pmod{p}$, where $1 \le i \le \frac{1}{2}(p-1)$. This implies that $n \le \frac{1}{2}(p-1)$.

It remains to show that $n \geq \frac{1}{2}(p - 1)$. To do this, we must verify that the least positive residues (mod p) of the quantities $1^2, 2^2, 3^2, \cdots ,$ $[(p - 1)/2]^2$ are all distinct (mod p). Suppose that $1 \leq k \leq i \leq \frac{1}{2}(p - 1)$ and $i^2 \equiv k^2 \pmod{p}$. We wish to show that $i = k$. Theorem 4.9 implies that $i \equiv \pm k \pmod{p}$. However, $2 \leq i + k \leq p - 1$, so $i \not\equiv -k \pmod{p}$. Therefore, $i \equiv k \pmod{p}$, so $|i - k| \equiv 0 \pmod{p}$. Now $0 \leq |i - k| \leq i + k \leq p - 1$. Therefore, $|i - k| = 0$, so $i = k$, and we are done. ∎

For example, if $p = 7$, then the quadratic residues (mod 7) are the least positive residues (mod 7) of $1^2, 2^2$, and 3^2, namely, 1, 4, and 2. The quadratic nonresidues (mod 7) are 3, 5, and 6.

Also, if $p = 11$, then the quadratic residues (mod 11) are the least positives residues (mod 11) of $1^2, 2^2, 3^2, 4^2$, and 5^2, namely, 1, 4, 9, 5, and 3. The quadratic nonresidues (mod 11) are 2, 6, 7, 8, and 10.

If p is an odd prime and $(a, p) = 1$, the Legendre symbol, which is defined below, provides a convenient way to indicate whether a is a quadratic residue or a quadratic nonresidue (mod p).

Definition 7.3   *Legendre Symbol*
If p is an odd prime and $(a, p) = 1$, let

$$\left(\frac{a}{p}\right) = \begin{cases} 1 & \text{if } a \text{ is a quadratic residue (mod } p) \\ -1 & \text{if } a \text{ is a quadratic nonresidue (mod } p) \end{cases}$$

For example, $(3/11) = 1$, because 3 is a quadratic residue (mod 11); also, $(2/5) = -1$, because 2 is a quadratic nonresidue (mod 5). Note that evaluating (a/p) is equivalent to determining whether congruence 7.5 has solutions.

In this chapter, the following lemma will be useful.

Lemma 7.1   *If p is an odd prime, $x \equiv y \pmod{p}$, and $|x| = |y| = 1$, then $x = y$.*

PROOF: Since $|x| = |y|$ by hypothesis, it follows that $x = \pm y$. Therefore, to prove that $x = y$, it suffices to prove that $x \neq -y$. If $x = -y$, then $|x - y| = |2x| = 2|x| = 2$. Since $x \equiv y \pmod{p}$ by hypothesis, we have $p \mid (x - y)$, so $p \mid (|x - y|)$. Therefore, $p \mid 2$, an impossibility because p is odd by hypothesis. ∎

Euler's criterion, which follows, provides a method of evaluating (a/p), where p is an odd prime and $0 < a < p$, but at the expense of approximately $\log_2 p$ multiplications (mod p).

Theorem 7.2   *(Euler's Criterion)*
If p is an odd prime and $(a, p) = 1$, then

$$\left(\frac{a}{p}\right) \equiv a^{1/2(p-1)} \quad (\bmod\ p)$$

PROOF: It suffices to show that $(a/p) = 1$ if and only if $a^{1/2(p-1)} \equiv 1 \pmod p$. Let g be a primitive root (mod p). If $(a/p) = 1$, then $a \equiv b^2 \pmod p$ for some b. Theorem 6.6 implies that $b \equiv g^k \pmod p$ for some k. Therefore, $a \equiv (g^k)^2 \equiv g^{2k} \pmod p$. Since $(g, p) = 1$, Fermat's theorem (Theorem 4.8) implies that $g^{p-1} \equiv 1 \pmod p$. Therefore, $a^{1/2(p-1)} \equiv (g^{2k})^{1/2(p-1)} \equiv (g^{p-1})^k \equiv 1^k \equiv 1 \pmod p$.

Conversely, suppose that $a^{1/2(p-1)} \equiv 1 \pmod p$. Theorem 6.6 implies that $a \equiv g^j \pmod p$ for some j. Therefore, $(g^j)^{1/2(p-1)} \equiv 1 \pmod p$; that is, $(g^{1/2(p-1)})^j \equiv 1 \pmod p$. However, by the result of Exercise 7 in Chapter 6, $g^{1/2(p-1)} \equiv -1 \pmod p$. Therefore, $(-1)^j \equiv 1 \pmod p$. Now Lemma 7.1 implies that $(-1)^j = 1$, so $j = 2r$ for some r. Therefore, $a \equiv g^{2r} \equiv (g^r)^2 \pmod p$, so that $(a/p) = 1$.   ■

For example, $2^{1/2(7-1)} = 2^3 = 8 \equiv 1 \pmod 7$, so $(2/7) = 1$; also, $3^{1/2(7-1)} = 3^3 = 27 \equiv -1 \pmod 7$, so $(3/7) = -1$. Again, $2^{1/2(11-1)} = 2^5 = 32 \equiv -1 \pmod{11}$, so $(2/11) = -1$; also, $3^{1/2(11-1)} = 3^5 = 243 \equiv 1 \pmod{11}$, so $(3/11) = 1$.

Theorem 7.3   *(Properties of the Legendre Symbol)*
Let p be an odd prime. Then

(i)   *If $a \equiv b \pmod p$, then $\left(\dfrac{a}{p}\right) = \left(\dfrac{b}{p}\right)$.*

(ii) $\left(\dfrac{ab}{p}\right) = \left(\dfrac{a}{p}\right)\left(\dfrac{b}{p}\right)$.

(iii) $\left(\dfrac{a^2}{p}\right) = 1$.

(iv) $\left(\dfrac{-1}{p}\right) = (-1)^{1/2(p-1)}$.

PROOF:

(i) It suffices to prove that $(a/p) = 1$ if and only if $(b/p) = 1$. Let $(a/p) = 1$. Then $a \equiv x^2 \pmod{p}$ for some x. Since $b \equiv a \pmod{p}$, we have $b \equiv x^2 \pmod{p}$, so $(b/p) = 1$. Similarly, if $(b/p) = 1$, then $(a/p) = 1$.

(ii) Invoking Euler's criterion (Theorem 7.2), we have $(ab/p) \equiv (ab)^{1/2(p-1)} \equiv a^{1/2(p-1)}b^{1/2(p-1)} \equiv (a/p)(b/p) \pmod{p}$. Since p is odd by hypothesis, and $|(ab/p)| = |(a/p)(b/p)| = 1$, Lemma 7.1 implies that $(ab/p) = (a/p)(b/p)$.

(iii) This follows directly from (ii).

(iv) This follows from Euler's criterion (Theorem 7.2) and Lemma 7.1.
∎

By using Theorem 7.3, one can reduce the problem of evaluating (a/p) to the problem of evaluating all Legendre symbols (q/p), where q is a prime factor of a. For example, suppose we wish to evaluate $(120/131)$. Since $120 = 2^3 3^1 5^1$, we have $(120/131) = (4/131)(2/131)(3/131)(5/131)$. Also, $(77/89) = (7/89)(11/89)$.

More generally, let p be an odd prime, $(n, p) = 1$, and $n = \prod_{i=1}^{r} q_i^{e_i}$, where the q_i are distinct odd primes. Then Theorem 7.3 implies that

$$\left(\frac{n}{p}\right) = \prod\left\{\left(\frac{q_i}{p}\right) : e_i \equiv 1 \pmod 2\right\}$$

For example, since $360 = 2^3 3^2 5^1$ and 401 is prime, it follows that $(360/401) = (2/401)(5/401)$. In the next section we will learn how to evaluate (q/p) when p and q are distinct primes and p is odd.

SECTION 7.3 EXERCISES

In each of exercises 1 through 5, use a change of variable to transform the given congruence to a pure congruence: $x^2 \equiv a \pmod{p}$.

1. $2y^2 + 3y + 4 \equiv 0 \pmod{11}$

2. $5y^2 - 3y - 7 \equiv 0 \pmod{13}$

3. $6y^2 + y - 13 \equiv 0 \pmod{17}$

4. $4y^2 + 4y - 6 \equiv 0 \pmod{19}$

5. $8y^2 + 7y + 6 \equiv 0 \pmod{23}$

6. Find all the quadratic residues and nonresidues $\pmod{p}$ for (a) $p = 13$, (b) $p = 17$, and (c) $p = 19$.

7. Prove that if p is an odd prime, then every primitive root $\pmod{p}$ is a quadratic nonresidue $\pmod{p}$.

*8. Prove that every quadratic nonresidue $\pmod{p}$ is a primitive root $\pmod{p}$ if and only if p is a Fermat prime.

*9. Find necessary and sufficient conditions on the odd prime p such that all but one of the quadratic nonresidues is also a primitive root (mod p).

10. Use Euler's criterion to evaluate each of the following Legendre symbols.

(a) $\left(\frac{7}{11}\right)$ (b) $\left(\frac{5}{17}\right)$

(c) $\left(\frac{3}{13}\right)$ (d) $\left(\frac{11}{19}\right)$

(e) $\left(\frac{6}{23}\right)$

11. Prove that if the prime $p \geq 7$, then there are two consecutive quadratic residues (mod p).

SECTION 7.3 COMPUTER EXERCISE

12. Write a computer program to evaluate (a/p) using Euler's criterion if p is an odd prime and $0 < a < p$.

7.3   GAUSS'S LEMMA AND THE LAW OF QUADRATIC RECIPROCITY

We have seen that if $a > 1$, p is an odd prime, and $(a, p) = 1$, then by virtue of the properties of the Legendre symbol, the problem of evaluating (a/p) reduces to the problem of evaluating all (q/p) where q is a prime factor of n and q carries an odd exponent. Our next item, Gauss's lemma, will enable us to obtain a formula for $(2/p)$.

Theorem 7.4   *(Gauss's Lemma)*
If p is an odd prime and $(a, p) = 1$, consider the least positive residues (mod p) of the integers: $a, 2a, 3a, \cdots, [(p-1)/2]a$. Let n be the number of these least positive residues that exceed $\frac{1}{2}p$. Then $(a/p) = (-1)^n$.

PROOF: Consider the first $[(p-1)/2]$ positive multiples of a, namely, a, $2a, 3a, \cdots, [(p-1)/2]a$. First we show that these integers are all distinct (mod p). Suppose that $ja \equiv ka \pmod{p}$, where $1 \leq k \leq j \leq (p-1)/2$, so that $0 \leq j - k \leq (p-3)/2$. Since $(a, p) = 1$ by hypothesis, it follows that $j \equiv k \pmod{p}$; that is, $p \mid (j - k)$. Therefore, $j - k = 0$; that is, $j = k$. Now we look at the least positive residues (mod p) of $a, 2a, 3a, \cdots$, $[(p-1)/2]a$. Let $r_1, r_2, r_3, \cdots, r_n$ be those least positive residues that exceed $\frac{1}{2}p$, while $s_1, s_2, s_3, \cdots, s_m$ are those least positive residues which are less than $\frac{1}{2}p$. Therefore, $n + m = \frac{1}{2}(p - 1)$. By the preceding argument, the r_i and the s_j are all distinct (mod p). For each index i, we have $\frac{1}{2}p < r_i < p$, so $0 < p - r_i < \frac{1}{2}p$. We claim that the $p - r_i$ are distinct from the s_j. Indeed, if $p - r_i = s_j$ for some i and j, then $r_i + s_j \equiv 0 \pmod{p}$. However, by definition of r_i and s_j, we have $r_i \equiv ua \pmod{p}$, $s_j \equiv va \pmod{p}$ for some u and v such that $1 \le u, v \le \frac{1}{2}(p - 1)$. Therefore, $ua + va \equiv r_i + s_j \equiv 0 \pmod{p}$, or $(u + v)a \equiv 0 \pmod{p}$. Again, since $(a, p) = 1$, we have $u + v \equiv 0 \pmod{p}$. However, this is impossible because $2 \le u + v \le p - 1$.

To recapitulate, the integers $p - r_1, p - r_2, \cdots, p - r_n$ and $s_1, s_2, \cdots, s_m$ are all distinct, are $\frac{1}{2}(p - 1)$ in number, and are all between 1 and $\frac{1}{2}(p - 1)$. Therefore, these integers are a permutation of 1, 2, 3, $\cdots, \frac{1}{2}(p - 1)$. It follows that

$$(p - r_1)(p - r_2) \cdots (p - r_n)s_1 s_2 \cdots s_m = 1 \cdot 2 \cdot 3 \cdot \cdots \cdot \left(\frac{p-1}{2}\right)$$

That is,

$$\prod_{i=1}^{n}(p - r_i)\prod_{j=1}^{m}s_j = \left(\frac{p-1}{2}\right)!$$

Therefore,

$$\prod_{i=1}^{n}(p - r_i)\prod_{j=1}^{m}s_j \equiv \left(\frac{p-1}{2}\right)! \pmod{p}$$

so

$$(-1)^n\prod_{i=1}^{n}r_i\prod_{j=1}^{m}s_j \equiv \left(\frac{p-1}{2}\right)! \pmod{p}$$

In addition, however,

$$\prod_{i=1}^{n}r_i\prod_{j=1}^{m}s_j \equiv \prod_{k=1}^{\frac{1}{2}(p-1)} ka \equiv a^{\frac{1}{2}(p-1)}\left(\frac{p-1}{2}\right)! \pmod{p}$$

Therefore,

$$(-1)^n a^{\frac{1}{2}(p-1)}\left(\frac{p-1}{2}\right)! \equiv \left(\frac{p-1}{2}\right)! \pmod{p}$$

which implies that

$$a^{\frac{1}{2}(p-1)} \equiv (-1)^n \pmod{p}$$

Finally, Euler's criterion (Theorem 7.2) and Lemma 7.1 imply that $(a/p) = (-1)^n$. ∎

For example, let us use Gauss's lemma to evaluate $(3/11)$. We look at the first $\frac{1}{2}(11 - 1) = 5$ multiples of 3, namely, 3, 6, 9, 12, and 15 and then reduce (mod 11) to obtain 3, 6, 9, 1, and 4. Exactly 2 of these residues exceed 11/2, namely, 6 and 9. Therefore, $(3/11) = (-1)^2 = 1$.

As a second example, let us use Gauss's lemma to evaluate (7/13). We look at the first $\frac{1}{2}(13 - 1) = 6$ multiples of 7, namely, 7, 14, 21, 28, 35, and 42 and then reduce (mod 13) to obtain 7, 1, 8, 2, 9, and 3. Exactly 3 of these residues exceed 13/2, namely, 7, 8, and 9. Therefore, $(7/13) = (-1)^3 = -1$.

If we apply Gauss's lemma to the case $a = 2$, we obtain an explicit formula for $(2/p)$, which is given by Theorem 7.5 below.

Theorem 7.5   *If p is an odd prime, then*

$$\left(\frac{2}{p}\right) = \begin{cases} 1 & \text{if } p \equiv \pm 1 \pmod 8 \\ -1 & \text{if } p \equiv \pm 3 \pmod 8 \end{cases}$$

PROOF: Applying Gauss's lemma, we look at 2, 2(2), 3(2), $\cdots$, $[(p-1)/2]2$; that is, we look at 2, 4, 6, $\cdots$, $p - 1$. These positive integers are all less than p, so they are their own least positive residues (mod p). If $p = 8k + 1$, then the residues that exceed $\frac{1}{2}p = 4k + \frac{1}{2}$ are $4k + 2$, $4k + 4$, $\cdots$, $8k$. If $p = 8k - 1$, then the residues that exceed $\frac{1}{2}p = 4k - \frac{1}{2}$ are $4k$, $4k + 2$, $\cdots$, $8k - 2$. In either case, these residues are $2k$ in number, so $(2/p) = (-1)^{2k} = 1$.

If $p = 8k + 3$, then the residues that exceed $\frac{1}{2}p = 4k + 1.5$ are $4k + 2, 4k + 4, \cdots, 8k + 2$. These residues are $2k + 1$ in number, so $(2/p) = (-1)^{2k+1} = -1$. Finally, if $p = 8k - 3$, then the residues that exceed $\frac{1}{2}p = 4k - 1.5$ are $4k, 4k + 2, \cdots, 8k - 4$. These residues are $2k - 1$ in number, so $(2/p) = (-1)^{2k-1} = -1$.   ∎

For example, $13 \equiv -3 \pmod 8$, so $(2/13) = -1$. Also, $17 \equiv 1 \pmod 8$, so $(2/17) = 1$.

Recall that a Mersenne prime is a prime of the form $2^p - 1$, where p itself is prime (see Definition 5.8). Usually, $2^p - 1$ is composite. By virtue of Theorem 7.5, we can prove that $2^p - 1$ is composite for certain primes.

Theorem 7.6   *If p is prime, $p > 3$, $p \equiv 3 \pmod 4$, and $q = 2p + 1$ is prime, then $2^p - 1$ is composite.*

PROOF: Since $p \equiv 3 \pmod 4$ and $q = 2p + 1$ by hypothesis, it follows that $q \equiv 7 \pmod 8$. Therefore, Theorem 7.5 implies that $(2/q) = 1$. Now Euler's criterion (Theorem 7.2) implies that $2^{\frac{1}{2}(q-1)} \equiv 1 \pmod q$;

that is, $q \mid (2^p - 1)$. If $n > 3$, one can prove by induction that $2^n - 1 > 2n + 1$. Since $p > 3$ by hypothesis, we have $2^p - 1 > q$, so $2^p - 1$ is composite and has q as a factor. ■

REMARK: If $2p + 1$ is prime, then $3 \nmid (2p + 1)$, so $p \equiv 2 \pmod 3$. If also $p \equiv 3 \pmod 4$, then $p \equiv 11 \pmod{12}$. The five smallest primes for which Theorem 7.6 holds are 11, 23, 83, 131, and 179.

Now that we know how to evaluate $(2/p)$, we can restrict our attention to the evaluation of (a/p), where p is prime, a and p are odd, and $(a, p) = 1$. The following highly technical theorem is the second of three items needed in the proof of the law of quadratic reciprocity.

Theorem 7.7   *If p is prime, a and p are odd, and $(a, p) = 1$, then*

$$\left(\frac{a}{p}\right) = (-1)^t \qquad \text{where } t = \sum_{k=1}^{\frac{p-1}{2}} \left[\frac{ka}{p}\right]$$

PROOF: We use the same notation as in the proof of Gauss's lemma (Theorem 7.4). We look at the least positive residues (mod p) of the integers ka, where $1 \le k \le \frac{1}{2}(p - 1)$. That is to say, we look at the remainders that are obtained when each of the integers ka is divided by p. The r_i (of which there are n) are the remainders that exceed $\frac{1}{2}p$, and the s_j (of which there are m) are the remainders that are less than $\frac{1}{2}p$. We know by Gauss's lemma that $(a/p) = (-1)^n$. Therefore, it suffices to show that $(-1)^t = (-1)^n$ or, equivalently, that $t \equiv n \pmod 2$.

By Theorem 1.12, part I_7 we have $ka = p\left[\dfrac{ka}{p}\right] + u_k$, where $u_k = r_i$ or s_j for some i or j. Therefore,

$$\sum_{k=1}^{\frac{p-1}{2}} ka = \sum_{k=1}^{\frac{p-1}{2}} \left(p\left[\frac{ka}{p}\right] + u_k\right)$$

$$= \sum_{k=1}^{\frac{p-1}{2}} p\left[\frac{ka}{p}\right] + \sum_{k=1}^{\frac{p-1}{2}} u_k$$

$$= p\sum_{k=1}^{\frac{p-1}{2}} \left[\frac{ka}{p}\right] + \sum_{k=1}^{\frac{p-1}{2}} u_k$$

This yields

7.14
$$a \sum_{k=1}^{\frac{p-1}{2}} k = pt + \sum_{i=1}^{n} r_i + \sum_{j=1}^{m} s_j$$

Now recall from the proof of Theorem 7.4 that the integers $p - r_i$ and s_j are precisely all the integers from 1 to $\frac{1}{2}(p - 1)$. Therefore, we have

7.15
$$\sum_{k=1}^{\frac{p-1}{2}} k = \sum_{i=1}^{n} (p - r_i) + \sum_{j=1}^{m} s_j$$

so

7.16
$$\sum_{k=1}^{\frac{p-1}{2}} k = np - \sum_{i=1}^{n} r_i + \sum_{j=1}^{m} s_j$$

Subtracting equation 7.16 from equation 7.14, we get

7.17
$$(a - 1) \sum_{k=1}^{\frac{p-1}{2}} k = p(t - n) + 2 \sum_{i=1}^{n} r_i$$

Therefore,

7.18
$$(a - 1) \sum_{k=1}^{\frac{p-1}{2}} k \equiv p(t - n) + 2 \sum_{i=1}^{n} r_i \quad (\bmod\ 2)$$

However, since a and p are odd by hypothesis, congruence 7.18 reduces (mod 2) to

7.19
$$0 \equiv t - n \quad (\bmod\ 2)$$

so

$$t \equiv n \quad (\bmod\ 2) \qquad \blacksquare$$

For example, let us use Theorem 7.7 to evaluate $(3/11)$. Here $p = 11$ and $\frac{1}{2}(p - 1) = 5$. Therefore,

$$t = \sum_{k=1}^{5} \left[\frac{3k}{11} \right]$$
$$= \left[\frac{3}{11} \right] + \left[\frac{6}{11} \right] + \left[\frac{9}{11} \right] + \left[\frac{12}{11} \right] + \left[\frac{15}{11} \right]$$
$$= 0 + 0 + 0 + 1 + 1$$
$$= 2$$

so $(3/11) = (-1)^2 = 1$.

As a second example, let us use Theorem 7.7 to evaluate (7/13). Here $p = 13$ and $\frac{1}{2}(p - 1) = 6$. Therefore,

$$t = \sum_{k=1}^{6}\left[\frac{7k}{13}\right]$$

$$= \left[\frac{7}{13}\right] + \left[\frac{14}{13}\right] + \left[\frac{21}{13}\right] + \left[\frac{28}{13}\right] + \left[\frac{35}{13}\right] + \left[\frac{42}{13}\right]$$

$$= 0 + 1 + 1 + 2 + 2 + 3$$

$$= 9$$

so $(7/13) = (-1)^9 = -1$.

The next theorem, which despite its algebraic appearance is really geometric in nature, is the last of three items needed to prove the law of quadratic reciprocity.

Theorem 7.8   *If p and q are distinct odd primes, then*

$$\sum_{i=1}^{\frac{p-1}{2}}\left[\frac{qi}{p}\right] + \sum_{j=1}^{\frac{q-1}{2}}\left[\frac{pj}{q}\right] = \left(\frac{p-1}{2}\right)\left(\frac{q-1}{2}\right)$$

PROOF: Consider the set of points with integer coordinates (i, j), where $1 \le i \le \frac{1}{2}(p - 1)$ and $1 \le j \le \frac{1}{2}(q - 1)$. The total number of these points is $\left(\frac{p-1}{2}\right)\left(\frac{q-1}{2}\right)$. Let line L have the equation $py = qx$. (Figure 7.1 illustrates the case where $p = 11$ and $q = 7$.) If a point (i, j) lies on L, then $pj = qi$. Since $(p, q) = 1$ by hypothesis, this implies that $p \mid i$, which is impossible because $1 \le i \le \frac{1}{2}(p - 1)$. Therefore, each point (i, j) is either below line L or above line L. Furthermore, the total number points, namely, $\left(\frac{p-1}{2}\right)\left(\frac{q-1}{2}\right)$ is the sum of the number of points below line L and the number of points above line L. Let m be the number of points (i, j) below line L. For each such point (i, j), we have $pj < qi$, so $j < qi/p$. Therefore, for each fixed i, if (i, j) is below L, then $1 \le j \le \left[\frac{qi}{p}\right]$. This implies that $m = \sum_{i=1}^{\frac{p-1}{2}}\left[\frac{qi}{p}\right]$. Similarly, let n be the number of points (i, j) above line L. For each point (i, j), we have $pj > qi$, so $i < pj/q$. Therefore, for each fixed j if (i, j) is above L, $1 \le i \le \left(\frac{pj}{q}\right)$. This implies that $n = \sum_{j=1}^{\frac{q-1}{2}}\left[\frac{pj}{q}\right]$. However, $m + n = \left(\frac{p-1}{2}\right)\left(\frac{q-1}{2}\right)$, from which the conclusion follows.   ∎

For example, in Figure 7.1, $p = 11$ and $q = 7$, so $\frac{1}{2}(p - 1) = 5$ and $\frac{1}{2}(q - 1) = 3$. The equation of line L is $11y = 7x$. Furthermore, $\left(\frac{p-1}{2}\right)\left(\frac{q-1}{2}\right) = 5 \cdot 3 = 15$. Now

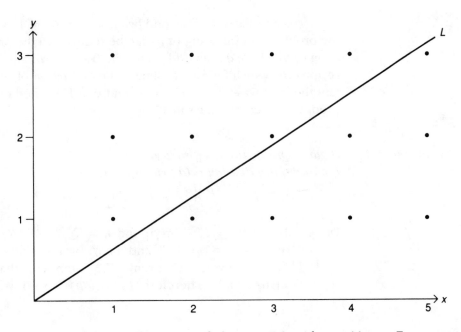

Figure 7.1. An Illustration of Theorem 7.8, with $p = 11$, $q = 7$.

$$m = \sum_{i=1}^{5}\left[\frac{7i}{11}\right]$$

$$= \left[\frac{7}{11}\right] + \left[\frac{14}{11}\right] + \left[\frac{21}{11}\right] + \left[\frac{28}{11}\right] + \left[\frac{35}{11}\right]$$

$$= 0 + 1 + 1 + 2 + 3$$

$$= 7$$

and

$$n = \sum_{j=1}^{3}\left[\frac{11j}{7}\right]$$

$$= \left[\frac{11}{7}\right] + \left[\frac{22}{7}\right] + \left[\frac{33}{7}\right]$$

$$= 1 + 3 + 4$$

$$= 8$$

So

$$m + n = 8 + 7 = 15$$

To recapitulate, if p is an odd prime and $(a, p) = 1$, we have seen that the problem of evaluating (a/p) can be reduced to the problem of evaluating (q/p) where q is an odd prime and $0 < q < p$. The law of quadratic reciprocity, which follows, facilitates the evaluation of (q/p). This important theorem was first stated by Legendre in 1785, and the first complete proof was given by Gauss in 1799.

Theorem 7.9   *(Law of Quadratic Reciprocity)*
If p and q are distinct odd primes, then

$$\left(\frac{p}{q}\right)\left(\frac{q}{p}\right) = (-1)^{[(p-1)/2][(q-1)/2]}$$

PROOF: Let $m = \sum_{i=1}^{\frac{1}{2}(p-1)}[qi/p]$ and $n = \sum_{j=1}^{\frac{1}{2}(q-1)}[pj/q]$. Theorem 7.6 implies that $(q/p) = (-1)^m$ and $(p/q) = (-1)^n$, so $(p/q)(q/p) = (-1)^{m+n}$. However, Theorem 7.7 implies that $m + n = [(p-1)/2][(q-1)/2]$. Therefore, $(p/q)(q/p) = (-1)^{[(p-1)/2][(q-1)/2]}$.

∎

It is more convenient in practice to use the following alternative form of the law of quadratic reciprocity.

Theorem 7.9a   *(Law of Quadratic Reciprocity, Alternative Form)*
Let p and q be distinct odd primes. If $p \equiv q \equiv 3$ (mod 4), then $(p/q) = -(q/p)$; if $p \equiv 1$ (mod 4) or $q \equiv 1$ (mod 4), then $(p/q) = (q/p)$.

PROOF: Exercise.

By using Theorem 7.9a, as well as Theorems 7.5 and 7.3, we can now evaluate (a/p) for any a such that $(a, p) = 1$, where p is an odd prime. For example, let us evaluate $(159/43)$. Since $159 > 43$, we note that $159 = 3(43) + 30$, so $(159/43) = (30/43)$ by Theorem 7.3, part (i). Now $30 = 2 \cdot 3 \cdot 5$, so Theorem 7.3, part (ii) implies that $(30/43) = (2/43)(3/43)(5/43)$. $(2/43) = -1$ by Theorem 7.5, because $43 \equiv 3$ (mod 8), and $(3/43) = -(43/3) = -(1/3) = -1$ by Theorem 7.9a and Theorem 7.3, parts (i) and (iii). $(5/43) = (43/5) = (3/5) = (5/3) = (2/3) = -1$ by Theorem 7.9a, Theorem 7.3, part (i), and Theorem 7.5. Therefore, $(159/43) = (30/43) = (-1)(-1)(-1) = -1$.

As a second example, let us evaluate $(105/139)$. Since $105 = 3 \cdot 5 \cdot 7$, we have $(105/139) = (3/139)(5/139)(7/139)$. Now $(3/139) = -(139/3) =$

$-(1/3) = -1, (5/139) = (139/5) = (4/5) = 1,$ and $(7/139) = -(139/7) = -(6/7) = -(2/7)(3/7).$ However, $(2/7) = 1,$ because $7 \equiv -1 \pmod 8$; also, $(3/7) = -(7/3) = -(1/3) = -1.$ Therefore, $(7/139) = -(-1) = 1.$ Finally, $(105/139) = (-1)(1)(1) = 1.$

In section 4.8 we defined the Fermat numbers $f_n = 2^{2^n} + 1.$ The law of quadratic reciprocity provides a test for the primality of Fermat numbers.

Theorem 7.10   *(Pepin's Test for Primality of Fermat Numbers)*
Let $f_n = 2^{2^n} + 1,$ where $n \geq 1.$ Then f_n is prime if and only if $3^{\frac{1}{2}(f_n - 1)} \equiv -1 \pmod{f_n}.$

PROOF: (Sufficiency)   If $n \geq 1,$ then $2^{2^n} \equiv 1 \pmod 3,$ so $f_n \equiv 2 \pmod 3.$ Also, $f_n \equiv 1 \pmod 4.$ Therefore, $(3/f_n) = (f_n/3) = (2/3) = -1.$ Now Euler's criterion implies that $3^{\frac{1}{2}(f_n - 1)} \equiv -1 \pmod{f_n}.$

(Necessity)   Suppose that $3^{\frac{1}{2}(f_n - 1)} \equiv -1 \pmod{f_n}.$ If q is a prime factor of $f_n,$ then $3^{\frac{1}{2}(f_n - 1)} \equiv -1 \pmod q,$ so $3^{f_n - 1} \equiv 1 \pmod q.$ Let 3 have order $m \pmod q.$ Then, by Theorem 6.1, $m \mid (f_n - 1);$' that is, $m \mid 2^{2^n}.$ Therefore, $m = 2^k$ for some k such that $0 \leq k \leq 2^n.$ Let $k = 2^n - j,$ where $j \geq 0.$ If $j > 0,$ then $3^{\frac{1}{2}(f_n - 1)} = 3^{2^{2^n - 1}} = (3^{2^k})^{2^{2^n - 1 - k}} \equiv 1 \pmod q$ because $3^{2^k} \equiv 1 \pmod q.$ However, then $-1 \equiv 1 \pmod q,$ an impossibility since q is odd. Therefore, 3 has order $2^{2^n} \pmod q.$ Now Theorem 6.2 implies that $2^{2^n} \mid (q - 1),$ so $2^{2^n} \leq q - 1;$ thus $f_n \leq q.$ Since $q \mid f_n,$ we must have $q = f_n,$ so f_n is prime.   ∎

REMARKS: We use the base 3 because 3 is a quadratic nonresidue $(\bmod f_n)$ for all $f_n,$ not just those f_n which are prime.

For example, let us use Pepin's test to verify that $f_4 = 65,537$ is prime. For each n such that $0 \leq n \leq 15,$ we must evaluate $3^{2^n} \pmod{65,537}.$ We list the results of this computation in the following table:

n	3^{2^n} (mod 65,537)	n	3^{2^n} (mod 65,537)
0	3	8	282
1	9	9	13,987
2	81	10	8,224
3	6,561	11	65,229
4	54,449	12	64
5	61,869	13	4,096
6	19,139	14	65,281
7	15,028	15	65,536

Since $3^{2^{15}} \equiv 65{,}536 \equiv -1 \pmod{65{,}537}$, we conclude that 65,537 is prime.

Carl Friedrich Gauss

Carl Friedrich Gauss was born in Braunschweig, Germany on April 30, 1777. His father was a casual laborer. A child prodigy, Gauss corrected his father's errors in addition when he was not yet 3 years old. Young Gauss's talents were made known by his mother to the Duke of Braunschweig, who became Gauss's patron, paying for his education and later supporting his research.

Originally, Gauss had intended to study philology, since his philology teacher was more stimulating than his mathematics teacher. In 1796, however, Gauss managed to solve a 2000-year-old mathematical problem: how to construct a regular 17-sided polygon using only a straight edge and a compass. This great success inspired him to study higher mathematics.

In 1799 Gauss received a doctorate from the University of Goettingen. In 1801, after returning to Braunschweig, he published a major work, *Disquisitiones Arithmeticae*, in which he introduced congruences and proved the law of quadratic reciprocity. In the same year he calculated the orbit of the newly discovered asteroid Ceres. Gauss's reputation grew rapidly in Europe, and in 1802 he received an offer from the St. Petersburg Academy. He remained in Braunschweig, where the duke increased his stipend.

In 1806 the Duke of Braunschweig died while leading the German forces in battle against the French. The death of his patron left Gauss in need of a source of income. In 1807 Gauss became Professor of Astronomy and Director of the Observatory at the University of Goettingen, where he spent the rest of his life. Gauss married twice; both wives suffered from ill health and died at an early age. The two sons from the second marriage emigrated to the United States.

Gauss made major contributions to several branches of mathematics, including number theory, complex analysis, and numerical analysis. He was so admired and respected for his achievements that he came to be known as the "prince of mathematics." Outside mathematics, Gauss did research in geodesy (surveying), magnetism, and dioptrics. His motto was "*pauca sed matura*" (few but ripe). He is known also for the quotation: "Mathematics is the queen of the sciences, and number theory is the queen of mathematics."

Gauss became rich late in life as a result of successful financial speculation. He died on February 23, 1855.

SECTION 7.3 EXERCISES

13. Use Gauss's lemma to evaluate each of the following.

(a) $\left(\frac{5}{11}\right)$ (b) $\left(\frac{5}{13}\right)$

(c) $\left(\frac{6}{17}\right)$ (d) $\left(\frac{19}{23}\right)$

(e) $\left(\frac{11}{29}\right)$

14. Let f_n be the nth Fermat number. Prove that if p is prime, $n \geq 2$, and $p \mid f_n$, then $p \equiv 1 \pmod{2^{n+2}}$. (*Hint:* Use Theorem 7.5.)

15. Use Theorems 7.9a and 7.5 to evaluate each of the following.

(a) $\left(\frac{3}{17}\right)$ (b) $\left(\frac{20}{29}\right)$

(c) $\left(\frac{105}{113}\right)$ (d) $\left(\frac{6}{23}\right)$

(e) $\left(\frac{11}{37}\right)$ (f) $\left(\frac{10}{79}\right)$

(g) $\left(\frac{226}{229}\right)$ (h) $\left(\frac{79}{97}\right)$

(i) $\left(\frac{60}{83}\right)$ (j) $\left(\frac{182}{263}\right)$

16. Prove that $2^n - 15 = x^2$ if and only if $n = 4$ or 6.

17. Let p and q be Fermat primes such that $p < q$. Prove that

$$\left(\frac{p}{q}\right) = \begin{cases} -1 & \text{if } p \leq 5 \\ 1 & \text{if } p \geq 17 \end{cases}$$

18. Prove that if p and q are odd primes and $q = 2p + 1$, then $(p/q) = (-1/p)$.

19. Prove that for each $k \geq 0$, there exist infinitely many primes p such that $p \equiv 1 \pmod{2^k}$.

20. If p is an odd prime, find necessary and sufficient conditions on p such that each of the following holds (not simultaneously).

(a) $\left(\frac{5}{p}\right) = 1$ (b) $\left(\frac{3}{p}\right) = 1$

(c) $\left(\frac{-3}{p}\right) = 1$ (d) $\left(\frac{-2}{p}\right) = 1$

(e) $\left(\frac{15}{p}\right) = 1$

SECTION 7.3 COMPUTER EXERCISES

21. Write a computer program to evaluate (a/p) using Gauss's lemma, where p is an odd prime and $0 < a < p$.

22. Given that $f_6 = 2^{64} + 1$ is composite, write a program to find the least prime factor of f_6, making use of the result of Exercise 14.

7.4   SOLUTION OF QUADRATIC CONGRUENCES (mod *p*)

Let $0 < a < p$, where p is an odd prime. Once we have verified that $(a/p) = 1$, that is, that the congruence

7.5 $$x^2 \equiv a \pmod{p}$$

has solutions, it remains to find these solutions. The algorithm given below, which is due to Daniel Shanks (see Bibliography), provides a method of obtaining the solutions.

Algorithm for Solving Quadratic Congruences (mod p)

Initialization

I_1:   Determine the integers k and m such that $p - 1 = 2^k m$, where $k \geq 1$, $m \geq 1$, and m is odd.
I_2:   Find q, a quadratic nonresidue (mod p).
I_3:   Set $b \equiv q^m$ (mod p).
I_4:   Set $t \equiv a^{\frac{1}{2}(m+1)}$ (mod p).
I_5:   Set $n \equiv a^m$ (mod p).

Main Routine

Step 1:   Find the least $j \geq 0$ such that $n^{2^j} \equiv 1$ (mod p).
Step 2:   If $j > 0$, go to Step 3; if $j = 0$, go to Step 7.
Step 3:   Set $c \equiv b^{2^{k-j-1}}$ (mod p).
Step 4:   Obtain a new value of t by multiplying the current value of t by c.
Step 5:   Obtain a new value of n by multiplying the current value of n by c^2.
Step 6:   Go to Step 1.
Step 7:   Terminate; the solutions of congruence 7.5 are $x \equiv \pm t$ (mod p).

For example, let us solve the congruence $x^2 \equiv 2$ (mod 41). [By Theorem 7.5, solutions exist, since $41 \equiv 1$ (mod 8).] Initializing, we have $41 - 1 = 40 = 2^3 \cdot 5$, so $k = 3$ and $m = 5$. Now $(3/41) = (41/3) = (2/3) = -1$, so let $q = 3$. We have $b \equiv q^m \equiv 3^5 \equiv 243 \equiv 38$ (mod 41). Also, $t \equiv a^{\frac{1}{2}(m+1)} \equiv 2^3 \equiv 8$ (mod 41), and $n \equiv a^m \equiv 2^5 \equiv 32$ (mod 41).

In the main routine, we have $n \equiv 32 \not\equiv 1$ (mod 41), $n^2 \equiv 32^2 \equiv 1024 \equiv -1 \not\equiv 1$ (mod 41), and $n^4 \equiv (-1)^2 \equiv 1$ (mod 41), so $j = 2$. Now we set $c \equiv b^{2^{k-j-1}} \equiv 38^{2^{3-2-1}} \equiv 38^{2^0} \equiv 38^1 \equiv 38$ (mod 41); also, $t \equiv 8(38) \equiv 304 \equiv 17$ (mod 41), and $n \equiv 32(38)^2 \equiv 1$ (mod 41). Therefore, the solutions are $x \equiv \pm 17$ (mod 41).

REMARKS: We now discuss why the preceding algorithm is guaranteed to work. After initialization, we have $t^2 \equiv a^{m+1} \equiv a(a^m) \equiv an$ (mod p). If $n \equiv 1$ (mod p), then $t^2 \equiv a$ (mod p), so the solutions of congruence 7.5 are $x \equiv \pm t$ (mod p). If $n \not\equiv 1$ (mod p), then further steps are necessary. When the values of t and n are altered, the congruence $t^2 \equiv an$ (mod p) is preserved, since $(tc)^2 \equiv a(nc^2)$ (mod p). Initially, we have $n^{2^{k-1}} \equiv (a^m)^{2^{k-1}} \equiv a^{2^{k-1}m} \equiv a^{1/2(p-1)} \equiv 1$ (mod p), since $(a/p) = 1$ by hypothesis. Therefore, if n has order d (mod p), then $d \mid 2^{k-1}$; that is, $d = 2^j$ for some j such that $0 \le j \le k - 1$. We seek an integer r such that r also has order 2^j (mod p). Then, by the result of Exercise 15 in Chapter 6, nr has order 2^i (mod p), where $0 \le i \le j - 1$.

Let $r \equiv b^{2^{k-j}}$ (mod p), where $b \equiv q^m$ (mod p). That is, $r \equiv (q^m)^{2^{k-j}} \equiv q^{2^{k-j}m}$ (mod p). Now $r^j \equiv (q^{2^{k-j}m})^{2^j} \equiv q^{2^k m} \equiv q^{p-1} \equiv 1$ (mod p) by Fermat's theorem, so r has order 2^b (mod p), where $0 \le b \le j$. Therefore, $q^{2^{k-j+b}m} \equiv (q^{2^{k-j}m})^{2^b} \equiv r^{2^b} \equiv 1$ (mod p). If $0 \le b \le j - 1$, then $q^{1/2(p-1)} \equiv q^{2^{k-1}m} \equiv (q^{2^{k-j+b}m})^{j-b-1} \equiv 1^{j-b-1} \equiv 1$ (mod p). However, this is impossible because q has been chosen so that $(q/p) = -1$. Therefore, r has order 2^j (mod p). Note that $r \equiv c^2$ (mod p), where $c \equiv b^{2^{k-j-1}}$ (mod p). (The exponent on 2 is a nonnegative integer because $j \le k - 1$.) Note also that j is reduced by at least 1 in each iteration of the main routine. Therefore, at most $k - 1$ iterations are needed to produce the solutions.

If $p \equiv 3$ (mod 4), then $k = 1$ and $m = \frac{1}{2}(p - 1)$, so $\frac{1}{2}(m + 1) = \frac{1}{4}(p + 1)$. In this case, $j = 0$ immediately, so the solutions of congruence 7.5 are given by $x \equiv \pm a^{1/4(p+1)}$ (mod p).

For example, in order to solve the congruence

$$x^2 \equiv 6 \quad (\text{mod } 43)$$

first we verify that $(6/43) = (2/43)(3/43) = (-1)(-1)(43/3) = (1/3) = 1$. Since $43 \equiv 3$ (mod 4), the solutions are given by $x \equiv \pm 6^{1/4(43+1)} \equiv \pm 6^{11} \equiv \pm 7$ (mod 43).

SECTION 7.4 EXERCISES

Find all solutions, if any, of each of the following congruences.

23. $x^2 \equiv 26$ (mod 37)

24. $x^2 \equiv 15$ (mod 47)

25. $x^2 \equiv 2$ (mod 53)

26. $x^2 \equiv 2$ (mod 73)

27. $x^2 \equiv 6$ (mod 97)

28. $x^2 \equiv 21$ (mod 113)

29. $x^2 \equiv 59$ (mod 137)

30. $x^2 \equiv 67$ (mod 257)

31. $x^2 \equiv 62$ (mod 577)

32. Prove that if $p \equiv 1$ (mod 4) and Q is a quadratic nonresidue (mod p), then the solutions of the congruence $x^2 \equiv -1$ (mod p) are given by $x \equiv \pm Q^{\frac{1}{4}(p-1)}$.

SECTION 7.4 COMPUTER EXERCISES

33. Write a computer program to solve quadratic congruences (mod p), where p is an odd prime, using the Shanks algorithm described at the beginning of this section.

7.5 QUADRATIC CONGRUENCES TO COMPOSITE MODULI

Up to this point we have studied the quadratic congruence

7.20 $$x^2 \equiv a \quad (\text{mod } m)$$

where $m = p$, a prime. Let us now consider the case where m is composite. To begin with, suppose that m is a power of an odd prime p; that is, $m = p^n$ for some $n \geq 2$.

We saw in Chapter 4 (Section 4.6) that if $f(x)$ is a nonconstant polynomial with integer coefficients, p is prime, $n \geq 2$, and the congruence

7.21 $$f(x) \equiv 0 \quad (\text{mod } p^n)$$

has a solution, then so does the congruence

7.22 $$f(x) \equiv 0 \quad (\text{mod } p^k)$$

for every k such that $1 \leq k \leq n - 1$. In particular, the congruence

7.23 $$f(x) \equiv 0 \quad (\text{mod } p)$$

must have a solution. Furthermore, if y is a solution of congruence 7.21, then

7.24 $$y \equiv x_i + tp^{n-1} \quad (\text{mod } p^n)$$

where x_i is a solution of

7.25 $$f(x) \equiv 0 \quad (\text{mod } p^{n-1})$$

and t satisfies the congruence

7.26 $tf'(x_i) \equiv \dfrac{-f(x_i)}{p^{n-1}}$ $(\bmod\ p)$

(See Theorem 4.7.)

   In this context, we now consider the congruence

7.27 $x^2 \equiv a$ $(\bmod\ p^n)$

where p is an odd prime, $n \geq 2$, $0 < a < p$, and $(a/p) = 1$. We shall see that congruence 7.27 always has exactly two solutions:

7.28 $x \equiv \pm b_n$ $(\bmod\ p^n)$

which are obtained by first solving congruence 7.27 where $n = 1$ and then "working one's way up."

Theorem 7.11   *Let p be an odd prime, $0 < a < p$, $(a/p) = 1$, and $n \geq 1$. Then the congruence*

7.27 $x^2 \equiv a$ $(\bmod\ p^n)$

always has exactly two solutions

7.28 $x \equiv \pm b_n$ $(\bmod\ p^n)$

where $(b_n, p) = 1$.

PROOF: (Induction on n)   By hypothesis, $(a/p) = 1$, so Theorem 7.10 holds for $n = 1$; that is, the congruence

7.29 $x^2 \equiv a$ $(\bmod\ p)$

has two solutions $x \equiv \pm b_1$ $(\bmod\ p)$, where $(b_1, p) = 1$. By induction hypothesis, suppose that the congruence

7.30 $x^2 \equiv a$ $(\bmod\ p^{n-1})$

has only the two solutions $x \equiv \pm b_{n-1}$ $(\bmod\ p^{n-1})$, where $(b_{n-1}, p) = 1$. According to Theorem 4.7, we obtain corresponding solutions of congruence 7.27, namely,

7.31 $x \equiv \pm(b_{n-1} + tp^{n-1})$ $(\bmod\ p^n)$

for each t such that

7.32 $tf'(b_{n-1}) \equiv \dfrac{-f(b_{n-1})}{p^{n-1}}$ $(\bmod\ p)$

Here $f(x) = x^2 - a$ and $f'(x) = 2x$, so congruence 7.32 becomes

7.33 $$2b_{n-1}t \equiv \frac{(b_{n-1}^2 - a)}{p^{n-1}} \pmod{p}$$

Now $(b_{n-1}, p) = 1$ by induction hypothesis, so Theorem 4.5 implies that congruence 7.33 has a unique solution

7.34 $$t \equiv t_{n-1} \pmod{p}$$

Therefore, if we let $b_n = b_{n-1} + t_{n-1}p^{n-1}$, we conclude that congruence 7.27 has exactly two solutions, namely,

7.28 $$x \equiv \pm b_n \pmod{p^n}$$

Since $(b_{n-1}, p) = 1$, it follows that $(b_n, p) = 1$. ∎

For example, suppose we wish to solve the congruence

7.35 $$x^2 \equiv 2 \pmod{7^3}$$

Since $7 \equiv -1 \pmod 8$, $(2/7) = 1$; that is, the congruence

7.36 $$x^2 \equiv 2 \pmod 7$$

has two solutions, namely,

7.37 $$x \equiv \pm 3 \pmod 7$$

So $b_1 = 3$.

Next, we solve the congruence

7.38 $$x^2 \equiv 2 \pmod{7^2}$$

According to Theorem 7.10, the solutions of congruence 7.38 are

7.39 $$x \equiv \pm b_2 \pmod{7^2}$$

where $b_2 \equiv b_1 + 7t_1 \pmod{7^2}$ and $2b_1t_1 \equiv -(b_1^2 - 2)/7 \pmod 7$. That is, $6t_1 \equiv -7/7 \pmod 7$. This yields $t_1 \equiv 1 \pmod 7$, so $b_2 \equiv 3 + 7(1) \equiv 10 \pmod{7^2}$. According to Theorem 7.10, the solutions of congruence 7.35 are

7.40 $$x \equiv \pm b_3 \pmod{7^3}$$

where $b_3 \equiv b_2 + 49t_2 \pmod{7^3}$ and $2b_2t_2 \equiv -(b_2^2 - 2)/7^2 \pmod 7$. That is, $20t_2 \equiv -98/49 \pmod 7$. This yields $t_2 \equiv 2 \pmod 7$, so $b_3 \equiv 10 + 49(2) \equiv 108 \pmod{7^3}$. Therefore, the solutions of congruence 7.35 are

7.41 $x \equiv \pm 108 \pmod{7^3}$

Next, we consider the congruence

7.20 $x^2 \equiv a \pmod{m}$

where $m = \prod_{i=1}^{r} p_i^{e_i}$, the p_i are distinct odd primes, and each $e_i \geq 1$. In fact, let $m_i = p_i^{e_i}$ for each i, so that $m = \prod_{i=1}^{r} m_i$ and the m_i are pairwise relatively prime. If congruence 7.20 has a solution, then so does each congruence in the system

7.42
$$\begin{cases} x^2 \equiv a \pmod{m_1} \\ x^2 \equiv a \pmod{m_2} \\ \quad\vdots \\ x^2 \equiv a \pmod{m_r} \end{cases}$$

Conversely, if each of the r congruences in the system 7.42 has a solution, then, since the m_i are pairwise relatively prime, congruence 7.20 has a solution. Now let each congruence

7.43 $x^2 \equiv a \pmod{m_i}$

have a solution $x \equiv c_i \pmod{m_i}$. [Theorem 7.10 guarantees the existence of such solutions, provided that $(a/p_i) = 1$ for each i.] By the Chinese remainder theorem (Theorem 4.6), the system of simultaneous linear congruences

7.44
$$\begin{cases} x \equiv c_1 \pmod{m_1} \\ x \equiv c_2 \pmod{m_2} \\ \quad\vdots \\ x \equiv c_r \pmod{m_r} \end{cases}$$

has a unique solution (mod m) for each set of parameters $c_1, c_2, \cdots, c_r$. According to Theorem 7.10, each of the r congruences in system 7.44 has two solutions. Therefore, we obtain 2^r sets of values of the parameters c_1, $c_2, \cdots, c_r$ and hence 2^r solutions of congruence 7.20.

For example, consider the congruence

7.45 $x^2 \equiv 2 \pmod{119}$

Since $119 = 7 \cdot 17$, congruence 7.45 is equivalent to the system of simultaneous congruences

7.46
$$\begin{cases} x^2 \equiv 2 \pmod{7} \\ x^2 \equiv 2 \pmod{17} \end{cases}$$

Since $7 \equiv -1$ (mod 8) and $17 \equiv 1$ (mod 8), Theorem 7.5 implies that $(2/7) = (2/17) = 1$, so each of the congruences in system 7.46 has solutions. In fact,

$$x^2 \equiv 2 \quad (\text{mod } 7) \qquad \text{if } x \equiv \pm 3 \quad (\text{mod } 7)$$
$$x^2 \equiv 2 \quad (\text{mod } 17) \qquad \text{if } x \equiv \pm 6 \quad (\text{mod } 17)$$

We therefore obtain four systems of simultaneous linear congruences:

7.47
$$\begin{cases} x \equiv 3 \quad (\text{mod } 7) \\ x \equiv 6 \quad (\text{mod } 17) \end{cases}$$

7.48
$$\begin{cases} x \equiv -3 \quad (\text{mod } 7) \\ x \equiv -6 \quad (\text{mod } 17) \end{cases}$$

7.49
$$\begin{cases} x \equiv 3 \quad (\text{mod } 7) \\ x \equiv -6 \quad (\text{mod } 17) \end{cases}$$

7.50
$$\begin{cases} x \equiv -3 \quad (\text{mod } 7) \\ x \equiv 6 \quad (\text{mod } 17) \end{cases}$$

The solution of system 7.47 is $x \equiv -11$ (mod 119). Therefore, the solution of the system 7.48 is $x \equiv 11$ (mod 119). The solution of the system 7.49 is $x \equiv 45$ (mod 119). Therefore, the solution of the system 7.50 is $x \equiv -45$ (mod 119). In conclusion, the solutions of congruence 7.45 are

7.51
$$x \equiv \pm 11, \pm 45 \quad (\text{mod } 119)$$

As a second example, consider the congruence

7.52
$$x^2 \equiv 3 \quad (\text{mod } 55)$$

Since $55 = 5 \cdot 11$, congruence 7.52 is equivalent to the system of simultaneous congruences

7.53
$$\begin{cases} x^2 \equiv 3 \quad (\text{mod } 5) \\ x^2 \equiv 3 \quad (\text{mod } 11) \end{cases}$$

Since $(3/5) = -1$, the first congruence in system 7.53 has no solution. Therefore, congruence 7.52 has no solution.

SECTION 7.5 EXERCISES

Find all solutions, if any, to each of the following congruences.

34. $x^2 \equiv 3$ (mod 11^2)

35. $x^2 \equiv 5$ (mod 19^2)

36. $x^2 \equiv 3 \pmod{13^3}$

37. $x^2 \equiv 2 \pmod{17^3}$

38. $x^2 \equiv 7 \pmod{33}$

39. $x^2 \equiv 3 \pmod{143}$

40. $x^2 \equiv 7 \pmod{57}$

41. $x^2 \equiv 10 \pmod{403}$

42. Prove the following generalization of Theorem 7.10. Let p be an odd prime. Let $n \geq 2$. Let a and m be integers such that $(a, p) = 1$ and $m \geq 2$. If the congruence $x^m \equiv a \pmod{p}$ has a solution, then so does the congruence $x^m \equiv a \pmod{p^n}$.

43. Let $f(x)$ be a nonconstant polynomial with integer coefficients. For $i = 1, 2, 3, \cdots, r$, suppose that the congruence $f(x) \equiv 0 \pmod{m_i}$ has $n_i \geq 0$ solutions. Suppose that $m_1, m_2, \cdots, m_r$ are pairwise relatively prime natural numbers whose product is m. Let n be the number of solutions of the congruence $f(x) \equiv 0 \pmod{m}$. Prove that $n = \prod_{i=1}^r n_i$.

7.6   JACOBI SYMBOL

Consider the congruence

7.54 $x^2 \equiv a \pmod{P}$

where P is odd and composite and $(a, P) = 1$. In the preceding section, we learned (1) how to determine whether congruence 7.54 has solutions and (2) how to find the solutions when they exist. The Jacobi symbol, which generalizes the Legendre symbol, sheds some additional light on how to determine whether congruence 7.54 has solutions.

Definition 7.4   *Jacobi Symbol*
Let $P = \prod_{i=1}^r p_i$, where the p_i are odd primes, not necessarily distinct. Let $(a, P) = 1$. let (a/p_i) denote a Legendre symbol for each i such that $1 \leq i \leq r$. Then $(a/P) = \prod_{i=1}^r (a/p_i)$ is called a *Jacobi symbol*.

For example, $(2/15) = (2/3)(2/5) = (-1)^2 = 1$ and $(3/77) = (3/7)(3/11) = (-1)1 = -1$.

We now discuss the relevance of the value of the Jacobi symbol (a/P) to congruence 7.54 above. If $(a/P) = -1$, then necessarily $(a/p_i) = -1$ for some i such that $1 \leq i \leq r$. Therefore, the congruence

7.55 $x^2 \equiv a \pmod{p_i}$

has no solutions, from which it follows that congruence 7.54 has no solutions. Unfortunately, if $(a/P) = 1$, it does not necessarily follow that congruence 7.54 has solutions. This is so because it may be that two or more factors $(a/p_i) = -1$. For example, although $(2/15) = 1$, as we saw above, nevertheless, the congruence:

$$x^2 \equiv 2 \pmod{15}$$

has no solution. In this sense, the Jacobi symbol provides less information than the Legendre symbol.

The Jacobi symbol has many properties in common with the Legendre symbol. Before establishing the properties of the Jacobi symbol, we need the following lemma.

Lemma 7.2   *Let $A_r = \prod_{i=1}^{r} a_i$, where all the a_i are odd. Let $B_r = \frac{1}{2}(A_r - 1)$ and $b_i = \frac{1}{2}(a_i - 1)$ for each i. Then $B_r \equiv \sum_{i=1}^{r} b_i \pmod{2}$.*

PROOF: (Induction on r)   If $r = 1$, then $A_1 = a_1$, from which it follows that $B_1 = b_1$ and hence $B_1 \equiv b_1 \pmod{2}$. Now $B_{r+1} = \frac{1}{2}(A_{r+1} - 1) = \frac{1}{2}(a_{r+1}A_r - 1) = \frac{1}{2}[(2b_{r+1} + 1)A_r - 1] = b_{r+1}A_r + \frac{1}{2}(A_r - 1) = b_{r+1}A_r + B_r$. By induction hypothesis, $B_r \equiv \sum_{i=1}^{r} b_i \pmod{2}$. Therefore,

$$B_{r+1} \equiv b_{r+1}A_r + \sum_{i=1}^{r} b_i \equiv \sum_{i=1}^{r+1} b_i \pmod{2} \qquad \blacksquare$$

For example, let $a_1 = 3, a_2 = 5$, and $r = 2$, so $A_2 = 15, B_2 = 7, b_1 = 1$, and $b_2 = 2$. Now $b_1 + b_2 = 1 + 2 = 3 \equiv 1 \pmod{2}$ and $B_2 = 7 \equiv 1 \pmod{2}$, so $B_2 \equiv b_1 + b_2 \pmod{2}$.

Theorem 7.12   *(Properties of the Jacobi Symbol)*

(i) $\left(\dfrac{ab}{P}\right) = \left(\dfrac{a}{P}\right)\left(\dfrac{b}{P}\right)$

(ii)   *If $a \equiv b \pmod{P}$, then $\left(\dfrac{a}{P}\right) = \left(\dfrac{b}{P}\right)$.*

(iii) $\left(\dfrac{a}{P_1 P_2}\right) = \left(\dfrac{a}{P_1}\right)\left(\dfrac{a}{P_2}\right)$.

(iv) $\left(\dfrac{a}{P^2}\right) = \left(\dfrac{a^2}{P}\right) = 1$.

(v) $\left(\dfrac{-1}{P}\right) = (-1)^{\frac{1}{2}(P-1)}$.

(vi) $\left(\dfrac{2}{P}\right) = (-1)^{(P^2-1)/8}$.

PROOF: Exercise. (Use Lemma 7.2 and the corresponding properties of the Legendre symbol.)

For   example, $(55/21) =$ $(13/21) =$ $(13/3)(13/7) =$ $(1/3)(6/7) =$ $1(2/7)(3/7) = -(7/3) = -(1/3) = -1$. Also, $(90/93) = (2/93)(9/93)(5/93) =$ $(-1)(1)(5/3)(5/31) = -(2/3)(31/5) = -(-1)(1/5) = 1$.

The following theorem states that the law of quadratic reciprocity (Theorem 7.9) remains valid when applied to the Jacobi symbol.

Theorem 7.13   *If each of P and Q is odd, positive, and at least 3, then*

$$\left(\frac{P}{Q}\right)\left(\frac{Q}{P}\right) = (-1)^{[(P-1)/2][(Q-1)/2]}.$$

PROOF: Exercise. (Use Theorem 7.9 and Lemma 7.2.)

For   example, $(13/21) =$ $(21/13) =$ $(8/13) =$ $(2/13) = -1$. Also, $(35/39) = -(39/35) = -(4/35) = -1$.

The Jacobi symbol sometimes provides a quicker way to evaluate a Legendre symbol. For example, suppose we wish to evaluate $(1155/1187)$. This is a Legendre symbol, since 1187 is prime. Now $1155 = 3 \cdot 5 \cdot 7 \cdot 11$, so using Theorem 7.3, part (ii), we have

$$\left(\frac{1155}{1187}\right) = \left(\frac{3}{1187}\right)\left(\frac{5}{1187}\right)\left(\frac{7}{1187}\right)\left(\frac{11}{1187}\right)$$

Now

$$\left(\frac{3}{1187}\right) = -\left(\frac{1187}{3}\right) = -\left(\frac{2}{3}\right) = -(-1) = 1$$

$$\left(\frac{5}{1187}\right) = \left(\frac{1187}{5}\right) = \left(\frac{2}{5}\right) = -1$$

$$\left(\frac{7}{1187}\right) = -\left(\frac{1187}{7}\right) = -\left(\frac{4}{7}\right) = -1$$

$$\left(\frac{11}{1187}\right) = -\left(\frac{1187}{11}\right) = -\left(\frac{10}{11}\right) = -\left(\frac{2}{11}\right)\left(\frac{5}{11}\right) = -(-1)\left(\frac{11}{5}\right) = \left(\frac{1}{5}\right) = 1$$

Therefore, $(1155/1187) = 1(-1)(-1)1 = 1$.

If we treat $(1155/1187)$ as a Jacobi symbol, we obtain

$$\left(\frac{1155}{1187}\right) = -\left(\frac{1187}{1155}\right) = -\left(\frac{32}{1155}\right) = -\left(\frac{2}{1155}\right) = -(-1) = 1$$

Since (1155/1187) is a Legendre symbol as well as a Jacobi symbol, we may therefore conclude, in this case, that the congruence $x^2 \equiv$ 1155   (mod 1187) has solutions.

Suppose we wish to find a quadratic nonresidue t (mod m) where either (1) it is not known whether m is prime or composite or (2) m is known to be composite but the prime factors of m are unknown. Using Theorem 7.12 and 7.13, we can nevertheless evaluate the Jacobi symbol (t/m). If $(t/m) = -1$, then we know that t is a quadratic nonresidue (mod m). If $(t/m) = 1$, then no conclusion may be drawn. In this case, one replaces t by another candidate, such as $s = t + 1$, and computes (s/m).

For example, suppose we wish to find a quadratic nonresidue (mod 11,111) without determining whether 11,111 is prime or composite. Now (2/11,111) = 1, since 11,111 $\equiv$ 7   (mod 8). Also, (3/11,111) = $-(11,111/3) = -(2/3) = -(-1) = 1$. In addition, (5/11,111) = (11,111/5) = (1/5) = 1. However, (7/11,111) = $-(11,111/7) = -(2/7) = -1$. Therefore, 7 is the least quadratic nonresidue (mod 11,111).

SECTION 7.6 EXERCISES

44. Evaluate each of the following Jacobi symbols.

   (a) $\left(\frac{30}{1001}\right)$ (b) $\left(\frac{91}{385}\right)$

   (c) $\left(\frac{130}{231}\right)$ (d) $\left(\frac{70}{141}\right)$

   (e) $\left(\frac{136}{455}\right)$

45. Evaluate (5/77). Does the congruence $x^2 \equiv$ 5   (mod 77) have solutions? (Explain.)

46. Evaluate (3/133). Does the congruence $x^2 \equiv$ 3   (mod 133) have solutions? (Explain.)

47. Use the Jacobi symbol to evaluate each of the following Legendre symbols.

   (a) $\left(\frac{105}{113}\right)$ (b) $\left(\frac{385}{401}\right)$

   (c) $\left(\frac{91}{101}\right)$ (d) $\left(\frac{77}{103}\right)$

   (e) $\left(\frac{65}{79}\right)$

48. Prove Theorem 7.12

49. Prove Theorem 7.14

50. Without factoring 1,111,111, find a quadratic nonresidue (mod 1,111,111).

SECTION 7.6 COMPUTER EXERCISE

51. Write a computer program that makes use of the Jacobi symbol to find the least quadratic nonresidue (mod m) where m is a given odd natural number.

CHAPTER EIGHT

SUMS OF SQUARES

8.1 INTRODUCTION

In this chapter we will give necessary and sufficient conditions on the natural number n so that n can be represented as a sum of two squares of nonnegative integers. (For example, $34 = 5^2 + 3^2$, $53 = 7^2 + 2^2$, and $64 = 8^2 + 0^2$.) First, we will prove a key result originally stated by Fermat: If p is prime and $p \not\equiv 3 \pmod{4}$, then p is a sum of two squares. (For example, $29 = 5^2 + 2^2$.)

Later we will show that every natural number can be represented as a sum of four squares of nonnegative integers. For example, $39 = 5^2 + 3^2 + 2^2 + 1^2$, and $71 = 7^2 + 3^2 + 3^2 + 2^2$.) The latter result is due to Lagrange.

8.2 SUMS OF TWO SQUARES

Let us look at the integers from 1 to 50 and determine which of them can be represented as a sum of two squares, counting 0 as a square. That is, for each integer n such that $1 \leq n \leq 50$, we seek integers x and y such that $x \geq y \geq 0$ and $n = x^2 + y^2$. The results are given in the following table:

n	x	y	n	x	y	n	x	y
1	1	0	16	4	0	34	5	3
2	1	1	17	4	1	36	6	0
4	2	0	18	3	3	37	6	1
5	2	1	20	4	2	40	6	2
8	2	2	25	4	3	41	5	4
9	3	0	26	5	1	45	6	3
10	3	1	29	5	2	49	7	0
13	3	2	32	4	4	50	7	1

Note that we could also write $25 = 5^2 + 0^2$ and $50 = 5^2 + 5^2$. None of the integers from 1 to 50 that are not listed in the table, that is, none of 3, 6, 7, 11, 12, 14, 15, 19, 21, 22, 23, 24, 27, 28, 30, 31, 33, 35, 38, 39, 42, 43, 44, 46, 47, and 48, can be represented as a sum of two squares.

Since $1 = 1^2 + 0^2$, we henceforth consider natural numbers n such that $n \geq 2$. We wish to develop criteria to determine whether n is a sum of two squares.

Let $n = \prod_{i=1}^{r} p_i^{e_i}$, where the p_i are distinct primes. We claim that (1) $n = x^2 + y^2$ if and only if (2) $e_i \equiv 0 \pmod 2$ for all i (if any) such that $p_i \equiv 3 \pmod 4$.

First, we show that (1) implies (2).

Theorem 8.1 Let $n \geq 2$. If $n = x^2 + y^2$, p is a prime such that $p^e \parallel n$, and $p \equiv 3 \pmod 4$, then $2 \mid e$.

PROOF: Let p be a prime such that $p^e \parallel n$, where $n = x^2 + y^2 \geq 2$. It suffices to show that if p and e are both odd, then $p \equiv 1 \pmod 4$. Let $(x, y) = d$. Then $x = du$ and $y = dv$, where $(u, v) = 1$. Now $n = d^2 m$, where $m = u^2 + v^2$. Let $p^j \parallel d$, where $j \geq 0$. Therefore, $p^{e-2j} \parallel m$. Since e is odd by hypothesis, $e - 2j \geq 1$, so $p \mid m$; that is, $p \mid (u^2 + v^2)$. Since $(u, v) = 1$, Theorem 4.12 implies that $p \equiv 1 \pmod 4$. ∎

For example, $6^2 + 3^2 = 45 = 3^2 5^1$. Also, $21^2 + 14^2 = 637 = 7^2 13^1$.

In order to show that (2) implies (1), some preliminaries are needed. The first of these preliminaries, namely, Lemma 8.1 below, appears in the *Arithmetica* of Diophantus, about 250 A.D.

Lemma 8.1   *If each of the natural numbers u and v is a sum of two squares, then so is uv.*

PROOF: Let $u = a^2 + b^2$ and $v = c^2 + d^2$. Then $uv = (a^2 + b^2)(c^2 + d^2) = a^2c^2 + b^2c^2 + a^2d^2 + b^2d^2 = a^2c^2 + 2abcd + b^2d^2 + a^2d^2 - 2abcd + b^2c^2 = (ac + bd)^2 + (ad - bc)^2$. ∎

REMARKS: Similarly, $uv = (ac - bd)^2 + (ad + bc)^2$. These two representations of uv as a sum of two squares are easily seen to be distinct unless $a = b$ or $c = d$. For example, since $5 = 2^2 + 1^2$, $13 = 3^2 + 2^2$, and $65 = 5 \cdot 13$, it follows that

$$65 = (6 + 2)^2 + (4 - 3)^2 = 8^2 + 1^2$$

Also,

$$65 = (6 - 2)^2 + (4 + 3)^2 = 4^2 + 7^2$$

The second necessary preliminary is Theorem 8.2 below, which was first stated by Fermat. To prove Theorem 8.2, we first show that if the prime $p \equiv 1 \pmod 4$, then a multiple of p, say $k_1 p$, is a sum of two squares. We then show that if $k_1 > 1$, then we can find a smaller multiple of p which is a sum of two squares. By iterating our argument, we obtain our conclusion.

Theorem 8.2   *If the prime $p \equiv 1 \pmod 4$, then there exist unique integers x and y such that $x > y > 0$ and $p = x^2 + y^2$.*

PROOF: Let q be any quadratic nonresidue $\pmod p$. (According to Theorem 7.1, these are not hard to find.) Let b be an integer such that $0 < b < p$ and $b \equiv q^{1/4(p-1)} \pmod p$. Let $x_1 = \text{Min}\{b, p - b\}$, so $x_1 < \frac{1}{2}p$. Then $x_1^2 \equiv b^2 \equiv q^{1/2(p-1)} \equiv -1 \pmod p$ by Euler's criterion. If we let $y_1 = 1$, then $x_1^2 + y_1^2 \equiv 0 \pmod p$, so $x_1^2 + y_1^2 = k_1 p$ for some $k_1 \geq 1$. Since $x_1^2 < \frac{1}{4}p^2$, we have $k_1 p < \frac{1}{4}p^2 + 1$, so $k_1 \leq \frac{1}{4}(p - 1)$. If $k_1 = 1$, then letting $x = x_1$ and $y = y_1$, we are done. Now for $i \geq 1$, assume that $x_i^2 + y_i^2 = k_i p$ for some $k_i > 1$. Choose integers a_i and b_i such that $|a_i| \leq \frac{1}{2}k_i$, $|b_i| \leq \frac{1}{2}k_i$, $a_i \equiv x_i \pmod{k_i}$, and $b_i \equiv y_i \pmod{k_i}$. Now $a_i^2 + b_i^2 \equiv x_i^2 + y_i^2 \equiv 0 \pmod{k_i}$, so $a_i^2 + b_i^2 = k_{i+1}k_i$ for some $k_{i+1} \geq 0$. If $k_{i+1} = 0$, then $a_i = b_i = 0$, so $k_i \mid x_i$ and $k_i \mid y_i$. However, then $k_i^2 \mid (x_i^2 + y_i^2)$, that is, $k_i^2 \mid k_i p$, so $k_i \mid p$. However, $k_{i+1}k_i = a_i^2 + b_i^2 \leq (\frac{1}{2}k_i)^2 + (\frac{1}{2}k_i)^2$, so $k_{i+1} \leq \frac{1}{2}k_i$. Since $k_1 \leq \frac{1}{4}(p - 1)$, it follows that $k_i \leq \frac{1}{4}(p - 1)$ for all i. Since $2 \leq k_i \leq \frac{1}{4}(p - 1)$, it is impossible that $k_i \mid p$. Therefore, $k_{i+1} \neq 0$, so $1 \leq k_{i+1} \leq \frac{1}{2}k_i$. Lemma 8.1 implies that $(a_i x_i + b_i y_i)^2 + (a_i y_i - b_i x_i)^2 = (a_i^2 + b_i^2)(x_i^2 + y_i^2) = k_{i+1}k_i^2 p$. Now

$a_i x_i + b_i y_i \equiv x_i^2 + y_i^2 \equiv 0 \pmod{k_i}$ and $a_i y_i - b_i x_i \equiv x_i y_i - y_i x_i \equiv 0 \pmod{k_i}$. Let $x_{i+1} = |(a_i x_i + b_i y_i)/k_i|$ and $y_{i+1} = |(a_i y_i - b_i x_i)/k_i|$. Now x_{i+1} and y_{i+1} are nonnegative integers such that $x_{i+1}^2 + y_{i+1}^2 = k_{i+1} p$, with $1 \leq k_{i+1} \leq \frac{1}{2} k_i$. Therefore, we obtain a sequence of equations $x_i^2 + y_i^2 = k_i p$ with $k_1 > k_2 > k_3 > \cdots$. Theorem 1.6 implies that this sequence must terminate, so there exists $n \geq 1$ such that $k_n = 1$; that is, $x_n^2 + y_n^2 = p$. Let $x = \operatorname{Max}\{x_n, y_n\}$ and $y = \operatorname{Min}\{x_n, y_n\}$. Then $p = x^2 + y^2$, with $x \geq y \geq 0$. If $y = 0$ or $y = x$, then p is not prime. Therefore, $x > y > 0$.

Finally, we show that the representation of p as a sum of two squares is unique. If also $p = v^2 + w^2$, with $v > w > 0$, then $x^2 \equiv -y^2 \pmod{p}$ and $w^2 \equiv -v^2 \pmod{p}$, so $x^2 w^2 \equiv y^2 v^2 \pmod{p}$, and hence $xw \equiv \pm yv \pmod{p}$. Now v, w, x, and y are all between 0 and $\sqrt{p}$, so $-p < xw - yv < p$ and $0 < xw + yv < 2p$. If $xw \equiv yv \pmod{p}$, then $xw - yv = 0$, so $w = yv/x$. Then $x^2 + y^2 = p = v^2 + w^2 = v^2 + y^2 v^2/x^2 = (v^2/x^2)(x^2 + y^2)$. This implies that $v = x$ and $w = y$. If $xw \equiv -yv \pmod{p}$, then $xw + yv = p$. However, then $p^2 + (xv - yw)^2 = (xw + yv)^2 + (xv - yw)^2 = (x^2 + y^2)(v^2 + w^2) = p^2$, so $xv - yw = 0$. However, this implies that $w = (x/y)v > v$, contrary to hypothesis.   ∎

For example, let us represent the prime 1013 as a sum of two squares. Since $1013 \equiv 5 \pmod{8}$, $(2/1013) = -1$. Now $2^{\frac{1}{4}(1013-1)} \equiv 2^{253} \equiv 45 \pmod{1013}$, so $x_1 = 45$ and $y_1 = 1$. Now $x_1^2 + y_1^2 = 45^2 + 1^2 = 2026 = 2(1013)$, so $k_1 = 2$. Next, we choose a_1 and b_1 such that $|a_1| \leq 1$, $|b_1| \leq 1$, $a_1 \equiv 45 \pmod{2}$, and $b_1 \equiv 1 \pmod{2}$; that is, $a_1 = b_1 = 1$. Now $x_2 = |(a_1 x_1 + b_1 y_1)/k_1| = |(45 + 1)/2| = 23$, while $y_2 = |(a_1 y_1 - b_1 x_1)/k_1| = |(1 - 45)/2| = 22$. Since $x_2^2 + y_2^2 = 23^2 + 22^2 = 1013$, we are done.

Similarly, or by inspection, one obtains the following representations as sums of two squares of the five smallest primes p such that $p \equiv 1 \pmod{4}$:

$$5 = 2^2 + 1^2$$
$$13 = 3^2 + 2^2$$
$$17 = 4^2 + 1^2$$
$$29 = 5^2 + 2^2$$
$$37 = 6^2 + 1^2$$

REMARK: The number of iterations needed to represent p as a sum of two squares is at most $\log_2 [\frac{1}{4}(p - 1)]$.

Now we are ready to prove that (2) implies (1), by way of the following theorem.

Theorem 8.3   *Let $n \geq 2$. Then $n = x^2 + y^2$ if and only if for every prime p such that $p \equiv 3 \pmod 4$ and $p^e \parallel n$ it is true that $2 \mid e$.*

PROOF: Theorem 8.1 implies necessity, so it suffices to prove sufficiency. By hypothesis, $n = AB^2$, where $A = 1$ or $A = \prod_{i=1}^{r} p_i^{e_i}$ with no $p_i \equiv 3 \pmod 4$, and $B^2 = 1$ or $B^2 = \prod_{j=1}^{s} q_j^{2f_j}$ with all $q_j \equiv 3 \pmod 4$. Noting that $2 = 1^2 + 1^2$ and using Theorem 8.2 and Lemma 8.1 repeatedly, we see that A is a sum of two squares. Since $B^2 = B^2 + 0^2$, Lemma 8.1 implies that n is a sum of two squares. ∎

Let the natural number $n \geq 2$. To determine whether n is a sum of two squares, we first factor n as a product of primes. If either (1) n has no prime factors p such that $p \equiv 3 \pmod 4$ or (2) each such prime factor carries an even exponent, then n is a sum of two squares. Otherwise, n is not a sum of two squares.

For example, let us determine which of the integers 116, 117, 118, 119, and 120 is a sum of two squares.

$$116 = 2^2 29^1 = 10^2 + 4^2$$
$$117 = 3^2 13^1 = 9^2 + 6^2$$
$$118 = 2^1 59^1 \neq x^2 + y^2 \qquad [59 \equiv 3 \pmod 4, \ 59^1 \parallel 118]$$
$$119 = 7^1 17^1 \neq x^2 + y^2 \qquad [7 \equiv 3 \pmod 4, \ 7^1 \parallel 119]$$
$$120 = 2^3 3^1 5^1 \neq x^2 + y^2 \qquad [3 \equiv 3 \pmod 4, \ 3^1 \parallel 120]$$

A composite integer that satisfies the hypothesis of Theorem 8.3 may have one or several distinct representations as a sum of two squares. For example, $441 = 3^2 7^2 = 21^2 + 0^2$; on the other hand, $325 = 5^2 13^1 = 18^2 + 1^2 = 17^2 + 6^2 = 15^2 + 10^2$. Let $r_2(n)$ be the number of distinct representations of the natural number n as a sum of two squares. The following theorem, whose proof we omit, tells us how to compute $r_2(n)$.

Theorem 8.4   *Let n be a natural number such that if p is a prime such that $p \equiv 3 \pmod 4$ and $p^e \parallel n$, then $2 \mid e$. Let $r_2(n)$ be the number of distinct representations $n = x^2 + y^2$, where $x \geq y \geq 0$. Let $n = lm$, where*

$$l = 2^a \prod_{i=1}^{s} q_j^{2f_j} \qquad m = \prod_{i=1}^{r} p_i^{e_i}$$

all $q_j \equiv 3$ (mod 4), all $p_i \equiv 1$ (mod 4), $a \geq 0$, and each product may assume the value 1. *Then $r_2(n) = \left[\frac{1}{2}(1 + \tau(m))\right]$.*

PROOF: Omitted, but see Hardy and Wright (1979, p. 241–243) (see Bibliography).

REMARKS: $\tau(m)$ is the number-of-divisors function that we studied in Chapter 5, so

$$\tau(m) = \prod_{i=1}^{r}(e_i + 1)$$

For example, let $n = 441$. Then $l = 3^2 7^2$ and $m = 1$, so $\tau(m) = 1$. Therefore, $r_2(441) = \left[\frac{1}{2}(1 + 1)\right] = 1$. Also, if $n = 325$, then $l = 1$ and $m = 5^2 13^1$, so $\tau(m) = 3 \cdot 2 = 6$. Therefore, $r_2(325) = \left[\frac{1}{2}(1 + 6)\right] = 3$.

One can obtain the various representations of n as a sum of two squares by first representing each prime factor of m as a sum of two squares and then applying Lemma 8.1.

For example, $5 = 2^2 + 1^2$, so $5^2 = 5^2 + 0^2 = 4^2 + 3^2$. Also, $13 = 3^2 + 2^2$. Recall that $325 = 5^2 13^1$. From $5^2 = 5^2 + 0^2$ and $13 = 3^2 + 2^2$, we obtain $325 = 15^2 + 10^2$; from $5^2 = 4^2 + 3^2$ and $13 = 3^2 + 2^2$, we obtain $325 = 18^2 + 1^2$ and $325 = 17^2 + 6^2$.

Joseph Louis Lagrange

Joseph Louis Lagrange was born in Turin, Italy on January 25, 1736. The paternal side of his family had French roots. His father, a local government official, wanted Lagrange to study law. Lagrange, however, preferred mathematics, for which he showed an early aptitude. He began publishing his research in 1754. In 1755 Lagrange became Professor of Mathematics at the Royal Artillery School in Turin. He did research on the calculus of variations, fluid mechanics, differential equations, and planetary motion. In 1765 Lagrange won a prize offered by the Paris Academy of Sciences for his research on the moons of Jupiter. Despite his growing reputation and his desire to remain in Turin, Lagrange never got a raise in salary while he remained there.

In 1766, through the intercession of d'Alembert, Lagrange obtained a post at the Berlin Academy of Sciences. (Euler had just left Berlin to return to St. Petersburg.) Lagrange remained in Berlin for over 20 years, working mostly on mechanics and number theory. In 1768 he gave the solution of Pell's equation by continued fractions and proved that only quadratic irrationals have periodic continued fractions. In 1770 he proved that every natural number is a sum of four squares. In 1771 he gave the first proof of Wilson's theorem and its converse.

In 1786 King Frederick II of Prussia died; in 1787 Lagrange moved to Paris, where he became a member of the Paris Academy of Sciences. The French Revolution, with all its upheavals, soon followed. In 1790 Lagrange was named chair of a commission charged with the standardization of weights and measures. This commission was retained even when the Paris Academy of Sciences was abolished, and Lagrange was kept as chair, even though other members of the commission were purged. In 1793, through the intervention of Lavoisier, Lagrange was exempted from a government order to arrest foreigners and confiscate their property. Lagrange survived the French Revolution unscathed, perhaps because of the following credo: "I believe that, in general, one of the first principles of every wise man is to conform strictly to the laws of the country in which he is living, even if they are unreasonable."

Lagrange taught at the newly formed Ecole Polytechnique from 1794 to 1799. Under Napoleon's rule, he became senator and, in 1808, Count of the Empire. He died on April 10, 1813.

SECTION 8.2 EXERCISES

1. Represent each of the following primes as a sum of two squares.

   (a) 233   (b) 317

   (c) 613   (d) 1009

   (e) 1409

2. For each integer from 131 to 140, (a) determine whether it can be represented as a sum of two squares and (b) if so, find such a representation.

3. Represent each of the following integers as a sum of two squares in as many ways as possible.

   (a) 65   (b) 85

   (c) 221   (d) 1073

   (e) 1105   (f) 1885

   (g) 5525

4. Suppose that p is an odd prime and $p = x^2 + y^2$, with $x > y > 0$. (a) Find integers u and v such that $2p = u^2 + v^2$, with $u > v > 0$. (b) Prove that this representation of $2p$ as a sum of two squares is unique.

5. Prove that if p is a prime such that $2p = u^2 + 1$, then p is the sum of the squares of two consecutive integers.

6. Prove that if the prime $p = x^2 + y^2$ and $p \equiv \pm 1 \pmod{10}$, then $5 \mid xy$.

7. Find the least three consecutive natural numbers such that none is a square but each is a sum of two squares.

*8. Prove that there are arbitrarily large gaps between consecutive integers each of which is a sum of two squares. (*Hint:* The Chinese remainder theorem is helpful.)

9. Let p be prime. Prove that there is a right triangle with integer sides and hypotenuse p if and only if $p \equiv 1 \pmod{4}$.

SECTION 8.2 COMPUTER EXERCISE

10. Write a computer program that, given a
    prime p, such that $p \equiv 1 \pmod 4$, finds the
    integers x and y such that $x > y > 0$ and
    $p = x^2 + y^2$.

8.3   SUMS OF FOUR SQUARES

Let us try to represent each of the integers from 1 to 50 as a sum of four
squares, counting 0 as a square. That is, for each integer n such that $1 \leq
n \leq 50$, we seek integers w, x, y, and z such that $w \geq x \geq y \geq z$ and $n =
w^2 + x^2 + y^2 + z^2$. The results are given in Table 8.1. We seek to
represent each n as a sum of a minimal number of nonzero squares.
Several integers have more than one such representation.

We shall see that every natural number can be represented as the sum
of four squares. First, we need the following lemma.

Table 8.1. Integers as Sums of Four Squares

n	w	x	y	z		n	w	x	y	z		n	w	x	y	z
1	1	0	0	0		18	3	3	0	0		35	5	3	1	0
2	1	1	0	0		19	3	3	1	0		36	6	0	0	0
3	1	1	1	0		20	4	2	0	0		37	6	1	0	0
4	2	0	0	0		21	4	2	1	0		38	6	1	1	0, 5 3 2 0
5	2	1	0	0		22	3	3	2	0		39	6	1	1	1, 5 3 2 1
6	2	1	1	0		23	3	3	2	1		40	6	2	0	0
7	2	1	1	1		24	4	2	2	0		41	5	4	0	0
8	2	2	0	0		25	5	0	0	0		42	5	4	1	0
9	3	0	0	0		26	5	1	0	0		43	5	3	3	0
10	3	1	0	0		27	5	1	1	0, 3 3 3 0		44	6	2	2	0
11	3	1	1	0		28	5	1	1	1, 3 3 3 1		45	6	3	0	0
12	2	2	2	0		29	5	2	0	0		46	6	3	1	0
13	3	2	0	0		30	5	2	1	0		47	6	3	1	1, 5 3 3 2
14	3	2	1	0		31	5	2	1	1, 3 3 3 2		48	4	4	4	0
15	3	2	1	1		32	4	4	0	0		49	7	0	0	0
16	4	0	0	0		33	5	2	2	0, 4 4 1 0		50	7	1	0	0, 5 5 0 0
17	4	1	0	0		34	5	3	0	0						

Lemma 8.2 *(Lagrange, 1770)*
If each of two natural numbers is a sum of four squares, then so is their product.

PROOF: Let $u = a^2 + b^2 + c^2 + d^2$ and $v = w^2 + x^2 + y^2 + z^2$. Then

$$uv = (a^2 + b^2 + c^2 + d^2)(w^2 + x^2 + y^2 + z^2)$$
$$= a^2w^2 + a^2x^2 + a^2y^2 + a^2z^2 + b^2w^2 + b^2x^2 + b^2y^2 + b^2z^2$$
$$+ c^2w^2 + c^2x^2 + c^2y^2 + c^2z^2 + d^2w^2 + d^2x^2 + d^2y^2 + d^2z^2$$
$$= (aw + bx + cy + dz)^2 + (ax - by + cz - dy)^2$$
$$+ (ay - bz - cw + dx)^2 + (az + by - cx - dw)^2 \qquad \blacksquare$$

For example, if $u = 3$ and $v = 5$, then $a = 1, b = 1, c = 1$, and $d = 0$; also, $w = 2, x = 1, y = 0$, and $z = 0$. Therefore, as in the proof of Lemma 8.2, we have $uv = 15 = (2 + 1)^2 + (1 - 2)^2 + (-2)^2 + (-1)^2 = 3^2 + 1^2 + 2^2 + 1^2$.

Therefore, once it is shown that every prime is a sum of four squares, it will follow from Lemma 8.2 and Theorem 3.10 that every natural number is a sum of four squares. Before proving that every prime is a sum of four squares, we need a couple of preparatory lemmas.

Lemma 8.3 *Let the prime $p \equiv 3 \pmod 4$. Then there exists a prime q such that $q \not\equiv 3 \pmod 4$ and $(q/p) = -1$.*

PROOF: By hypothesis, either $p \equiv 3 \pmod 8$ or $p \equiv 7 \pmod 8$. If $p \equiv 3 \pmod 8$, then $(2/p) = -1$, so let $q = 2$. If $p \equiv 7 \pmod 8$, let q be the least prime such that $q \equiv 2p - 1 \pmod{4p}$. Since $(2p - 1, 4p) = 1$, such a q exists by Dirichlet's theorem (Theorem 3.12). Therefore, $q \equiv 1 \pmod 4$ and $q \equiv -1 \pmod p$, so $(q/p) = (-1/p) = -1$. $\blacksquare$

For example, if $p = 7$, the least such q is 5; also, if $p = 31$, the least such q is 13.

Lemma 8.4 *If p is an odd prime, then there exists k such that $0 < k < p$ and kp is a sum of four squares.*

PROOF: If $p \equiv 1 \pmod 4$, then Theorem 8.2 implies that $p = x^2 + y^2 + 0^2 + 0^2$. If $p \equiv 3 \pmod 4$, let q be the least prime such that $q \not\equiv 3 \pmod 4$ and $(q/p) = -1$. (Such a q exists, according to Lemma 8.3.)

Theorem 8.2 implies that $q = s^2 + t^2$ for some s and t. Now choose integers a_1, a_2, and a_3 such that $0 \leq a_i < \frac{1}{2}p$ for each i and $a_1 \equiv \pm q^{\frac{1}{4}(p+1)}$ (mod p), $a_2 \equiv \pm s$ (mod p), and $a_3 \equiv \pm t$ (mod p). Then $a_1^2 + a_2^2 + a_3^2 + 0^2 = q^{\frac{1}{2}(p+1)} + s^2 + t^2 \equiv q(q^{\frac{1}{2}(p-1)}) + q \equiv q(-1) + q \equiv 0$ (mod p), by Euler's criterion. Therefore, $a_1^2 + a_2^2 + a_3^2 + 0^2 = kp$ for some $k > 0$. However, $a_1^2 + a_2^2 + a_3^2 < 3(\frac{1}{2}p)^2 = \frac{3}{4}p^2$, which implies that $k < \frac{3}{4}p$. ∎

For example, if $p = 7$, then $q = 5 = 2^2 + 1^2$ and $q^{\frac{1}{4}(p+1)} \equiv 5^2 \equiv 4$ (mod 7). Now $a_1 \equiv \pm 4$ (mod 7) and $0 < a_1 < 3.5$, so $a_1 = 3$; also, $a_2 = 2$ and $a_3 = 1$, so we have $a_1^2 + a_2^2 + a_3^2 + 0^2 = 3^2 + 2^2 + 1^2 + 0^2 = 14 = 2(7)$.

REMARKS: Let the primes q such that $q \not\equiv 3$ (mod 4) be listed as $q_1 = 2$, $q_2 = 5$, $q_3 = 13$, etc. If p is a randomly chosen prime such that $p \equiv 3$ (mod 4), then for any q_i, the probability that $(q_i/p) = -1$ is $\frac{1}{2}$. Therefore, the probability that n is the least integer such that $(q_n/p) = -1$ is $1/2^n$. Thus one can usually quickly find an appropriate q_i that is small enough that it is easily represented as a sum of two squares.

Lemma 8.5   *If p is an odd prime, then there exists an odd integer m such that $0 < m < p$ and mp is a sum of four squares.*

PROOF: By Lemma 8.4, there exists k such that $kp = a^2 + b^2 + c^2 + d^2$ and $0 < k < p$. If k is odd, then $m = k$ and we are done. If $k = 2^j m$, where $j \geq 1$ and m is odd, then the set $\{a, b, c, d\}$ must contain an even number of odd integers, so $\{a, b, c, d\}$ can be split into two pairs of integers having the same parity. Suppose $a \equiv b$ (mod 2) and $c \equiv d$ (mod 2). Then

$$\frac{1}{2}kp = 2^{j-1}m = \left(\frac{a+b}{2}\right)^2 + \left(\frac{a-b}{2}\right)^2 + \left(\frac{c+d}{2}\right)^2 + \left(\frac{c-d}{2}\right)^2$$

Iterating this procedure j times, we obtain the desired result: a representation of mp as a sum of four squares, where m is odd and $0 < m < k < p$. ∎

For example, we saw earlier that $3^2 + 2^2 + 1^2 + 0^2 = 2(7)$. Let $a = 3$, $b = 1$, $c = 2$, and $d = 0$. Therefore,

$$\left(\frac{3+1}{2}\right)^2 + \left(\frac{3-1}{2}\right)^2 + \left(\frac{2+0}{2}\right)^2 + \left(\frac{2-0}{2}\right)^2 = 2^2 + 1^2 + 1^2 + 1^2$$
$$= 1(7) = 7$$

We have just seen that if an even multiple of p is a sum of four squares, then a smaller odd multiple of p is also a sum of four squares. We are about to see that if an odd multiple of p (other than p itself) is a sum of four squares, then so is a smaller odd multiple of p.

Lemma 8.6   *If p is an odd prime, m is odd, $1 < m < p$, and $mp = a_1^2 + a_2^2 + a_3^2 + a_4^2$, then there exists odd k such that $1 \le k < m$ and $kp = b_1^2 + b_2^2 + b_3^2 + b_4^2$.*

PROOF: Choose integers A_i such that $A_i \equiv a_i \pmod{m}$ and $|A_i| < \frac{1}{2}m$ for $1 \le i \le 4$. Now $\sum_{i=1}^{4} A_i^2 \equiv \sum_{i=1}^{4} a_i^2 \equiv 0 \pmod{m}$, so $\sum_{i=1}^{4} A_i^2 = km$ for some k. Also, $\sum_{i=1}^{4} A_i^2 < 4(\frac{1}{2}m)^2 = m^2$, so $k < m$. If $k = 0$, then each $A_i = 0$, in which case $m \mid a_i$ for each i. However, then $m^2 \mid \sum_{i=1}^{4} a_i^2$; that is, $m^2 \mid mp$. This implies that $m \mid p$, contrary to hypothesis. Therefore, $0 < k < m$. Now let

$$
\begin{aligned}
b_1 &= |(a_1A_1 + a_2A_2 + a_3A_3 + a_4A_4)/m| \\
b_2 &= |(a_1A_2 - a_2A_1 + a_3A_4 - a_4A_3)/m| \\
b_3 &= |(a_1A_3 - a_2A_4 - a_3A_1 + a_4A_2)/m| \\
b_4 &= |(a_1A_4 + a_2A_3 - a_3A_2 - a_4A_1)/m|
\end{aligned}
$$

Each b_i is an integer, since

$$
\sum_{i=1}^{4} a_iA_i \equiv \sum_{i=1}^{4} a_i^2 \equiv 0 \pmod{m}
$$

Also, $a_iA_j - a_jA_i + a_rA_s - a_sA_r \equiv a_ia_j - a_ja_i + a_ra_s - a_sa_r \equiv 0 \pmod{m}$ for any $i, j, r,$ and s. By applying Lemma 8.3, we obtain

$$
\begin{aligned}
\sum_{i=1}^{4} b_i^2 &= \left(\sum_{i=1}^{4} A_i^2 \right)\left(\sum_{i=1}^{4} a_i^2 \right)/m^2 \\
&= (km)(mp)/m^2 \\
&= kp
\end{aligned}
$$

If k is odd, then we are done. If k is even, then applying Lemma 8.5 yields the desired result.   ∎

For example, if $p = 47$, then, following the method of Lemma 8.4, we have $q = 5, r \equiv 5^{1/4(47+1)} \equiv 5^{12} \equiv 18 \pmod{47}$, and $5 = 2^2 + 1^2$, so $18^2 + 2^2 + 1^2 + 0^2 = 329 = 7 \cdot 47$. Let $A_1 \equiv 18 \pmod{7}, A_2 \equiv 2 \pmod{7}, A_3 \equiv 1 \pmod{7}, A_4 \equiv 0 \pmod{7}$, and $|A_i| < 3.5$. Therefore, $A_1 = -3, A_2 = 2, A_3 = 1,$ and $A_4 = 0$. Now

$$b_1 = |[18(-3) + 2(2) + 1(1) + 0(0)]/7| = 7$$
$$b_2 = |[18(2) - (-3)2 + 1(0) - 0(1)]/7| = 6$$
$$b_3 = |[18(1) - 2(0) - 1(-3) + 0(2)]/7| = 3$$
$$b_4 = |[18(0) + 2(1) - 1(2) - 0(-3)]/7| = 0$$

so $7^2 + 6^2 + 3^2 + 0^2 = 94 = 2(47)$. Finally,

$$\left(\frac{7+3}{2}\right)^2 + \left(\frac{7-3}{2}\right)^2 + \left(\frac{6+0}{2}\right)^2 + \left(\frac{6-0}{2}\right)^2 = 5^2 + 2^2 + 3^2 + 3^2 = 47$$

Now the most difficult work has been done, and it remains only to put the pieces together.

Lemma 8.7   *If p is prime, then p can be represented as a sum of four squares.*

PROOF: First, $2 = 1^2 + 1^2 + 0^2 + 0^2$. If p is odd, let m be the least natural number such that mp is a sum of four squares. If $m > 1$, then Lemma 8.5 implies that m is not even; also, Lemma 8.6 implies that m is not odd. Therefore, $m = 1$.   ∎

REMARKS: Lemmas 8.4 through 8.6 provide a constructive way to obtain a representation of a given prime as a sum of four squares. The reader can verify that if $p = 71$, then one obtains

$$22^2 + 3^2 + 2^2 + 0^2 = 7 \cdot 71$$
$$5^2 + 9^2 + 6^2 + 0^2 = 2(71)$$
$$7^2 + 2^2 + 3^2 + 3^2 = 71$$

We now come to the main result of this section.

Theorem 8.4   *(Lagrange)*
Every natural number can be represented as a sum of four squares.

PROOF: First, $1 = 1^2 + 0^2 + 0^2 + 0^2$. If $n \geq 2$, then n is a product of primes, so the conclusion follows from the repeated application of Lemmas 8.2 and 8.7.   ∎

For example, let us find a representation of 55 as a sum of four squares. First, we note that $55 = 5 \cdot 11$. Now $5 = 2^2 + 1^2 + 0^2 + 0^2$ and $11 = 3^2 + 1^2 + 1^2 + 0^2$, so we get

$$55 = (6 + 1 + 0 + 0)^2 + (2 - 3 + 0 - 0)^2 + (2 - 0 - 0 + 0)^2$$
$$+ (0 + 1 - 0 - 0)^2 = 7^2 + 1^2 + 2^2 + 1^2.$$

Theorem 8.4 is best possible in the sense that there are infinitely many natural numbers that cannot be represented as a sum of three of fewer squares, as the following theorem shows.

Theorem 8.5   *If $n = 4^k m$ with $k \geq 0$ and $m \equiv 7 \pmod 8$, then $n \neq a^2 + b^2 + c^2$.*

PROOF: Suppose that $4^k m = a^2 + b^2 + c^2$, where $k \geq 0$ and $m \equiv 7 \pmod 8$. If $k \geq 1$, then $a^2 + b^2 + c^2 \equiv 0 \pmod 4$. Therefore, either a, b, and c are all even or two are odd and one is even. In the latter case, we would have $a^2 + b^2 + c^2 \equiv 2 \pmod 4$. Therefore, $a = 2a_1$, $b = 2b_1$, and $c = 2c_1$, which yields $4^{k-1} = a_1^2 + b_1^2 + c_1^2$. If $k \geq 2$, then we may continue in like fashion, ultimately obtaining $m = a_k^2 + b_k^2 + c_k^2$. Without loss of generality, assume that $a_k^2 \geq b_k^2 \geq c_k^2 \pmod 4$. However, each of a_k^2, b_k^2, and c_k^2 is congruent to each 0, 1, or 4 $\pmod 8$. Therefore, $a_k^2 + b_k^2 + c_k^2 \equiv 0 + 0 + 0, 1 + 0 + 0, 1 + 1 + 0, 1 + 1 + 1, 4 + 0 + 0, 4 + 1 + 0, 4 + 1 + 1, 4 + 4 + 0, 4 + 4 + 1,$ or $4 + 4 + 4 \pmod 8$; that is, $a_k^2 + b_k^2 + c_k^2 \equiv 0, 1, 2, 3, 4, 5,$ or $6 \pmod 8$. Therefore, $m \not\equiv 7 \pmod 8$, contrary to hypothesis.   ∎

The converse of Theorem 8.5 is also true; that is, every natural number n such that $n \neq 4^k m$ where $k \geq 0$ and $m \equiv 7 \pmod 8$ can be represented as a sum of three squares. The proof of the converse is beyond the scope of this text and is therefore omitted. (See Gauss, Article 291.)

Let S_k denote the set of natural numbers that can be represented as a sum of at most k squares. According to Lemma 8.1, if each of u and v belongs to S_2, then so does uv. According to Lemma 8.2, if each of u and v belongs to S_4, then so does uv. We could say that each of the sets S_2 and S_4 is closed under multiplication. This property is not enjoyed by S_3. For example, $3 = 1^2 + 1^2 + 1^2$ and $21 = 4^2 + 2^2 + 1^2$, so each of 3 and 21 belongs to S_3. Now $3 \cdot 21 = 63$ and $63 \equiv 7 \pmod 8$, so by Theorem 8.5, 63 does not belong to S_3.

A natural number may have several representations as a sum of four squares. For example, $31 = 5^2 + 2^2 + 1^2 + 1^2 = 3^2 + 3^2 + 3^2 + 2^2$. Also, $55 = 7^2 + 2^2 + 1^2 + 1^2 = 6^2 + 3^2 + 3^2 + 1^2 = 5^2 + 5^2 + 2^2 + 1^2$. Suppose that we allow squares of negative as well as positive integers and that we count two representations as distinct if they differ merely in the order of their terms. Then 2 has 24 representations as a sum of four squares, namely, $(\pm 1)^2 + (\pm 1)^2 + 0^2 + 0^2$, and the six permutations of each of these four representations.

We state without proof the following remarkable theorem of Jacobi, which counts the number of representations of a natural number as a sum of four squares.

Theorem 8.6   *If n is a natural number let $r_4(n)$ denote the number of representations of n as a sum of four squares of integers. (Representations that differ only in the order of the terms are considered distinct.) Then $r_4(n) = 8\sigma^*(n)$, where $\sigma^*(n)$ denotes the sum of those divisors d of n such that $4 \nmid d$.*

PROOF: Omitted (See Hardy and Wright.)

For example, $55 = 7^2 + 2^2 + 1^2 + 1^2 = 6^2 + 3^2 + 3^2 + 1^2 = 5^2 + 5^2 + 2^2 + 1^2$. These representations give rise to $3 \cdot 16 = 48$ representations if we allow negative as well as positive integers. Also, the four terms of each representation can be permuted in $4!/2! = 12$ ways, so that it appears that $r_4(55) = 3 \cdot 16 \cdot 12 = 576$. Using the formula of Theorem 8.5, we have $r_4(55) = 8\sigma^*(55) = 8\sigma(55) = 8\sigma(5 \cdot 11) = 8 \cdot 6 \cdot 12 = 576$. Therefore, there are no additional representations of 55 as a sum of four squares.

SECTION 8.3 EXERCISES

11. Prove that if $n = a^2 + b^2 + c^2 + d^2$, with $a \ge b \ge c \ge d \ge 0$, then $\frac{1}{2}\sqrt{n} \le a \le \sqrt{n}$.

12. Prove that if $n \equiv 7 \pmod 8$, then $4n$ can be represented as a sum of four odd squares.

13. Find the smallest pair of consecutive integers such that each is the sum of not fewer than four squares.

14. Represent each of 60, 92, and 124 as a sum of four odd squares.

15. Prove that there do not exist three consecutive integers such that each is the sum of not fewer than four squares.

16. Find a representation of each of the following primes as a sum of four squares.

   (a) 167   (b) 191

   (c) 263   (d) 431

   (e) 479   (f) 1031

   (g) 8831

17. For each of the following integers, find all representations $n = a^2 + b^2 + c^2 + d^2$, with $a \ge b \ge c \ge d \ge 0$: 63, 71, 95.

18. Let $n \equiv 3 \pmod 8$. Prove that if $n = a^2 + b^2 + c^2 + d^2$, then $abcd \equiv 0 \pmod 4$.

19. Let $n \equiv 7 \pmod 8$. Prove that if $n = a^2 + b^2 + c^2 + d^2$, then $abcd \equiv 2 \pmod 4$.

20. Let $n \equiv 7 \pmod 8$. If $n = a^2 + b^2 + c^2 + d^2$, then by Theorem 8.5, $abcd \neq 0$. Furthermore, since n is odd, $\sigma^*(n) = \sigma(n)$. Therefore, if we count the number of representations of n as a sum of four squares of positive integers, where representations that differ only in the order of terms are considered distinct, by Theorem 8.6 we expect $\frac{1}{2}\sigma(n)$

representations. Find all representations of 127 as a sum of the squares of four positive integers.

21. Let $p = a^2 + b^2 + c^2 + d^2$, where p is prime and $a \geq b \geq c \geq d \geq 0$. Prove that if $p \equiv 23 \pmod{24}$, then at most 2 of a, b, c, and d are equal.

SECTION 8.3 COMPUTER EXERCISES

22. Write a computer program that, given an odd prime p, finds a representation of p as a sum of four squares.

23. Let p be a prime such that $p \equiv 7 \pmod 8$. In the proofs of Lemmas 8.3 and 8.4, we saw that starting with a prime q such that $q \not\equiv 3 \pmod 4$ and $(q/p) = -1$, we can obtain a

representation of p as a sum of four squares. Distinct values of q may or may not lead to distinct representations of p as a sum of four squares. Write a computer program to test the conjecture that all representations of p as a sum of four squares may be obtained by picking sufficiently many such q.

CHAPTER NINE

FIBONACCI AND LUCAS NUMBERS AND BINARY LINEAR RECURRENCES

9.1 INTRODUCTION

Fibonacci numbers $(1, 1, 2, 3, 5, 8, 13, \cdots)$ were defined recursively in Chapter 1 (Definition 1.2, page 9). In this chapter we will learn more about these remarkable numbers, which occur in nature and have many applications in mathematics and computer science. For example, if F_n denotes the nth Fibonacci number where $n \geq 1$ and $\alpha = \frac{1}{2}(1 + \sqrt{5})$, recall that

1.8 $$F_n \geq \alpha^{n-2}$$

This inequality is used in deriving a formula that links the number of iterations needed to compute the greatest common divisor of two given integers using the Euclidean algorithm to the size of the given integers (see Theorem 9.2 below).

We will also study the so-called companion sequence to the Fibonacci numbers, namely, the *Lucas numbers*. Finally, we will study a generalization of the Fibonacci numbers known as *binary linear recurrences*.

Fibonacci and Lucas Numbers

Fibonacci numbers are named for their discoverer, Leonardo Fibonacci (1170–1250). Fibonacci, also known as Leonardo Pisano, was born in Pisa, Italy, then an important city-state. During Fibonacci's lifetime (the high Middle Ages), European mathematics was largely dormant. Integers were represented by Roman numerals, which made computation rather cumbersome.

Fibonacci's father, who was involved with foreign trade, sent young Leonardo to study in the city of Bugia on the north coast of Africa. There, he learned the Hindu-Arabic numeral system, which he later introduced to Europe. Fibonacci wrote several books, including *Liber Abbaci*, of which the first edition appeared in 1202, the second in 1228. Donald Knuth calls Fibonacci "The most important European mathematician before the Renaissance." Fibonacci was so highly regarded that he was introduced to the Holy Roman Emperor, Frederick Barbarossa, on the occasion of the latter's visit to Pisa about 1225. A statue of Fibonacci is located in the Giardino Scotto, a park in Pisa.

In *Liber Abbaci*, Fibonacci poses the following problem regarding rabbits. Suppose a pair of rabbits, upon reaching sexual maturity at age 1 month, gives birth each month to a new pair of rabbits (one of each sex) and that all of their descendants do the same. Suppose furthermore that none of these rabbits ever dies or runs away. How many pairs of rabbits will there be after n months? (See Figure 9.1.)

In Figure 9.1, an empty circle represents a newborn pair of rabbits, whereas a darkened circle represents a mature pair of rabbits. If we let F_n denote the number of pairs of rabbits at time n, then $F_1 = 1, F_2 = 1, F_3 = 2, F_4 = 3, F_5 = 5, F_6 = 8, F_7 = 13$, etc. At time n, let M_n denote the number of pairs of mature rabbits, and let I_n denote the number of pairs of immature rabbits. Then $F_n = M_n + I_n$. If $n \geq 2$, then $M_n = F_{n-1}$, since every pair of rabbits that is mature at time n must have existed at time $n - 1$. Also, if $n \geq 3$, then $I_n = M_{n-1}$, since every newborn pair at time n is the offspring of a pair that was mature at time $n - 1$. However, $M_{n-1} = F_{n-2}$, so $I_n = F_{n-2}$. We therefore obtain the recurrence relation

9.1 $$F_n = F_{n-1} + F_{n-2} \qquad (n \geq 3)$$

with $F_1 = F_2 = 1$. This recurrence relation generates a sequence of integers known as the Fibonacci sequence. F_n denotes the nth Fibonacci number.

Fibonacci numbers occur in nature. For example, the surface of a pineapple consists of diamond-shaped scales that occur in spirals about the trunk. Some spirals are clockwise; others are counterclockwise. The number of spirals in each of the two directions are consecutive Fibonacci

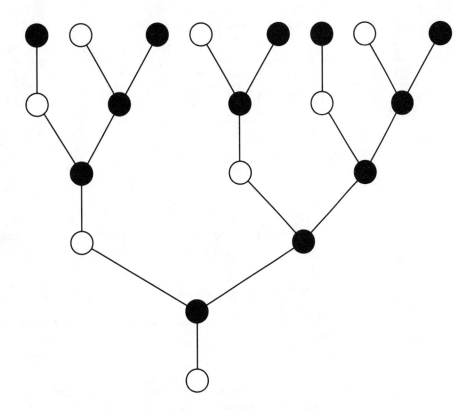

Figure 9.1. The Fibonacci Tree

numbers, usually 8 and 13. Similar patterns occur on pine cones and on the heads of sunflowers.

Fibonacci numbers have several applications in computer science. They are used in (1) file merging, (2) searching of ordered arrays, (3) construction of search trees with a minimal number of nodes, and (4) optimal procedures for finding extreme values of a unimodular function defined on a closed interval.

Associated with the Fibonacci sequence is another sequence known as the *Lucas sequence*, which is defined by the recurrence

9.2 $$L_n = L_{n-1} + L_{n-2}   (n \geq 3)$$

with $L_1 = 1$ and $L_2 = 3$. Here L_n denotes the *n*th *Lucas* number. Table 3 in Appendix C (p. 299) lists F_n and L_n for all n from 1 to 50.

The Fibonacci and Lucas sequences are closely interrelated, as is shown by the following three identities:

Table 9.1. The First 16 Fibonacci and Lucas Numbers

n	1	2	3	4	5	6	7	8	9	10	11	12	13	14	15	16
F_n	1	1	2	3	5	8	13	21	34	55	89	144	233	377	610	987
L_n	1	3	4	7	11	18	29	47	76	123	199	322	521	843	1364	2207

9.9 $$F_{2n} = F_n L_n$$

9.13 $$L_n = F_{n+1} + F_{n-1}$$

9.8 $$L_n^2 - 5F_n^2 = 4(-1)^n$$

The Fibonacci and Lucas sequences have been defined recursively. They may be defined explicitly as follows. Let the roots of the equation $t^2 - t - 1 = 0$ be

$$\alpha = \frac{1}{2}(1 + \sqrt{5}) \quad \text{and} \quad \beta = \frac{1}{2}(1 - \sqrt{5})$$

In 1843, the French mathematician Jacques Binet showed that

9.3 $$F_n = \frac{\alpha^n - \beta^n}{\alpha - \beta} \quad \text{and} \quad L_n = \alpha^n + \beta^n$$

These equations, known as the *Binet equations*, are logically equivalent to the recursive definitions 9.1 and 9.2 above. That is, the recursive definitions imply the Binet equations (see Theorem 9.1 below) and vice versa. The proof of the converse of Theorem 9.1 is left as an exercise.

Theorem 9.1   *If $F_1 = F_2 = 1$ and $F_n = F_{n-1} + F_{n-2}$ for $n \geq 3$, then*

$$F_n = \frac{\alpha^n - \beta^n}{\alpha - \beta}$$

where $\alpha = \frac{1}{2}(1 + \sqrt{5})$ and $\beta = \frac{1}{2}(1 - \sqrt{5})$.

PROOF: (Induction on n) $F_1 = 1 = (\alpha^1 - \beta^1)/(\alpha - \beta)$ and $F_2 = 1 = \alpha + \beta = (\alpha^2 - \beta^2)/(\alpha - \beta)$. If $n \geq 3$, then

$$F_n = F_{n-1} + F_{n-2}$$

$$= \frac{\alpha^{n-1} - \beta^{n-1}}{\alpha - \beta} + \frac{\alpha^{n-2} - \beta^{n-2}}{\alpha - \beta}$$

$$= \frac{(\alpha^{n-1} + \alpha^{n-2}) - (\beta^{n-1} + \beta^{n-2})}{\alpha - \beta}$$

However, $\alpha^2 = \alpha + 1$, so $\alpha^n = \alpha^{n-1} + \alpha^{n-2}$; also, $\beta^2 = \beta + 1$, so $\beta^n = \beta^{n-1} + \beta^{n-2}$. Therefore, $F_n = (\alpha^n - \beta^n)/(\alpha - \beta)$.   ∎

Note that $\alpha + \beta = 1$ and $\alpha\beta = -1$. As an example of Theorem 9.1,

$$F_3 = \frac{\alpha^3 - \beta^3}{\alpha - \beta} = \alpha^2 + \alpha\beta + \beta^2 = (\alpha + \beta)^2 - \alpha\beta = 1^2 - (-1) = 2$$

Also,

$$F_4 = \frac{\alpha^4 - \beta^4}{\alpha - \beta} = (\alpha + \beta)(\alpha^2 + \beta^2) = 1[(\alpha + \beta)^2 - 2\alpha\beta]$$

$$= 1 - 2(-1) = 3$$

We now list some properties of Fibonacci and Lucas numbers, namely, explicit formulas and identities. The symbol $\langle x \rangle$ denotes the nearest integer to the real number x; that is, $\langle x \rangle = [x + \frac{1}{2}]$. Arbitrarily, $\langle n - \frac{1}{2} \rangle = n$ (not $n - 1$) for integers n.

Explicit Formulas

9.3 $F_n = \dfrac{\alpha^n - \beta^n}{\alpha - \beta}$ and $L_n = \alpha^n + \beta^n$ (Binet equations)

9.4 $F_n = \langle \dfrac{\alpha^n}{\sqrt{5}} \rangle$ for $n \geq 1$ and $L_n = \langle \alpha^n \rangle$ for $n \geq 2$

9.5 $F_n = \dfrac{1}{2^{n-1}} \displaystyle\sum_{k=0}^{[\frac{1}{2}(n-1)]} \binom{n}{2k+1} 5^k$ and $L_n = \dfrac{1}{2^{n-1}} \displaystyle\sum_{k=0}^{[\frac{1}{2}n]} \binom{n}{2k} 5^k$

9.6 $F_{n+1} = \displaystyle\sum_{k=0}^{[\frac{1}{2}n]} \binom{n-k}{k}$ and $L_n = \displaystyle\sum_{k=0}^{[\frac{1}{2}n]} \dfrac{n}{n-k}\binom{n-k}{k}$

$$F_{n+1} = \sum_{k=-\left[\frac{n+1}{5}\right]}^{[n/5]} (-1)^k \binom{n}{[\frac{1}{2}(n-5k)]}$$

9.7   and

$$L_n = \sum_{k=-\left[\frac{n+1}{5}\right]}^{[n/5]} (-1)^k \frac{n + [\frac{1}{2}(n-5k)]}{n} \binom{n}{[\frac{1}{2}(n-5k)]}$$

Identities

9.8 $L_n^2 - 5F_n^2 = 4(-1)^n$

9.9 $F_{2n} = F_n L_n$

9.10 $L_{2n} = L_n^2 - 2(-1)^n$

9.11
$$F_{2n+1} = F_{n+1}^2 + F_n^2$$

9.12
$$L_{2n+1} = L_n L_{n+1} - (-1)^n$$

9.13
$$F_{n+1} + F_{n-1} = L_n$$

9.14
$$L_{n+1} + L_{n-1} = 5F_n$$

9.15
$$F_{n-1}F_{n+1} - F_n^2 = (-1)^n$$

9.16
$$L_{n-1}L_{n+1} - L_n^2 = 5(-1)^{n-1}$$

9.17
$$F_{m+n} + (-1)^n F_{m-n} = F_m L_n$$

9.18
$$F_{m+n} - (-1)^n F_{m-n} = F_n L_m$$

9.19
$$\sum_{k=1}^{n} F_k = F_{n+2} - 1$$

9.20
$$\sum_{k=1}^{n} L_k = L_{n+2} - 3$$

9.21
$$\alpha^n = \alpha F_n + F_{n-1} \quad \text{and} \quad \beta^n = \beta F_n + F_{n-1}$$

9.22
$$F_{kn+r} = \sum_{j=0}^{k} \binom{k}{j} F_n^j F_{n-1}^{k-j} F_{r+j} \quad \text{and} \quad L_{kn+r} = \sum_{j=0}^{k} \binom{k}{j} F_n^j F_{n-1}^{k-j} L_{r+j}$$

9.23
$$(F_m, F_n) = F_{(m,n)}$$

9.24 $(L_m, L_n) = L_{m,n}$ if $m/(m, n) \equiv n/(m, n) \equiv 1 \pmod 2$

9.25
$$\lim_{n \to \infty} \frac{F_{n+1}}{F_n} = \lim_{n \to \infty} \frac{L_{n+1}}{L_n} = \alpha$$

9.26
$$\lim_{n \to \infty} \frac{L_n}{F_n} = \sqrt{5}$$

9.27
$$F_{-n} = (-1)^{n+1} F_n$$

9.28
$$L_{-n} = (-1)^n L_n$$

9.29 $p \mid F_{p-(5/p)}$ if $p \neq 5$ is prime

9.30
$$\begin{pmatrix} 0 & 1 \\ 1 & 1 \end{pmatrix}^n = \begin{pmatrix} F_{n-1} & F_n \\ F_n & F_{n+1} \end{pmatrix}$$

Most of these properties may be proved by means of one or more of the following: (1) Binet equations, (2) the binomial theorem, or (3) mathematical induction. The proof of 9.7 is more esoteric; see Andrews and Robbins (1991). Formula 9.6 establishes a connection between Fibonacci numbers and Pascal's triangle: If we write the entries in Pascal's triangle in left-justified fashion and then sum along the diagonals, we obtain the Fibonacci numbers (see Figure 9.2).

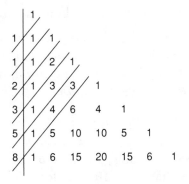

Figure 9.2. Fibonacci Numbers
and Pascal's Triangle

Suppose that a and b are given natural numbers, with $a > b$. In Chapter 2 we presented Euclid's algorithm for finding (a, b), the greatest common divisor of a and b. Generally speaking, if Euclid's algorithm requires n iterations, we would expect n to increase if a and b increase. The following theorem, which is useful in computation number theory, relates n to b. Its proof makes use of Fibonacci numbers.

Theorem 9.2   *(Lamé)*
If a and b are integers such that $a > b > 0$ and n iterations are needed to compute (a, b) using Euclid's algorithm, then $b > \alpha^{n-1}$, where $\alpha = \frac{1}{2}(1 + \sqrt{5})$.

PROOF: Let $a = r_0$ and $b = r_1$, with $0 < r_1 < r_0$. Then

$$\begin{aligned}
r_0 &= r_1 q_1 + r_2 & \text{with } 0 < r_2 < r_1 \\
r_1 &= r_2 q_2 + r_3 & \text{with } 0 < r_3 < r_2 \\
r_2 &= r_3 q_3 + r_4 & \text{with } 0 < r_4 < r_3 \\
&\;\;\vdots \\
r_{n-2} &= r_{n-1} q_{n-1} + r_n & \text{with } 0 < r_n < r_{n-1} \\
r_{n-1} &= r_n q_n & \text{with } q_n \geq 2
\end{aligned}$$

Now $r_n \geq 1$ but $F_2 = 1$, so $r_n \geq F_2$. Also, $r_{n-1} \geq 2r_n \geq 2$, but $F_3 = 2$, so $r_{n-1} \geq F_3$. If $1 \leq j \leq n - 2$, then $r_j = r_{j+1} q_{j+1} + r_{j+2} \geq r_{j+1} + r_{j+2}$. However, $r_{j+1} \geq F_{n+1-j}$ and $r_{j+2} \geq F_{n-j}$. Therefore, $r_j \geq F_{n+1-j} + F_{n-j}$; that is, $r_j \geq F_{n+2-j}$. In particular, $b = r_1 \geq F_{n+1}$. Theorem 1.12 implies that $F_{n+1} > \alpha^{n-1}$. Therefore, $b > \alpha^{n-1}$. ∎

As a corollary, we obtain the following theorem.

Theorem 9.3   *If a and b are integers such that $a > b > 0$, the decimal representation of b has k digits, and n iterations are needed to compute (a, b) using the Euclidean algorithm, then $n \leq 5k$.*

PROOF: By hypothesis and Theorem 9.2, we have $\alpha^{n-1} < b$, so $(n - 1) \log_{10} \alpha < \log_{10} b$. Since $1/5 < \log_{10} \alpha$, it follows that $(n-1)/5 < \log_{10} b$. Since b has k decimal digits by hypothesis, it follows that $b < 10^k$, so $\log_{10} b < k$. Therefore, $(n-1)/5 < k$. This implies that $n - 1 < 5k$, so $n \leq 5k$.   ∎

For example, to compute $(1597, 987)$ using the Euclidean algorithm should require at most $5(3) = 15$ iterations according to Theorem 9.3. The reader can verify that 15 iterations are indeed required.

SECTION 9.1 EXERCISES

1. Prove the converse of Theorem 9.1; that is, prove that if $F_n = (\alpha^n - \beta^n)/(\alpha - \beta)$, where $\alpha = \frac{1}{2}(1 + \sqrt{5})$ and $\beta = \frac{1}{2}(1 - \sqrt{5})$, then $F_1 = F_2 = 1$ and $F_n = F_{n-1} + F_{n-2}$ for $n \geq 3$.

2. Suppose that a fair coin is tossed until two consecutive heads appear. Let p_n be the probability that exactly n tosses are required, with $n \geq 2$. Prove that $P_n = F_{n-1}/2^n$.

3. A staircase has n steps. In climbing this staircase, one may advance either one or two steps at a time. Prove that the number of distinct ways one may climb the staircase is F_{n+1}.

*4. Suppose that n cities are located on the north bank of a straight river that flows from west to east. Each city may build its own sewage disposal plant, or the cities may agree to share costs by having some cities send their sewage through pipes to disposal plants in other cities. No city, however, will send its sewage partly to an eastern neighbor and partly to a western neighbor. Prove that the total number of possible sewage disposal arrangements among the n cities is F_{2n}.

5. Prove explicit formulas 9.5. (*Hint:* Use explicit formulas 9.3 and the binomial theorem.)

Prove each of the two following generating function formulas.

6. $\dfrac{x}{1-x-x^2} = \displaystyle\sum_{n=1}^{\infty} F_n x^n$

7. $\dfrac{e^{\alpha x} - e^{\beta x}}{\alpha - \beta} = \displaystyle\sum_{n=1}^{\infty} \dfrac{F_n x^n}{n!}$

8 through 30. Prove each of the identities 9.8 through 9.30.

31. Prove that $(F_n, F_{kn}/F_n) \mid k$.

32. Prove that $F_n \mid F_{kn}$ for all k and $L_n \mid L_{kn}$ for all odd k.

33. Prove that $(F_n, L_n) = \begin{cases} 2 & \text{if } 3 \mid n \\ 1 & \text{if } 3 \nmid n. \end{cases}$

34. Prove that $4 \mid \phi(F_n)$ for all $n \geq 5$.

35. Prove that every natural number has a representation as a sum of distinct Fibonacci numbers. (*Hint:* Use induction.) For example, $100 = 89 + 8 + 3$ and $50 = 34 + 13 + 3$.

36. Prove that if $n \geq 2$, then there is a right triangle with integer sides and hypotenuse F_{2n+1}. Also, express the legs of the right triangle in terms of Fibonacci numbers.

SECTION 9.1 COMPUTER EXERCISES

37. Write a computer program to represent a given natural number as a sum of distinct Fibonacci numbers.

38. Write a computer program to evaluate $\sum_{k=0}^{n} \binom{n}{k} F_k L_{n-k}$. Is there a pattern?

39. Write a computer program to evaluate $\sum_{k=0}^{n} \binom{n}{k} L_k$. Is there a pattern?

9.2   BINARY LINEAR RECURRENCES

Fibonacci and Lucas numbers can be generalized as follows: Let A and B be nonzero, relatively prime integers such that $D = A^2 - 4B \neq 0$. Now define sequences $\{u_n\}$ and $\{v_n\}$ by

9.31 $u_0 = 0, \quad u_1 = 1, \quad u_n = Au_{n-1} - Bu_{n-2} \quad$ for all $n \geq 2$

9.32 $v_0 = 2, \quad v_1 = A, \quad v_n = Av_{n-1} = Bv_{n-2} \quad$ for all $n \geq 2$

If $A = 1$ and $B = -1$, then $u_n = F_n$ (the nth Fibonacci number) and $v_n = L_n$ (the nth Lucas number).

An alternative is to let the roots of the equation $t^2 - At + B = 0$ be

9.33 $\qquad \alpha = \frac{1}{2}(A + \sqrt{D}) \qquad$ and $\qquad \beta = \frac{1}{2}(A - \sqrt{D})$

Then

9.34 $\quad u_n = \dfrac{\alpha^n - \beta^n}{\alpha - \beta} \qquad$ and $\qquad v_n = \alpha^n + \beta^n \qquad$ for $n \geq 0$

Equations 9.34 are known as the generalized Binet equations. One can show that the recursive definitions of the sequences $\{u_n\}$ and $\{v_n\}$ given by equations 9.31 and 9.32 is logically equivalent to the explicit definitions of these sequences given by equations 9.33 and 9.34. In any case, $\{u_n\}$ and $\{v_n\}$ are called *binary linear recurrences* or *Lucas-type sequences*.

Edouard Lucas (1842–1891), who taught mathematics at the Lycee Charlemagne in Paris, was the first to make a detailed investigation of such sequences (see Lucas). Lucas-type sequences can be used to test an integer n for primality if $n + 1$ can be easily factored (see Ribenboim).

Binary linear recurrences satisfy many identities that generalize the Fibonacci and Lucas identities 9.8 to 9.30. We now list several such identities.

9.35
$$u_n = \frac{1}{2^{n-1}} \sum_{k=0}^{[\frac{1}{2}(n-1)]} \binom{n}{2k+1} A^{n-(2k+1)} D^k \qquad \text{and}$$

$$v_n = \frac{1}{2^{n-1}} \sum_{k=0}^{[\frac{1}{2}n]} \binom{n}{2k} A^{n-2k} D^k$$

9.36
$$v_n^2 - Du_n^2 = 4B^n$$

9.37
$$u_{2n} = u_n v_n$$

9.38
$$v_{2n} = v_n^2 - 2B^n$$

9.39
$$u_{2n+1} = u_{n+1}^2 - Bu_n^2$$

9.40
$$v_{2n+1} = v_n v_{n+1} - AB^n$$

9.41
$$u_{n-1}u_{n+1} - u_n^2 = -B^{n-1}$$

9.42
$$v_{n-1}v_{n+1} - v_n^2 = DB^{n-1}$$

9.43
$$u_{m+n} + B^n u_{m-n} = u_m v_n$$

9.44
$$u_{m-n} - B^n u_{m-n} = u_n v_m$$

9.45 $\alpha^n = \alpha u_n - Bu_{n-1}$ and $\beta^n = \beta u_n - Bu_{n-1}$

9.46
$$(u_m, u_n) = u_{(m,n)}$$

9.47 $(v_m, v_n) = v_{(m,n)}$ if $m/(m, n) \equiv n/(m, n) \equiv 1 \pmod 2$

9.48 $p \mid u_{p-(D/p)}$ if p is prime and $p \nmid D$

9.49 $\begin{pmatrix} 0 & 1 \\ 1 & A \end{pmatrix}^n = \begin{pmatrix} u_{n-1} & u_n \\ u_n & u_{n+1} \end{pmatrix}$ if $B = 1$

Several binary linear recurrences other than the Fibonacci and Lucas numbers are of particular interest. If $A = 3$ and $B = 2$, then $D = 1$, so that we obtain $\alpha = 2$ and $\beta = 1$; hence

9.50 $u_n = 2^n - 1$ and $v_n = 2^n + 1$

Here $\{u_n\}$ may be called the *Mersenne sequence*, and $\{v_n\}$ may be called the *Fermat sequence*. If f_n denotes the nth Fermat number (recall Definition 4.9), then $f_n = v_{2^n}$. Earlier, we saw applications of the Mersenne and Fermat numbers.

If $A = 2$ and $B = 1$, then $D = \sqrt 8 = 2\sqrt 2$, so we obtain

9.51 $\alpha = 1 + \sqrt{2}$ and $\beta = 1 - \sqrt{2}$

Here the sequence $\{u_n\}$ satisfies

9.52 $u_0 = 0,\ \ u_1 = 1,\ u_n = 2u_{n-1} + u_{n-2}$ if $n \geq 2$

This sequence is known as the *Pell sequence*; that is, u_n is called the nth Pell number and is sometimes denoted as P_n. The first few Pell numbers are 0, 1, 2, 5, 12, 29, 70, 169, 408, and 985. The companion sequence $\{v_n\}$ satisfies

9.53 $v_0 = 2,\ \ v_1 = 2,\ \ v_n = 2v_{n-1} + v_{n-2}$ if $n \geq 2$

The terms of the companion sequence are called *associated Pell numbers* and are sometimes denoted R_n. The first few associated Pell numbers are 2, 2, 6, 14, 34, 82, 198, 478, 1154, and 2786.

One can show that if all sides of a right triangle have integer length, and if the legs differ in length by 1, then there exist $n \geq 1$ such that sides have lengths $2P_nP_{n+1}$, $\frac{1}{4}R_nR_{n+1}$, P_{2n+1}. For example, the smallest such triangle, corresponding to $n = 1$, is the familiar 3-4-5 right triangle (see Shanks, 1972).

SECTION 9.3 EXERCISES

40 through 54. Prove each of the identities 9.35 to 9.49.

55. Prove that (u_n, v_n) is determined as follows:

If A is odd and B is even, then $(u_n, v_n) = 1$ for all n.

If A is even and B is odd, then

$$(u_n, v_n) = \begin{cases} 2 & \text{if } 2 \mid n \\ 1 & \text{if } 2 \nmid n \end{cases}.$$

If A and B are both odd, then

$$(u_n, v_n) = \begin{cases} 2 & \text{if } 3 \mid n \\ 1 & \text{if } 3 \nmid n \end{cases}.$$

56. Let p_n denote the nth Pell number. Prove that $2^k \mid P_n$ if and only if $2^k \mid n$.

57. Let u_n be the nth term of the sequence 0, 1, 1, 3, 5, 11, 21, etc. Find a formula for u_n. (The initial term is u_0.)

SECTION 9.3 COMPUTER EXERCISES

58. Write a computer program to compute the first 100 Pell numbers. How many digits do you expect P_{100} to have?

59. Which Pell numbers seem to be divisible by 5? Care to conjecture?

CHAPTER TEN

CONTINUED FRACTIONS

10.1 INTRODUCTION

The awkward-looking expression:

$$4 + \cfrac{1}{7 + \cfrac{1}{2 + \cfrac{1}{3}}}$$

is an example of a *simple finite continued fraction*. Despite their uninviting appearance, soon to be remedied by appropriate notation, continued fractions provide (1) the best possible approximations of irrational numbers by rational numbers, (2) the solutions of the Diophantine equation known as Pell's equation and some generalizations thereof, as we shall see in Chapter 11, and (3) a method of factoring large integers, as we shall see in Chapter 12.

10.2 FINITE CONTINUED FRACTIONS

Let $a_1, a_2, a_3, \cdots, a_n$ be positive real numbers, and let a_0 be a real nonnegative number. We define finite continued fractions in the following inductive manner.

Definition 10.1   *Finite Continued Fraction*

$$[a_0] = a_0 \quad \text{and}$$

$$[a_0, a_1, a_2, \cdots, a_n] = a_0 + \frac{1}{[a_1 a_2, \cdots, a_n]} \quad \text{for } n \geq 1$$

For example,

$$[2] = 2$$
$$[7,2] = 7 + \frac{1}{[2]} = 7 + \frac{1}{2} = \frac{15}{2}$$
$$[3, 7, 2] = 3 + \frac{1}{[7,2]} = 3 + \frac{2}{15} = \frac{47}{15}$$
$$[8, 3, 7, 2] = 8 + \frac{1}{[3,7,2]} = 8 + \frac{15}{47} = \frac{391}{47}$$

Note that, in general,

$$[a_0, a_1] = a_0 + \frac{1}{a_1}$$

$$[a_0, a_1, a_2] = a_0 + \frac{1}{[a_1,a_2]} = a_0 + \frac{1}{a_1 + \frac{1}{a_2}}$$

and

$$[a_0, a_1, a_2, \cdots, a_{n-1}, a_n] = a_0 + \cfrac{1}{a_1 + \cfrac{1}{a_2 + \cdots}} + \cfrac{1}{a_{n-1} + \cfrac{1}{a_n}}$$

Our first theorem regarding continued fractions shows that a given finite continued fraction can be represented as another finite continued fraction with one less term.

Theorem 10.1 *Let $n \geq 1$. Let the real numbers a_i, where $0 \leq i \leq n$ satisfy $a_0 \geq 0$ and $a_i > 0$ for all $i > 0$. Then*

$$[a_0, a_1, a_2, \cdots, a_{n-2}, a_{n-1}, a_n] = \left[a_0, a_1, a_2, \cdots, a_{n-2}, a_{n-1} + \frac{1}{a_n}\right]$$

PROOF: (Induction on n)

$$[a_0, a_1] = a_0 + \frac{1}{a_1} = \left[a_0 + \frac{1}{a_1}\right]$$

so the theorem holds for $n = 1$. Then

$$[a_0, a_1, a_2, \cdots, a_n] = a_0 + \frac{1}{[a_1,a_2,\cdots,a_n]}$$

$$= a_0 + \frac{1}{[a_1,a_2,\cdots,a_{n-2},a_{n-1}+1/a_n]}$$

by induction hypothesis, and this equals $[a_0, a_1, a_2, \cdots, a_{n-2}, a_{n-1} + 1/a_n]$ by Definition 9.1.   ∎

An alternative statement of the conclusion of Theorem 10.1 is

$$[a_0, a_1, a_2, \cdots, a_{n-2}, a_{n-1}, a_n] = [a_0, a_1, a_2, \cdots, a_{n-2}, [a_{n-1}, a_n]]$$

It follows that for any k such that $0 \le k \le n - 1$, we have

$$[a_0, a_1, a_2, \cdots, a_n] = [a_0, a_1, a_2, \cdots, a_k, [a_{k+1}, a_{k+2}, \cdots, a_n]]$$

Repeated application of Theorem 10.1 provides one way of evaluating finite continued fractions. For example,

$$[5, 3, 2] = \left[5, 3 + \frac{1}{2}\right] = \left[5, \frac{7}{2}\right] = \left[5 + \frac{2}{7}\right] = \left[\frac{37}{7}\right] = \frac{37}{7}$$

Also,

$$[1, 2, 3, 4] = \left[1, 2, 3 + \frac{1}{4}\right] = \left[1, 2, \frac{13}{4}\right] = \left[1, 2 + \frac{4}{13}\right] = \left[1, \frac{30}{13}\right]$$

$$= \left[1 + \frac{13}{30}\right] = \left[\frac{43}{30}\right] = \frac{43}{30}$$

Later we will develop a more convenient way of evaluating finite continued fractions. As a corollary to Theorem 10.1, we have the following:

Theorem 10.2   *If n and the a_i are as in Theorem* 10.1, *then*

$$[a_0, a_1, a_2, \cdots, a_{n-1}, 1] = [a_0, a_1, a_2, \cdots, a_{n-2}, a_{n-1} + 1]$$

PROOF: This follows from Theorem 10.1 with $a_n = 1$.   ■

For example, $[7, 5, 3, 1] = [7, 5, 4]$; also, $[8, 5, 1] = [8, 6]$.

The a_i are called the *terms*, or *partial quotients*, of the continued fraction. Clearly, if all the a_i are rational numbers (or integers), then $[a_0, a_1, a_2, \cdots, a_n]$ is rational. Also, $[a_0, a_1, a_2, \cdots, a_n] > 0$ unless $a_0 = n = 0$. The case of greatest interest is where all the a_i are integers. We shall see that every positive rational number has an essentially unique representation as a finite continued fraction with integer terms. (By Theorem 10.2, any continued fraction whose last term is 1 can be converted to an equivalent continued fraction whose last term is greater than 1, and vice versa.) In this text we will consider only *simple* continued fractions, that is, continued fractions whose numerators are all 1s.

The history of continued fractions dates back to ancient times. In the fourth century B.C., Baudhayana, an Indian mathematician, used continued fractions to approximate $\frac{1}{2}\sqrt{\pi}$. Archimedes (287–212 B.C.) used continued fractions to obtain a well-known approximation to π: 22/7. During the Middle Ages, several Indian mathematicians were able to solve certain Diophantine equations by means of algorithms that are equivalent to the use of continued fractions. These mathematicians include Aryabhata (476–550), Brahmagupta (598–665), and Bhaskara (1114–1185). The first systematic treatment of continued fractions, based on efforts by Lord William Brouncker (1620–1684) and John Wallis (1616–1703), was given by the latter in his *Arithmetica Infinitorum*, published in 1655.

Given a positive rational number a/b, one may find its representation as a finite continued fraction with integer terms by employing Euclid's algorithm to find (a, b) and using the *quotients* that arise. For example, let us find the continued fraction representation of 100/13:

$$100 = 7(13) + 9 \qquad \text{that is,} \qquad \frac{100}{13} = 7 + \frac{9}{13}$$

$$13 = 1(9) + 4 \qquad\qquad\qquad \frac{13}{9} = 1 + \frac{4}{9}$$

$$9 = 2(4) + 1 \qquad\qquad\qquad \frac{9}{4} = 2 + \frac{1}{4}$$

$$4 = 4(1) \qquad\qquad\qquad\qquad \frac{4}{1} = 4$$

Therefore, $100/13 = [7, 1, 2, 4]$ (or $[7, 1, 2, 3, 1]$).

As a second example, let us find the continued fraction representation of 27/50:

$$27 = 0(50) + 27 \qquad \text{that is,} \qquad \frac{27}{50} = 0 + \frac{27}{50}$$

$$50 = 1(27) + 23 \qquad\qquad\qquad \frac{50}{27} = 1 + \frac{23}{27}$$

$$27 = 1(23) + 4 \qquad\qquad\qquad \frac{27}{23} = 1 + \frac{4}{23}$$

$$23 = 5(4) + 3 \qquad\qquad\qquad \frac{23}{4} = 5 + \frac{3}{4}$$

$$4 = 1(3) + 1 \qquad\qquad\qquad \frac{4}{3} = 1 + \frac{1}{3}$$

$$3 = 3(1) \qquad\qquad\qquad\qquad \frac{3}{1} = 3$$

Therefore, $27/50 = [0, 1, 1, 5, 1, 3]$ (or $[0, 1, 1, 5, 1, 2, 1]$).

This procedure may be formalized as follows:

Step 1: Set $x = a/b$ (the given rational number).
Step 2: Compute and store $[x]$ (the integer part of x).
Step 3: Compute $x - [x]$ (the fractional part of x).
Step 4: If $x - [x] \neq 0$, then replace x by $1/(x - [x])$ and go to Step 1; if $x - [x] = 0$, then terminate. (The answer is the sequence of integers stored when executing Step 2.)

Note that this procedure consists of alternately subtracting the integer part of a number that is greater than 1 and inverting the resulting fractional part of the number until eventually zero is generated.

In the work that follows, we prove that the finite continued fraction representation of a positive rational number is essentially unique.

Lemma 10.1   *If $n \geq 1$ and $a_0 \geq 1$, then $[a_0, a_1, a_2, \cdots, a_n] > 1$.*

PROOF: (Induction on n) $[a_0, a_1] = a_0 + \frac{1}{a_1} \geq 1 + \frac{1}{a_1} > 1$ by hypothesis. Similarly, $[a_{n-1}, a_n] > 1$. Now Theorem 10.1 implies that

$$[a_0, a_1, a_2, \cdots, a_{n-2}, a_{n-1}, a_n] = [a_0, a_1, a_2, \cdots, a_{n-2}, [a_{n-1}, a_n]]$$

so $[a_0, a_1, a_2, \cdots, a_{n-2}, a_{n-1}, a_n] > 1$ by induction hypothesis.   ∎

For example, $[1, 2, 3] = [1, [2, 3]] = \left[1, \frac{7}{3}\right] = \frac{10}{7} > 1$.

The following lemma shows that, with minor exceptions, a finite continued fraction with integer terms is never itself an integer.

Lemma 10.2   *If $n \geq 0$, let $a_0, a_1, a_2, \cdots, a_n$ be integers such that $a_0 \geq 0$ and $a_i > 0$ for all $i > 0$. Then $[a_0, a_1, a_2, \cdots, a_n] = k$, an integer, if and only if (i) $n = 0$ and $a_0 = k$ or (ii) $n = 1$ and $a_0 = k - 1$.*

PROOF: (Sufficiency) $k = [k] = [k - 1, 1]$ by Definition 10.1 and Theorem 10.2.   (Necessity)   If $n = 0$, then by hypothesis and Definition 10.1, we have $k = [a_0] = a_0$. If $n \geq 1$, then by hypothesis and Definition 10.1, we have $a_0 + 1/[a_1, a_2, \cdots, a_n] = k$, so $1 / [a_1, a_2, \cdots, a_n] = k - a_0$, integer. If $n \geq 2$, then by hypothesis and Lemma 10.1, we have $[a_1, a_2, \cdots, a_n] > 1$, so $0 < k - a_0 < 1$, an impossibility. Therefore, $n = 1$ and $1/a_1 = 1/[a_1] = k - a_0$, so $a_1 = 1$ and $a_0 = k - 1$.   ∎

For example, $[4] = 4$ and $[2, 1] = 3$.

Next we show that if two finite continued fractions with integer terms have the same value, and if neither has 1 as its last term, then they must be identical, that is, have exactly the same terms.

Theorem 10.3   *If $[a_0, a_1, a_2, \cdots, a_m] = [b_0, b_1, b_2, \cdots, b_n]$, where all a_i and b_j are integers, and if $a_m \neq 1$ and $b_n \neq 1$, then $m = n$ and $a_i = b_i$ for all i such that $0 \leq i \leq m$.*

PROOF: Without loss of generality, assume that $m \leq n$. If $m = 0$, then by hypothesis we have $a_0 = [a_0] = [b_0, b_1, b_2, \cdots, b_n]$. Now Lemma 10.2 implies that either $b_0 = a_0$ and $n = 0$ or $b_0 = a_0 - 1$ and $b_1 = n = 1$. However, the latter possibility is excluded by hypothesis. If $m \geq 1$, then by hypothesis and definition 10.1 we have

$$a_0 + \frac{1}{[a_1, a_2, \cdots, a_m]} = b_0 + \frac{1}{[b_1, b_2, \cdots, b_n]}$$

Now Lemma 10.1 implies that each of $[a_1, a_2, \cdots, a_m]$ and $[b_1, b_2, \cdots, b_n]$ exceeds 1. Therefore, $a_0 = b_0$ and $[a_1, a_2, \cdots, a_m] = [b_1, b_2, \cdots, b_n]$. By repeated application of this argument, we get $a_1 = b_1, a_2 = b_2, \cdots, a_{m-1} = b_{m-1}, a_m = [a_m] = [b_m, b_{m+1}, b_{m+2}, \cdots, b_n]$. Since $b_n > 1$ by hypothesis, Lemma 10.2 implies that $a_m = b_m$ and $m = n$.

∎

We are now ready to combine the results of Theorems 10.2 and 10.3 in Theorem 10.4.

Theorem 10.4   *Every positive rational number has exactly two representations as a finite continued fraction with integer terms: one whose last term exceeds* 1 *and one whose last term is* 1.

PROOF: Given a positive rational number c/d, let $c = r_{-1}$ and $d = r_0$. Then by the Euclidean algorithm, we have

$$\frac{c}{d} = \frac{r_{-1}}{r_0} = a_0 + \frac{r_1}{r_0} \qquad \text{with } 0 < r_1 < r_0$$

$$\frac{r_0}{r_1} = a_1 + \frac{r_2}{r_1} \qquad \text{with } 0 < r_2 < r_1$$

$$\frac{r_1}{r_2} = a_2 + \frac{r_3}{r_2} \qquad \text{with } 0 < r_3 < r_2$$

$$\vdots$$

$$\frac{r_{n-2}}{r_{n-1}} = a_{n-1} + \frac{r_n}{r_{n-1}} \qquad \text{with } 0 < r_n < r_{n-1}$$

$$\frac{r_{n-1}}{r_n} = a_n > 1$$

This sequence of $n + 1$ equations yields $c/d = [a_0, r_0/r_1] = [a_0, a_1, r_1/r_2] = [a_0, a_1, a_2, r_2/r_3] = \cdots = [a_0, a_1, a_2, \cdots, a_{n-1}, r_{n-1}/r_n] = [a_0, a_1, a_2, \cdots, a_{n-1}, a_n]$. By Theorem 10.2, we also have $c/d = [a_0, a_1, a_2, \cdots, a_{n-1}, a_n - 1, 1]$. Theorem 10.3 guarantees that c/d has no additional representations as a finite continued fraction with integer terms.

∎

Incidentally, the preceding sequence of $n + 1$ equations may seem familiar. In slightly different form (fractions cleared and q_{j+1} instead of a_j), these are just the equations that arise when computing (c, d) (see Chapter 2, pages 26–27). Earlier, using the Euclidean algorithm, we obtained the results: $100/13 = [7, 1, 2, 4]$ and $27/50 = [0, 1, 1, 5, 1, 3]$.

Now let us consider the reverse problem of computing the positive rational number that corresponds to a given finite continued fraction. One possibility is to work backwards, that is, to employ Theorem 10.1 repeatedly. For example,

$$[7, 1, 2, 4] = \left[7, 1, 2 + \frac{1}{4}\right] = \left[7, 1, \frac{9}{4}\right] = \left[7, 1 + \frac{4}{9}\right] = \left[7, \frac{13}{9}\right]$$

$$= \left[7 + \frac{9}{13}\right] = \left[\frac{100}{13}\right] = \frac{100}{13}$$

Also,

$$[0, 1, 1, 5, 1, 3] = \left[0, 1, 1, 5, 1 + \frac{1}{3}\right] = \left[0, 1, 1, 5, \frac{4}{3}\right]$$

$$= \left[0, 1, 1, 5 + \frac{3}{4}\right] = \left[0, 1, 1, \frac{23}{4}\right]$$

$$= \left[0, 1, 1 + \frac{4}{23}\right] = \left[0, 1, \frac{27}{23}\right]$$

$$= \left[0, 1 + \frac{23}{27}\right] = \left[0, \frac{50}{27}\right] = \left[0 + \frac{27}{50}\right]$$

$$= \left[\frac{27}{50}\right] = \frac{27}{50}$$

If we apply this method to $[a_0, a_1, a_2, \cdots, a_n]$, where all the a_i are integers, then n additions of fractions are required. A second method, which we are about to present, is more convenient, since it involves only operations on integers and allows us to work forwards, that is, from a_0 to a_n. We must first define what are known as the *convergents* to a continued fraction.

Definition 10.2   *Convergent*

Let $A = [a_0, a_1, a_2, \cdots, a_n]$, where the a_i are real and are all positive, except possibly a_0. If $0 \le k \le n$, let $C_k = [a_0, a_1, a_2, \cdots, a_k]$. C_k is called the kth *convergent* to A.

For example, let $A = [8, 3, 7, 2]$. Then

$$
\begin{aligned}
C_0 &= [8] = 8 \\
C_1 &= [8, 3] = 25/3 \\
C_2 &= [8, 3, 7] = 183/22 \\
C_3 &= [8, 3, 7, 2] = 391/47 = A
\end{aligned}
$$

Next we develop an iterative method to generate the convergents of a continued fraction, assuming that we have the terms a_k.

Theorem 10.5 *Given a finite continued fraction $A = [a_0, a_1, a_2, \cdots, a_n]$, where all the a_i are real and all except possibly a_0 are positive, define sequences $\{p_n\}$ and $\{q_n\}$ by*

$$
\begin{aligned}
p_{-2} &= 0, \quad p_{-1} = 1, \quad p_k = a_k p_{k-1} + p_{k-2} \quad \text{for } k \geq 0 \\
q_{-2} &= 1, \quad q_{-1} = 0, \quad q_k = a_k q_{k-1} + q_{k-2} \quad \text{for } k \geq 0
\end{aligned}
$$

If $0 \leq k \leq n$ and C_k is the kth convergent to A, that is, $C_k = [a_0, a_1, a_2, \cdots, a_k]$, then $C_k = p_k/q_k$.

PROOF: (Induction on k) If $k = 0$, then $p_0/q_0 = a_0/1 = a_0 = [a_0] = C_0$. If $k = 1$, then

$$
\frac{p_1}{q_1} = \frac{a_1 p_0 + p_{-1}}{a_1 q_0 + q_{-1}} = \frac{a_1 a_0 + 1}{a_1} = a_0 + \frac{1}{a_1} = [a_0, a_1] = C_1
$$

If $k \geq 2$, then

$$
\begin{aligned}
\frac{p_k}{q_k} &= \frac{a_k p_{k-1} + p_{k-2}}{a_k q_{k-1} + q_{k-2}} = \frac{a_k(a_{k-1}p_{k-2} + p_{k-3}) + p_{k-2}}{a_k(a_{k-1}q_{k-2} + q_{k-3}) + q_{k-2}} \\[2mm]
&= \frac{(a_k a_{k-1} + 1)p_{k-2} + a_k p_{k-3}}{(a_k a_{k-1} + 1)q_{k-2} + a_k q_{k-3}} = \frac{\left(a_{k-1} + \dfrac{1}{a_k}\right)p_{k-2} + p_{k-3}}{\left(a_{k-1} + \dfrac{1}{a_k}\right)q_{k-2} + q_{k-3}} \\[2mm]
&= \left[a_0, a_1, a_2, \cdots, a_{k-2}, a_{k-1} + \frac{1}{a_k}\right]
\end{aligned}
$$

by induction hypothesis. Therefore, Theorem 10.1 implies that $p_k/q_k = [a_0, a_1, a_2, \cdots, a_{k-1}, a_k] = C_k$. $\blacksquare$

Our first example of a continued fraction was $[8, 3, 7, 2]$. Let us now recompute this quantity using Theorem 10.5. Starting with $p_{-2} = 0$, $p_{-1} = 1$, $q_{-2} = 1$, and $q_{-1} = 0$, we get

$$p_0 = a_0 p_{-1} + p_{-2} = 8(1) + 0 = 8$$
$$p_1 = a_1 p_0 + p_{-1} = 3(8) + 1 = 25$$
$$p_2 = a_2 p_1 + p_0 = 7(25) + 8 = 183$$
$$p_3 = a_3 p_2 + p_1 = 2(183) + 25 = 391$$

$$q_0 = a_0 q_{-1} + q_{-2} = 8(0) + 1 = 1$$
$$q_1 = a_1 q_0 + q_{-1} = 3(1) + 0 = 3$$
$$q_2 = a_2 q_1 + q_0 = 7(3) + 1 = 22$$
$$q_3 = a_3 q_2 + q_1 = 2(22) + 3 = 47$$

Therefore, $[8, 3, 7, 2] = p_3/q_3 = 391/47$.

We now develop some properties of convergents. Recall that p_n/q_n is the nth convergent to a continued fraction and that F_n is the nth Fibonacci number.

Lemma 10.3 $q_n \geq F_{n+1}$ for all $n \geq 0$

PROOF: (Induction on n) $q_0 = 1 = F_1$, and $q_1 = a_1 q_0 + q_{-1} = a_1(1) + 0 = a_1 \geq 1 = F_2$. If $n \geq 2$, then $q_n = a_n q_{n-1} + q_{n-2} \geq q_{n-1} + q_{n-2} \geq F_n + F_{n-1}$ by induction hypothesis. However, $F_n + F_{n-1} = F_{n+1}$, so $q_n \geq F_{n+1}$. ∎

In the previous example, we had $q_0 = 1 = F_1, q_1 = 3 \geq F_2, q_2 = 22 \geq 2 = F_3$, and $q_3 = 47 \geq 3 = F_4$.

Lemma 10.4 $q_n \geq \alpha^{n-1}$ for all $n \geq 0$

PROOF: This follows from Lemma 10.3 and Theorem 1.12.

Lemma 10.5   *If $n \geq 2$, then $q_n > q_{n-1}$.*

PROOF: If $n \geq 2$, then $q_{n-2} \geq \alpha^{n-3} > 0$ by Lemma 10.4 Therefore, $q_{n-2} \geq 1$. Now $q_n = a_n q_{n-1} + q_{n-2} \geq q_{n-1} + 1$, so $q_n > q_{n-1}$. ∎

Theorem 10.6   *(a) If $n \geq -1$, then*

$$p_n q_{n-1} - p_{n-1} q_n = (-1)^{n-1}$$

(b) If $n \geq 1$, then

$$\frac{p_n}{q_n} - \frac{p_{n-1}}{q_{n-1}} = \frac{(-1)^{n-1}}{q_n q_{n-1}}$$

PROOF: If $n \geq 1$, then Lemma 10.3 implies that $q_{n-1} \geq 1$ and $q_n \geq 1$. Therefore (b) follows from (a), so it suffices to prove (a). We use induction on n. Now $p_{-1}q_{-2} - p_{-2}q_{-1} = 1(1) - 0(0) = 1 = (-1)^{-2}$. If $n \geq 0$, then

$$p_n q_{n-1} - p_{n-1}q_n = (a_n p_{n-1} + p_{n-2})q_{n-1} - p_{n-1}(a_n q_{n-1} + q_{n-2})$$
$$= -(p_{n-1}q_{n-2} - p_{n-2}q_{n-1}) = -(-1)^{n-2}$$

by induction hypothesis, and this equals $(-1)^{n-1}$.   ∎

For example, looking back at the convergents to $[8, 3, 7, 2]$, we have

$$p_0 q_{-1} - p_{-1}q_0 = 8(0) - 1(1) = -1$$
$$p_1 q_0 - p_0 q_1 = 25(1) - 8(3) = 1$$
$$p_2 q_1 - p_1 q_2 = 183(3) - 22(25) = -1$$
$$p_3 q_2 - p_2 q_3 = 391(22) - 183(47) = 1$$

Theorem 10.7   *If $n \geq 0$, then $(p_n, q_n) = (p_n, p_{n-1}) = (q_n, q_{n-1}) = 1$.*

PROOF: Theorem 10.6a implies that $(p_n, q_n) \mid 1$, $(p_n, p_{n-1}) \mid 1$, and $(q_n, q_{n-1}) \mid 1$, from which the conclusion follows.   ∎

For example, again referring to the convergents to $[8, 3, 7, 2]$, we have $(p_3, q_3) = (391, 47) = 1$, $(p_3, p_2) = (391, 183) = 1$, and $(q_3, q_2) = (47, 22) = 1$.

Theorem 10.8   *(a)   If $n \geq 0$, then*

$$p_n q_{n-2} - p_{n-2}q_n = (-1)^n a_n$$

(b) If $n \geq 2$, then

$$\frac{p_n}{q_n} - \frac{p_{n-2}}{q_{n-2}} = \frac{(-1)^n a_n}{q_n q_{n-2}}$$

PROOF: $q_2 > q_1 \geq F_2 = 1$ by Lemmas 10.3 and 10.5; also, $q_0 = 1$. If $n \geq 3$, then $q_n > q_{n-1} > q_{n-2} \geq \alpha^{n-3} \geq \alpha^0 = 1$ by Lemmas 10.4 and 10.5. Therefore, (b) follows from (a), so it suffices to prove (a). Now $p_0 q_{-2} - p_{-2}q_0 = a_0(1) - 0(1) = a_0 = (-1)^0 a_0$. If $n \geq 1$, then $p_n q_{n-2} - p_{n-2}q_n = (a_n p_{n-1} + p_{n-2})q_{n-2} - p_{n-2}(a_n q_{n-1} + q_{n-2}) = a_n(p_{n-1}q_{n-2} - p_{n-2}q_{n-1}) = (-1)^{n-2}a_n = (-1)^n a_n$ by Theorem 10.6a.   ∎

For example, again referring to the convergents to $[8, 3, 7, 2]$, we have

$$p_1q_{-1} - p_{-1}q_1 = 25(0) - 1(3) = -3 = -a_1$$
$$p_2q_0 - p_0q_2 = 183(1) - 8(22) = 7 = a_2$$
$$p_3q_1 - p_1q_3 = 391(3) - 25(47) = -2 = -a_3$$

As a consequence of Theorem 10.8b, we obtain two further results that will be useful when we deal with infinite continued fractions.

Theorem 10.9   Let $C_n = p_n/q_n$. Then (a) the sequence $\{C_{2n-1}\}$ is strictly decreasing and (b) the sequence $\{C_{2n}\}$ is strictly increasing.

PROOF: By Theorem 10.8b, we have

(a) $C_{2n+1} - C_{2n-1} = \dfrac{p_{2n+1}}{q_{2n+1}} - \dfrac{p_{2n-1}}{q_{2n-1}} = \dfrac{(-1)^{2n+1}a_{2n+1}}{q_{2n+1}q_{2n-1}} < 0$

(b) $C_{2n+2} - C_{2n} = \dfrac{p_{2n+2}}{q_{2n+2}} - \dfrac{p_{2n}}{q_{2n}} = \dfrac{(-1)^{2n+2}a_{2n+2}}{q_{2n+2}q_{2n}} > 0$ ∎

For example, returning again to the convergents to $[8, 3, 7, 2]$, we see that $p_0/q_0 = 8/1 < 183/22 = p_2/q_2$; also, $p_1/q_1 = 25/3 > 391/47 = p_3/q_3$. Even more can be said, namely:

Theorem 10.10   Let $C_k = [a_0, a_1, a_2, \cdots, a_k]$, where $0 \le k \le n$. Suppose that j is even, $j \le n$, and m is odd, $m \le n$. Then (a) $C_j < C_m$, (b) $C_j \le C_n < C_m$ if n is even, and (c) $C_j < C_n \le C_m$ if n is odd.

PROOF: Let $j = 2b$ and $m = 2t - 1$. Then $C_m - C_j = C_{2t-1} - C_{2b}$. If $t \le b$, then

$$C_{2t-1} - C_{2b} \ge C_{2b-1} - C_{2b} = \frac{(-1)^{2b}}{q_{2b}q_{2b-1}} > 0$$

by Theorems 10.9a and 10.6b. If $t > b$, then

$$C_{2t-1} - C_{2b} > C_{2t-1} - C_{2t} = \frac{(-1)^{2t}}{q_{2t}q_{2t-1}} > 0$$

by Theorems 10.9b and 10.6b. This proves (a). Now (b) and (c) follow from hypothesis and (a).   ∎

For example, returning once more to the convergents to $[8, 3, 7, 2]$, we have $p_0/q_0 < p_2/q_2 < p_3/q_3 < p_1/q_1$; that is, $8/1 < 183/22 < 391/47 < 25/3$.

SECTION 10.2 EXERCISES

1. Find finite continued fraction representations with integer terms (and $a_n > 1$) for each of the following.

    (a) $\frac{100}{37}$
    (b) $\frac{1001}{45}$
    (c) $\frac{21}{13}$
    (d) $\frac{13}{35}$
    (e) $\frac{1000}{301}$

2. Find rational numbers having the following continued fraction representations.

    (a) $[1, 2, 3, 4, 5]$
    (b) $[1, 1, 2, 2, 3, 3]$
    (c) $[1, 1, 1, 1, 1, 1]$
    (d) $[2, 1, 1, 1, 4]$
    (e) $[0, 1, 2, 3]$

3. If $0 < c < d$ and $d/c = [a_0, a_1, a_2, \cdots, a_n]$, prove that $c/d = [0, a_0, a_1, a_2, \cdots, a_n]$.

4. Use induction to prove that $[a_0, a_1, a_2, \cdots, a_n] = [a_0, a_1, a_2, \cdots, a_{m-1}, [a_m, a_{m+1}, \cdots, a_n]]$ for $0 \le m \le n$.

5. Prove that if $x = [a_0, a_1, a_2, \cdots, a_{n-1}, x]$, then x is a quadratic irrational.

6. Prove that the rational number whose continued fraction representation consists of n 1s is F_{n+1}/F_n.

7. Prove that the rational number whose continued fraction representation consists of n 2s is P_{n+1}/P_n. (Here P_n denotes the nth Pell number. See page 195.)

SECTION 10.2 COMPUTER EXERCISES

8. Write a computer program to evaluate $[1, 2, 3, \cdots, n - 1, n]$ for $1 \le n \le 50$ and to convert the resulting fraction to a decimal approximation. What seems to be happening?

9. Write a computer program to find the continued fraction representation of a positive rational number.

10.3   INFINITE CONTINUED FRACTIONS

The notion of an infinite continued fraction follows naturally from the notion of a finite continued fraction via the limit process.

Definition 10.3   *Infinite Continued Fractions*

Let $a_0 \ge 0$, and let $a_i > 0$ for all $i > 0$. Then we define the *infinite continued fraction*, denoted $[a_0, a_1, a_2, \cdots]$, as $\lim_{n \to \infty} [a_0, a_1, a_2, \cdots, a_n]$.

For example,

$$[1, 1, 1, \cdots] = \lim_{n \to \infty} [1, 1, 1, \cdots, 1] = \lim_{n \to \infty} \frac{F_{n+1}}{F_n} = \alpha$$
$$= \frac{1}{2}(1 + \sqrt{5})$$

by the result of Exercise 6 and the Fibonacci identity 9.25.

One might question whether the limit whose existence is implied by Definition 10.3 always exists. The following theorem answers this question affirmatively and gives an idea of how well the infinite continued fraction is approximated by the nth convergent.

Theorem 10.11   *If $a_0 \geq 0$ and $a_i > 0$ for all $i > 0$, then $L = \lim_{n \to \infty} [a_0, a_1, a_2, \cdots, a_n]$ exists. Furthermore, $|L - (p_n/q_n)| < 1/q_n q_{n+1}$ for all $n \geq 1$.*

PROOF: Let $C_{2n} = [a_0, a_1, a_2, \cdots, a_{2n}]$ and $C_{2n+1} = [a_0, a_1, a_2, \cdots, a_{2n+1}]$. Theorems 10.8b and 10.10a imply that the sequence $\{C_{2n}\}$ is monotone increasing and bounded from above by C_1. Therefore, $\{C_{2n}\}$ converges to a limit L_1. Likewise, Theorems 10.8a and 10.10a imply that the sequence $\{C_{2n+1}\}$ is monotone decreasing and bounded from below by C_0. Therefore, $\{C_{2n+1}\}$ converges to a limit L_2. For all n, we have $p_{2n}/q_{2n} = C_{2n} < L_1$ and $p_{2n+1}/q_{2n+1} = C_{2n+1} > L_2$, so

$$|L_2 - L_1| < \left|\frac{p_{2n+1}}{q_{2n+1}} - \frac{p_{2n}}{q_{2n}}\right| = \frac{1}{q_{2n+1}q_{2n}} \leq \frac{1}{\alpha^{4n-1}}$$

by Theorem 10.6b and Lemma 10.4. Since $\alpha > 1$ and n is arbitrary, we must have $|L_2 - L_1| = 0$, so $L_2 = L_1 = L$. Furthermore, since L falls between p_n/q_n and p_{n+1}/q_{n+1} for all n, we have

$$\left|L - \frac{p_n}{q_n}\right| < \left|\frac{p_n}{q_n} - \frac{p_{n+1}}{q_{n+1}}\right| = \frac{1}{q_n q_{n+1}}$$ ∎

For example, suppose we wish to estimate the golden ratio α to three decimal places using a convergent to its infinite continued fraction. That is, we seek the least n such that $|\alpha - (p_n/q_n)| < \frac{1}{2}(10^{-3})$. By Theorem 10.11, we may choose n so that $1/(q_n q_{n+1}) \leq \frac{1}{2}(10^{-3})$; that is, $q_n q_{n+1} \geq 2000$. However, in this case, by the result of Exercise 6, $q_n = F_n$, so we want $F_n F_{n+1} \geq 2000$. The least such n is 10. Since $p_n = F_{n+1}$, our estimate for α is $F_{11}/F_{10} = 89/55 = 1.618$.

We saw previously that every positive rational number has an essentially unique representation as a finite continued fraction. We now establish the connections between positive irrational numbers and infinite continued fractions.

Theorem 10.12   *If the real number x has a representation as an infinite continued fraction, then x is irrational.*

PROOF: Let $x = [a_0, a_1, a_2, \cdots]$. As in the proof of Theorem 10.11, we have $p_{2n}/q_{2n} < x < p_{2n+1}/q_{2n+1}$ for all n, so

$$0 < x - \frac{p_{2n}}{q_{2n}} < \frac{p_{2n+1}}{q_{2n+1}} - \frac{p_{2n}}{q_{2n}} = \frac{1}{q_{2n}q_{2n+1}}$$

If $x = a/b$, then

$$0 < \frac{a}{b} - \frac{p_{2n}}{q_{2n}} < \frac{1}{q_{2n}q_{2n+1}}$$

so $0 < aq_{2n} - bp_{2n} < b/q_{2n+1} \leq b/\alpha^{2n}$ by Lemma 10.4. However, if n is sufficiently large, then $b < \alpha^{2n}$, so that $0 < aq_{2n} - bp_{2n} < 1$, an impossibility. ∎

It remains to be shown that every positive irrational number has a unique representation as an infinite continued fraction with integer terms. First, some additional preliminaries are needed.

Lemma 10.6   *If u and v are nonnegative integers, $0 \leq f < 1$, $0 \leq g < 1$, and $u + f = v + g$, then $u = v$ and $f = g$.*

PROOF: Exercise

Theorem 10.13   *Every positive irrational number has at most one representation as an infinite continued fraction with integer terms.*

PROOF: Suppose $x_0 = y_0$, where $x_0 = [a_0, a_1, a_2, \cdots]$ and $y_0 = [b_0, b_1, b_2, \cdots]$. Then $\lim_{t \to \infty} [a_0, a_1, a_2, \cdots, a_t] = \lim_{t \to \infty} [b_0, b_1, b_2, \cdots, b_t]$, so $\lim_{t \to \infty} (a_0 + 1/[a_1, a_2, \cdots, a_t]) = \lim_{t \to \infty} (b_0 + 1/[b_1, b_2, \cdots, b_t])$; that is, $a_0 + 1 \lim_{t \to \infty} [a_1, a_2, \cdots, a_t] = b_0 + 1 \lim_{t \to \infty} [b_1, b_2, \cdots, b_t]$. Since $a_1 > 1$ and $b_1 > 1$, it follows that $\lim_{t \to \infty} [a_1, a_2, \cdots, a_t] > 1$ and $\lim_{t \to \infty} [b_1, b_2, \cdots, b_t] > 1$. Therefore, Lemma 10.6 implies that $a_0 = b_0$ and $\lim_{t \to \infty} [a_1, a_2, \cdots, a_t] = \lim_{t \to \infty} [b_1, b_2, \cdots, b_t]$. Similarly, $a_1 = b_1, a_2 = b_2, \cdots, a_n = b_n, \cdots$. ∎

The preceding theorem says that if an irrational positive number has a representation as an infinite continued fraction with integer terms, then that representation is unique. The following theorem guarantees the existence of such a representation and indicates how to compute it.

Theorem 10.14   *If x_0 is a positive irrational number, let $a_0 = [x_0]$. If $n \geq 1$, let $x_n = 1/(x_{n-1} - a_{n-1})$, and let $a_n = [x_n]$. Then $x_0 = [a_0, a_1, a_2, \cdots]$.*

PROOF: First, we show by induction on n that $x_0 = [a_0, a_1, \cdots, a_{n-1}, x_n]$ for all $n \geq 1$. By hypothesis, $x_{n-1} = a_{n-1} + 1/x_n$ for all $n \geq 1$, so $x_0 = a_0 + 1/x_1 = [a_0, x_1]$. Now, by induction hypothesis, we have $x_0 = [a_0, a_1, a_2, \cdots, a_{n-2}, x_{n-1}]$. However, recall that $x_{n-1} = a_{n-1} + 1/x_n$, so $x_0 = [a_0, a_1, a_2, \cdots, a_{n-2}, a_{n-1} + 1/x_n]$. Now Theorem 10.1 implies that $x_0 = [a_0, a_1, a_2, \cdots, a_{n-2}, a_{n-1}, x_n]$. Now let $p_n/q_n = C_n = [a_0, a_1, a_2, \cdots, a_n]$. Theorem 10.5 implies that $x_0 = (x_{n+1}p_n + p_{n-1})/(x_{n+1}q_n + q_{n-1})$. Therefore,

$$|x_0 - C_n| = |\frac{x_{n+1}p_n + p_{n-1}}{x_{n+1}q_n + q_{n-1}} - \frac{p_n}{q_n}| = |\frac{p_{n-1}q_n - p_nq_{n-1}}{q_n(x_{n+1}q_n + q_{n-1})}|$$

$$= \frac{1}{q_n(x_{n+1}q_n + q_{n-1})} < \frac{1}{q_n(a_{n+1}q_n + q_{n-1})}$$

$$= \frac{1}{q_n q_{n+1}} < \frac{1}{\alpha^{2n-1}}$$

by Lemma 10.4 and Theorem 10.6a. Therefore, $\lim_{n \to \infty} (x_0 - C_n) = 0$, so $x_0 = \lim_{n \to \infty} C_n = \lim_{n \to \infty} [a_0, a_1, a_2, \cdots, a_n] = [a_0, a_1, a_2, \cdots]$. ■

For example, let us compute the infinite continued fraction representation of $\sqrt{22}$.

$$x_0 = \sqrt{22} = 4 + (\sqrt{22} - 4) \qquad a_0 = 4$$

$$x_1 = \frac{1}{\sqrt{22} - 4} = \frac{\sqrt{22} + 4}{6} = 1 + \frac{\sqrt{22} - 2}{6} \qquad a_1 = 1$$

$$x_2 = \frac{6}{\sqrt{22} - 2} = \frac{\sqrt{22} + 2}{3} = 2 + \frac{\sqrt{22} - 4}{3} \qquad a_2 = 2$$

$$x_3 = \frac{3}{\sqrt{22} - 4} = \frac{\sqrt{22} + 4}{2} = 4 + \frac{\sqrt{22} - 4}{2} \qquad a_3 = 4$$

$$x_4 = \frac{2}{\sqrt{22} - 4} = \frac{\sqrt{22} + 4}{3} = 2 + \frac{\sqrt{22} - 2}{3} \qquad a_4 = 2$$

$$x_5 = \frac{3}{\sqrt{22} - 2} = \frac{\sqrt{22} + 2}{6} = 1 + \frac{\sqrt{22} - 4}{6} \qquad a_5 = 1$$

$$x_6 = \frac{6}{\sqrt{22} - 4} = \sqrt{22} + 4 = 8 + (\sqrt{22} - 4) \qquad a_6 = 8$$

At this point, repetition sets in. We have $x_6 - a_6 = \sqrt{22} - 4 = x_0 - a_0$. Now $x_7 = (x_6 - a_6)^{-1} = (x_0 - a_0)^{-1} = x_1$, so $a_7 = [x_7] = [x_1] = a_1$. More generally, for all $k \geq 1$, we have $x_{6+k} = (x_{5+k} - a_{5+k})^{-1} = (x_{k-1} - a_{k-1})^{-1} = x_k$, so $a_{6+k} = a_k$. Therefore, $\sqrt{22} = [4, 1, 2, 4, 2, 1, 8, 1, 2, 4, 2, 1, 8, 1, 2, 4, 2, 1, 8, \cdots]$.

As a consequence of Theorems 10.13 and 10.14, we are able to state the following theorem.

Theorem 10.15   *Every positive irrational number x_0 has a unique representation as an infinite continued fraction with integer terms, namely, $x_0 = [a_0, a_1, a_2, \cdots]$, where $a_n = [x_n]$ for all $n \geq 0$ and $x_n = 1/(x_{n-1} - a_{n-1})$ for all $n \geq 1$.*

PROOF: This follows from Theorems 10.13 and 10.14.   ∎

It appears from Theorem 10.15 that the infinite continued fraction representation of a positive irrational number may be obtained by an automatic procedure. In fact, the procedure is the same as that for obtaining the finite continued fraction representation of a rational number (see page 200), except that it never terminates. There is a catch, however. In the preceding example, x_1 through x_6 were presented in radical form, so no roundoff error was introduced in computing the x_i. Generally, however, this is not the case. If any x_i, which is irrational, is approximated by a rational number, then the error thus introduced will propagate larger errors in the x_k such that $k > i$. As a consequence, for sufficiently large k, the error in x_k will exceed 1, so a_k will be incorrect, as well as all a_n with $n > k$.

For example, it is known that

10.1 $\pi = [3, 7, 15, 1, 292, 1, 1, 1, 2, 1, 3, \cdots]$

In fact, the familiar approximation 22/7 is just the first convergent: [3, 7]. If you try to verify equation 10.1 with a hand calculator, using the approximation $\pi = 3.141592654$, you will not obtain $a_8 = 2$.

In the next section we will see that the best rational approximations to an irrational number are obtained from the convergents to the infinite continued fraction representation of that irrational number. In general, however, the infinite continued fraction representation of an irrational number is hard to obtain. A notable exception is the case where the irrational number is quadratic; that is, the irrational number is a real, positive root of a quadratic equation with integer coefficients. This case is investigated in Section 10.5.

SECTION 10.3 EXERCISES

10. Let x be the irrational number whose continued fraction representation is $[1, 2, 3, \cdots]$. Find the least n such that $C_n = p_n/q_n$ estimates x correctly to 6 decimal places.

11. Use Theorem 10.14 to obtain the continued fraction representation of $\sqrt{41}$. Using this result, estimate $\sqrt{41}$ to the nearest hundredth.

12. Prove that if x is the irrational number whose continued fraction representation is $[m, m, m, \cdots]$, then $x = \frac{1}{2}(m + \sqrt{m^2+4})$.

13. Prove that if $x = [a_0, a_1, a_2, \cdots]$, where $a_0 \geq 1$, then $1/x = [0, a_0, a_1, a_2, \cdots]$.

14. Let a family of polynomials be defined by

$$P_n(x) = \sum_{k=0}^{\left[\frac{n}{2}\right]} \binom{n-k}{k} x^{n-2k}$$

Let θ be the irrational number whose continued fraction representation is $[m, m, m, \cdots]$. Let $C_n(\theta)$ be the nth convergent to θ. Prove that $C_n(\theta) = P_{n+1}(m)/P_n(m)$.

10.4   APPROXIMATION BY CONTINUED FRACTIONS

We will now investigate just how well an irrational number is approximated by the convergents of its infinite continued fraction. First, we show that each convergent provides a better approximation than its immediate predecessor.

Theorem 10.16 *If x_0 is a positive irrational number, p_n/q_n is the nth convergent in the infinite continued fraction for x_0, and $n \geq 2$, then*

$$\left| x_0 - \frac{p_n}{q_n} \right| < \left| x_0 - \frac{p_{n-1}}{q_{n-1}} \right|$$

PROOF: As in the proof of Theorem 10.14, if we let $a_n = [x_n]$ for all $n \geq 0$ and $x_n = 1/(x_{n-1} - a_{n-1})$ for all $n \geq 1$, then $x_0 = [a_0, a_1, a_2, \cdots, a_n, x_{n+1}]$. Now Theorem 10.5 implies that

$$x_0 = \frac{x_{n+1}p_n + p_{n-1}}{x_{n+1}q_n + q_{n-1}}$$

so $x_{n+1}q_n x_0 + q_{n-1}x_0 = x_{n+1}p_n + p_{n-1}$. Therefore,

$$x_{n+1}(q_n x_0 - p_n) = -q_{n-1}x_0 + p_{n-1}$$

or

$$x_{n+1}(q_n x_0 - p_n) = -q_{n-1}\left(x_0 - \frac{p_{n-1}}{q_{n-1}} \right)$$

Dividing by $x_{n+1}q_n$, we get

$$x_0 - \frac{p_n}{q_n} = \frac{-q_{n-1}}{x_{n+1}q_n}\left(x_0 - \frac{p_{n-1}}{q_{n-1}} \right)$$

so

$$\left| x_0 - \frac{p_n}{q_n} \right| = \frac{q_{n-1}}{x_{n+1} q_n} \left| x_0 - \frac{p_{n-1}}{q_{n-1}} \right|$$

Now $0 < x_n - a_n < 1$, so $x_{n+1} = 1/(x_n - a_n) > 1$. Furthermore, Lemma 10.5 implies that $q_n > q_{n-1}$, since $n \geq 2$ by hypothesis. Therefore,

$$\left| x_0 - \frac{p_n}{q_n} \right| < \left| x_0 - \frac{p_{n-1}}{q_{n-1}} \right| \qquad \blacksquare$$

For example, if $x_0 = \sqrt{22}$, we saw previously that $\sqrt{22} = [4, 1, 2, 4, 2, 1, 8, \cdots]$. Therefore, we have

$$C_0 = [4] = 4 \qquad\qquad |\sqrt{22} - C_0| = |\sqrt{22} - 4| = 0.690$$
$$C_1 = [4, 1] = 5 \qquad\qquad |\sqrt{22} - C_1| = |\sqrt{22} - 5| = 0.310$$

$$C_2 = [4, 1, 2] = \frac{14}{3} \qquad\qquad |\sqrt{22} - C_2| = \left|\sqrt{22} - \frac{14}{3}\right| = 0.024$$

$$C_3 = [4, 1, 2, 4] = \frac{61}{13} \qquad |\sqrt{22} - C_3| = \left|\sqrt{22} - \frac{61}{13}\right| = 0.002$$

Let $y_n = |x_0 - C_n|$, where C_n is the nth convergent in the infinite continued fraction representation of x_0. By virtue of Theorems 10.14 and 10.15, we know that the sequence $\{y_n\}$ decreases monotonically as it approaches zero. But how rapidly do the convergents C_n approach x_0 in absolute value? This question is answered by the next several theorems.

Theorem 10.17   *If p_n/q_n is the nth convergent in the infinite continued fraction representation of the irrational x_0, then*

$$\frac{1}{2 q_n q_{n+1}} < \left| x_0 - \frac{p_n}{q_n} \right| < \frac{1}{q_n q_{n+1}}$$

PROOF: Since x_0 is between p_n/q_n and p_{n+1}/q_{n+1}, we have

$$\left| x_0 - \frac{p_n}{q_n} \right| + \left| x_0 - \frac{p_{n+1}}{q_{n+1}} \right| = \left| \frac{p_{n+1}}{q_{n+1}} - \frac{p_n}{q_n} \right| = \frac{1}{q_n q_{n+1}}$$

by Theorem 10.6b. By Theorem 10.15,

$$0 < \left| x_0 - \frac{p_{n+1}}{q_{n+1}} \right| < \left| x_0 - \frac{p_n}{q_n} \right|$$

Therefore,

$$\left| x_0 - \frac{p_n}{q_n} \right| < \frac{1}{q_n q_{n+1}} < 2 \left| x_0 - \frac{p_n}{q_n} \right|$$

from which the conclusion follows. $\blacksquare$

For example, we saw previously that if $x_0 = \sqrt{22}$, then $C_2 = 14/3$, so $q_2 = 3$; also, $q_3 = 13$. Theorem 9.16 says that $1/(2q_2q_3) < |\sqrt{22} - C_2| < 1/(q_2q_3)$; that is, $1/78 < |\sqrt{22} - (14/3)| < 1/39$, or $0.0128 < |\sqrt{22} - (14/3)| < 0.0256$. [Actually, $|\sqrt{22} - (14/3)| = 0.0239$.]

If p_n/q_n is the nth convergent in the infinite continued fraction expansion of x, then Theorem 10.17 and Lemma 10.5 imply that $|x - (p_n/q_n)| < 1/q_n^2$. As a result, we see that for every positive irrational number x, there are infinitely many rationals a/b such that $|x - (a/b)| < 1/b^2$. This result may be improved as follows:

Theorem 10.18 *If x is a positive irrational number and $n \geq 1$, then (a) at least one of every two consecutive convergents to x satisfies*

$$\left| x - \frac{p_n}{q_n} \right| < \frac{1}{2q_n^2}$$

and (b) at least one of every three consecutive convergents to x satisfies

$$\left| x - \frac{p_n}{q_n} \right| < \frac{1}{\sqrt{5}q_n^2}$$

PROOF OF (a):   Recall from the proof of Theorem 10.16 that

$$\left| x - \frac{p_n}{q_n} \right| + \left| x - \frac{p_{n+1}}{q_{n+1}} \right| = \frac{1}{q_nq_{n+1}}$$

If $|x - (p_k/q_k)| \geq 1/2q_k^2$ for $k = n$ and $k = n + 1$, then we have $(1/2q_n^2) + (1/2q_{n+1}^2) \leq 1/q_nq_{n+1}$ by Theorem 9.16, so $(q_{n+1}/q_n) + (q_n/q_{n+1}) \leq 2$. If we let $r = q_{n+1}/q_n$, we have $r + (1/r) \leq 2$. This implies that $(r - 1)^2 = 0$, so $r = 1$. However, since $n \geq 1$ by hypothesis, Lemma 10.5 implies $q_{n+1} > q_n$, hence $r > 1$, an impossibility.   ■

PROOF OF (b):   Let $r = q_{n+1}/q_n$ and $s = q_{n+2}/q_{n+1}$. If $|x - (p_k/q_k)| \geq 1/\sqrt{5}q_k^2$ for $k = n, n + 1$, and $n + 2$, then, as in the proof of (a), we get $r + (1/r) \leq \sqrt{5}$, so $r < \frac{1}{2}(1 + \sqrt{5})$. Similarly, $s < \frac{1}{2}(1 + \sqrt{5})$. However, $q_{n+2} = a_{n+1}q_{n+1} + q_n \geq q_{n+1} + q_n$, so $q_{n+2}/q_{n+1} \geq 1 + (q_n/q_{n+1})$; that is, $s \geq 1 + (1/r)$. Since $r < \frac{1}{2}(\sqrt{5} + 1)$, we have $1/r > \frac{1}{2}(\sqrt{5} - 1)$, so $s > \frac{1}{2}(\sqrt{5} + 1)$, an impossibility.   ■

For example, let $x = \frac{1}{2}(1 + \sqrt{5})$. By the results of Exercises 6 and 11, the nth convergent to x is F_{n+1}/F_n, where, as usual, F_n denotes the nth Fibonacci number. One can show that $|x - (F_{n+1}/F_n)| < 1/\sqrt{5}F_n^2$ if and only if n is even. We leave the proof of this assertion as an exercise.

As an immediate consequence of Theorem 10.17, we obtain the following:

Theorem 10.19   *(Hurwitz)*

If x is an irrational number, then there are infinitely many rationals a/b such that $|x - (a/b)| < 1/\sqrt{5}b^2$.

PROOF: Let x be irrational. If $x > 0$, then the conclusion follows from Theorem 10.17b. If $x < 0$, then $-x > 0$, so Theorem 10.17b implies that there exist infinitely many rationals $r = a/b$ such that $|-x-r| < 1/b^2$. For each such rational, we have $|x - (-r)| = |x + r| = |-x-r| < 1/b^2$, so we are done.   ∎

The preceding example, where $x = \frac{1}{2}(1 + \sqrt{5})$, also illustrates Hurwitz's theorem (Theorem 10.19).

It can be shown that $\sqrt{5}$ is the best possible, that is, largest, constant that can appear in the denominator on the right side of the inequality in the statement of Hurwitz's theorem. We leave the proof of this assertion as an exercise.

Theorem 10.17a has a "converse" of sorts that states that a sufficiently accurate rational approximation to an irrational number *must* be a convergent to the infinite continued fraction expansion of the irrational number. Before we can prove this surprising result, we need the following theorem.

Theorem 10.20   *If x is a positive irrational number, p_n/q_n is the nth convergent to the infinite continued fraction expansion of x, and $1 \le b < q_{n+1}$, then $|q_n x - p_n| \le |bx - a|$ for all a.*

PROOF: Let us seek integers u and v such that

$$\begin{cases} p_n u + p_{n+1} v = a \\ q_n u + q_{n+1} v = b \end{cases}$$

Since the determinant of this system of linear equations is $p_n q_{n+1} - p_{n+1} q_n = (-1)^{n+1} \neq 0$, the system has a unique solution, namely,

$$u = (-1)^{n+1}(a q_{n+1} - b p_{n+1})$$
$$v = (-1)^{n+1}(b p_n - a q_n)$$

If $u = 0$, the $aq_{n+1} = bp_{n+1}$. Since $(p_{n+1}, q_{n+1}) = 1$ by Theorem 10.7, Euclid's lemma implies that $q_{n+1} \mid b$, so $q_{n+1} \leq b$, contrary to hypothesis. If $u \neq 0$ but $v = 0$, then $b = q_n u$ and $a = p_n u$, so $|bx - a| = |q_n ux - p_n u| = |u| |q_n x - p_n| \geq |q_n x - p_n|$. If $u \neq 0$ and $v \neq 0$, we claim that u and v have opposite signs. Indeed, since $q_n u = b - q_{n+1} v$, if $v < 0$, then $q_n u > 0$; hence, $u > 0$. If $v > 0$, then $v \geq 1$, so $b - q_{n+1} v \leq b - q_{n+1}$. However, $b - q_{n+1} < 0$ by hypothesis, so $b - q_{n+1} v < 0$; that is, $q_n u < 0$; hence $u < 0$.

In addition, since x lies between the consecutive convergents p_n/q_n and p_{n+1}/q_{n+1}, the quantities $q_n x - p_n$ and $q_{n+1} x - p_{n+1}$ have opposite signs. Therefore, $u(q_n x - p_n)$ and $v(q_{n+1} x - p_{n+1})$ have the same sign. Now

$$
\begin{aligned}
|bx - a| &= |(q_n u + q_{n+1} v)x - (p_n u + p_{n+1} v)| \\
&= |u(q_n x - p_n) + v(q_{n+1} x - p_{n+1})| \\
&= |u(q_n x - p_n)| + |v(q_{n+1} x - p_{n+1})| \geq |u| |q_n x - p_n| \\
&\geq |q_n x - p_n| \qquad\blacksquare
\end{aligned}
$$

For example, if $x = \sqrt{22}$ and $n = 1$, then we have the following: If $1 \leq b < q_2$, then $|q_1 x - p_1| \leq |bx - a|$ for all a. Recall that $q_1 = 1$, $p_1 = 5$, and $q_2 = 3$, so $|\sqrt{22} - 5| \leq |b\sqrt{22} - a|$ for all $b < 3$ and all a. Let us verify this statement. If $b = 1$, then $|\sqrt{22} - 5| \leq |\sqrt{22} - a|$ for all a, since 5 is the integer closest to $\sqrt{22}$. If $b = 2$, then we expect $|\sqrt{22} - 5| = |2\sqrt{22} - a|$ for all a. Now $|\sqrt{22} - 5| = 0.31$ and $|2\sqrt{22} - a| = |\sqrt{88} - a|$, we assumes a minimum value of 0.38 at $a = 9$, so the statement holds.

Now we can show that the convergents of the continued fraction representation of positive irrational x provide the best rational approximations with bounded denominators.

Theorem 10.21   *If p_n/q_n is the nth convergent in the continued fraction expansion of positive irrational x, $1 \leq b \leq q_n$, and a is arbitrary, then*

$$\left| x - \frac{p_n}{q_n} \right| \leq \left| x - \frac{a}{b} \right|$$

PROOF: If $|x - (a/b)| < |x - (p_n/q_n)|$, then $|q_n x - p_n| = q_n |x - (p_n/q_n)| > q_n |x - (a/b)| \geq b |x - (a/b)| = |bx - a|$, so $|q_n x - p_n| > |bx - a|$, contrary to Theorem 10.20. $\blacksquare$

For example, if $x = \sqrt{22}$, and $n = 2$, we get $|x - (p_2/q_2)| \leq |x - (a/b)|$ for all $b \leq q_2$ and all a; that is, $|\sqrt{22} - (14/3)| \leq |\sqrt{22} - (a/b)|$ for all $b \leq 3$.

We are now in a position to state the following theorem.

Theorem 10.22 *If x is a positive irrational number, a and b are integers such that $(a, b) = 1$, and $|x - (a/b)| < 1/2b^2$, then a/b is a convergent in the continued fraction representation for x; that is, $a = p_n$ and $b = q_n$ for some n.*

PROOF: Assume the contrary. Lemma 10.5 implies that we can find an n such that $q_n \leq b < q_{n+1}$. Now $|q_n x - p_n| \leq |bx - a| = b|x - (a/b)| < 1/2b$ by Theorem 10.20 and hypothesis. Therefore, $|x - (p_n/q_n)| < 1/2bq_n$. If $a/b \neq p_n/q_n$, then $|bp_n - aq_n| \geq 1$. Therefore,

$$\frac{1}{bq_n} \leq \frac{|bq_n - aq_n|}{bq_n} = \left|\frac{p_n}{q_n} - \frac{a}{b}\right| \leq \left|\frac{p_n}{q_n} - x\right| + \left|x - \frac{a}{b}\right| < \frac{1}{2bq_n} + \frac{1}{2b^2}$$

This implies that $b < q_n$, an impossibility. Therefore, $a/b = p_n/q_n$. Since $(a, b) = 1$ by hypothesis and $(p_n, q_n) = 1$ by Theorem 10.7, it follows that $a = p_n$ and $b = q_n$. ∎

For example, since $|e - (8/3)| < 1/2(3^2)$, it follows that 8/3 is a convergent in the continued fraction expansion for e.

Note that if x is a positive irrational number, then Theorem 10.22 gives conditions that are sufficient, but not necessary, for a rational approximation to x to be a convergent.

For example, if $x = x_0 = \sqrt{13}$, then $a_0 = [\sqrt{13}] = 3$, and

$$x_1 = \frac{1}{x_0 - a_0} = \frac{1}{\sqrt{13} - 3} = \frac{\sqrt{13} + 3}{4} \qquad a_1 = [x_1] = 1$$

$$x_2 = \frac{1}{x_1 - a_1} = \frac{4}{\sqrt{13} - 1} = \frac{\sqrt{13} + 1}{3} \qquad a_2 = [x_2] = 1$$

$$x_3 = \frac{1}{x_2 - a_2} = \frac{3}{\sqrt{13} - 2} = \frac{\sqrt{13} + 2}{3} \qquad a_3 = [x_3] = 1$$

The following table lists a_n, p_n, and q_n for $0 \leq n \leq 3$:

n	a_n	p_n	q_n
0	3	3	1
1	1	4	1
2	1	7	2
3	1	11	3

Now $|x - (p_3/q_3)| = |\sqrt{13} - (11/3)| = 0.0611 > 0.0556 = 1/18 = 1/2q_3^2$, so the inequality of Theorem 10.21 is not satisfied, even though 11/3 is a convergent to $\sqrt{13}$. The following theorem, however, gives conditions that are both necessary and sufficient for a rational approximation to irrational x to be a convergent.

Theorem 10.23 *Let a and b be natural numbers such that $a/b = [a_0, a_1, a_2, \cdots, a_n] = p_n/q_n$. If x is a positive irrational number, then a/b is a convergent in the continued fraction expansion for x if and only if $|x - (a/b)| < 1/q_n(q_n + q_{n-1})$.*

PROOF: We omit the proof but refer the interested reader to Perron (1913, pp. 42–45).

If we apply Theorem 10.22 to the preceding example, we see that since $11/3 = [3, 1, 2]$, with $a_0 = 3, p_0 = 3, q_0 = 1, a_1 = 1, p_1 = 4, q_1 = 1$, $a_2 = 2$, $p_2 = 11$, and $q_2 = 3$, we have $11/3 = p_2/q_2$ and $1/q_2(q_2 + q_1) = 1/3(3 + 1) = 1/12 = 0.0833 > 0.0611 = |\sqrt{13} - (11/3)|$, and it follows that 11/3 is indeed a convergent to $\sqrt{13}$.

Theorem 10.23 enables us to obtain the following improvement to Theorem 10.22.

Theorem 10.24 *If a and b are natural numbers such that $(a, b) = 1$ and $b > 1$ and if x is a positive irrational number such that $|x - (a/b)| < 1/b(2b - 1)$, then a/b is a convergent in the continued fraction representation of x.*

PROOF: Let $a/b = [a_0, a_1, a_2, \cdots, a_n] = p_n/q_n$, with $a_n > 1$ if $n > 0$. If $n = 0$, then $a/b = [a_0] = a_0/1$, so $b = 1$, contrary to hypothesis. If $n = 1$, then $a/b = [a_0, a_1] = (a_0 a_1 + 1)/a_1 = p_1/q_1$, so $q_1 = a_1 = b > 1$. However, $q_0 = 1$, so $q_1 > q_0$ or $q_0 \leq q_1 - 1$. If $n \geq 2$, then $q_{n-2} \geq 1$, so $q_n = a_n q_{n-1} + q_{n-2} \geq q_{n-1} + 1$. Therefore, $q_{n-1} \leq q_n - 1$ for all $n \geq 1$. This implies that $q_n + q_{n-1} \leq 2q_n - 1$ for all $n \geq 1$. Now $1/b(2b - 1) = 1/q_n(2q_n - 1) \leq 1/q_n(q_n + q_{n-1})$, so the conclusion now follows from hypothesis and Theorem 10.23.   ∎

SECTION 10.4 EXERCISES

15. Find the least n such that $\sqrt{6}$ can be approximated to three decimal places by the nth convergent of its continued fraction.

16. Given that $(e+1)/(e-1) = [2, 6, 10, 14, \cdots]$, use a convergent to estimate $(e+1)/(e-1)$ to four decimal places.

17. Let p_n/q_n be the nth convergent in the continued fraction expansion of $x = \frac{1}{2}(1 + \sqrt{5})$.

Prove that $|x - (p_n/q_n)| < 1/5q_n^2$ if and only if n is even.

18. Use the properties of Fibonacci numbers to prove that if $x = \frac{1}{2}(1 + \sqrt{5})$ and $|x - (a/b)| < 1/bb^2$, then $b \le \sqrt{5}$.

SECTION 10.4 COMPUTER EXERCISE

19. Let p_n/q_n be the nth convergent in the infinite continued fraction representation of θ, where θ is a positive irrational number. Write a computer program to evaluate

$$\left|\frac{p_n}{q_n} - \theta\right|q_n^2$$

for $n = 1, 2, 3$, etc. (You might take $\theta = \sqrt{2}$ or $\sqrt{5}$, for example.) What does the numerical evidence suggest?

10.5 PERIODIC CONTINUED FRACTIONS

Previously, in Section 10.3, we saw that $\sqrt{22} = [4, 1, 2, 4, 2, 1, 8, 1, 2, 4, 2, 1, 8, \cdots]$. That is, the continued fraction representation of $\sqrt{22}$ consists of an initial term 4 followed by a sequence of six integers that repeats indefinitely, namely, 1, 2, 4, 2, 1, 8. A more convenient notation for this infinite continued fraction is

$$\sqrt{22} = [4, \overline{1, 2, 4, 2, 1, 8}]$$

We say that the continued faction representation for $\sqrt{22}$ is *periodic*. In general, an infinite continued fraction is called periodic if it ends with an indefinitely repeating finite sequence of integers. Some examples are

$$[2, 1, 3, 4, 5, 3, 4, 5, 3, 4, 5, \cdots] = [2, 1, \overline{3, 4, 5}]$$

and

$$[1, 4, 2, 3, 1, 4, 2, 3, 1, 4, 2, 3, \cdots] = [\overline{1, 4, 2, 3}]$$

The second example, where the continued fraction contains no initial nonperiodic terms, is called *purely periodic*.

It can be shown that if x is a positive irrational number, then the continued fraction representation for x is periodic if and only if x is a quadratic irrational, that is, if x is the real root of a quadratic equation with integer coefficients. For example, we saw earlier, in Section 10.3, that if $\alpha = \frac{1}{2}(1 + \sqrt{5})$, then $\alpha = [1, 1, 1, \cdots] = [\overline{1}]$. Now α is the positive real root of the equation $t^2 - t - 1 = 0$.

We will prove the slightly narrower result: If C and D are integers such that $0 < C < D$ and D/C is not a square, then

10.2 $$\sqrt{D/C} = [a_0, \overline{a_1, a_2, \cdots, a_t}]$$

for some $t \geq 1$. The smallest t for which equation 10.2 holds is called the *length of the period* of the continued fraction. The periodic infinite continued fractions presented above have periods of length 6, 3, 4, and 1, respectively.

In the work ahead, $[y]$ denotes the integer part of y, *not* a continued fraction with a single term.

Theorem 10.25 *If C and D are integers such that $0 < C < D$ and $D/C \neq s^2$, then there exists $t \geq 1$ such that $\sqrt{D/C} = [a_0, \overline{a_1, a_2, \cdots, a_t}]$.*

PROOF: Let $b_0 = 0, r_0 = C, x_0 = \sqrt{D/C}$, and $a_0 = [x_0] = [\sqrt{D/C}]$. For $k \geq 1$, let $b_k = a_{k-1}r_{k-1} - b_{k-1}, r_k = (CD - b_k^2)/r_{k-1}, x_k = (\sqrt{CD} + b_k)/r_k$, and $a_k = [x_k]$. In particular, $b_1 = C[\sqrt{D/C}]$. Then $x_k = a_k + (\sqrt{CD} + b_k - a_kr_k)/r_k = a_k + r_{k+1}/(\sqrt{CD} + b_{k+1}) = a_k + 1/x_{k+1}$. We will show by induction on k that each of b_k, r_k, and a_k is an integer. From this and from Theorem 10.15, it follows that $\sqrt{D/C} = [a_0, a_1, a_2, \cdots]$. The periodicity is a consequence of the fact (to be demonstrated) that the b_k and the r_k are bounded from above. First, we need several lemmas.

Lemma 10.7 r_k *and* b_k *are integers for all* $k \geq 0$.

PROOF: (Induction on k) $b_0 = 0, r_0 = C, b_1 = a_0r_0 - b_0 = C[\sqrt{D/C}]$, and $r_1 = (CD - b_1^2)/r_0 = (CD - C^2[\sqrt{D/C}]^2)/C = D - C[\sqrt{D/C}]^2$. By definition, a_k is an integer for all k. If $k \geq 1$, since $b_k = a_{k-1}r_{k-1} - b_{k-1}$, b_k is an integer by induction hypothesis. If $k \geq 2$, then $r_k = (CD - b_k^2)/r_{k-1} = [CD - (a_{k-1}r_{k-1} - b_{k-1})^2]/r_{k-1} = (CD - b_{k-1}^2)/r_{k-1} + a_{k-1}(2b_{k-1} - a_{k-1}r_{k-1}) = r_{k-2} + a_{k-1}(b_{k-1} - b_k)$. By induction hypothesis, each of r_{k-2} and b_{k-1} is an integer. We have just proved that b_k is an integer. Therefore, r_k is an integer.   ∎

Lemma 10.8 $1 \le r_k \le 2[\sqrt{CD}]$, $b_k \le [\sqrt{CD}]$, and $a_k \ge 1$ for all $k \ge 0$.

PROOF: (Induction on k)   Lemma 10.8 holds for $k = 0$, since $r_0 = C$, $b_0 = 0$, and $a_0 = [\sqrt{D/C}]$. Let $k \ge 1$. Since b_{k-1} and r_{k-1} are integers, it follows that $x_{k-1} = (\sqrt{CD} + b_{k-1})/r_{k-1}$ is irrational. Now

$$x_{k-1} - [x_{k-1}] = \frac{\sqrt{CD} - b_k}{r_{k-1}} = \frac{r_k}{\sqrt{CD} + b_k}$$

so we have

10.3
$$0 < \frac{\sqrt{CD} - b_k}{r_{k-1}} < 1$$

and

10.4
$$0 < \frac{r_k}{\sqrt{CD} + b_k} < 1$$

Since $r_{k-1} \ge 1$ by induction hypothesis, 10.3 implies that $b_k < \sqrt{CD}$, so $b_k \le [\sqrt{CD}]$ and $CD - b_k^2 > 0$. Since $a_k = [(\sqrt{CD} + b_k)/r_k]$, 10.4 implies that $a_k \ge 1$. Since $r_k = (CD - b_k^2)/r_{k-1}$, $r_k > 0$, so Lemma 10.7 implies that $r_k \ge 1$. Finally, 10.4 implies that $r_k < \sqrt{CD} + b_k \le \sqrt{CD} + [\sqrt{CD}]$. Since r_k is an integer, we have $r_k \le 2[\sqrt{CD}]$.   ∎

Lemma 10.9 $b_k \ge 1$ for all $k \ge 1$.

PROOF: Recall $b_1 = C[\sqrt{D/C}] \ge 1$. Now assume that $k \ge 1$ and $b_{k+1} \le 0$. This implies that $a_k r_k \le b_k$. Since $a_k \ge 1$, we have $r_k \le b_k$. Also, $a_k \le b_k/r_k$, so $a_k \le [b_k/r_k]$; that is, $[(\sqrt{CD} + b_k)/r_k] \le [b_k/r_k]$. Since $CD > 0$, the reverse inequality holds; thus $[(\sqrt{CD} + b_k)/r_k] = [b_k/r_k]$. This implies that $\sqrt{CD} < r_k$. However, then Lemma 10.8 implies that $b_k < r_k$, an impossibility. We therefore conclude that $b_k \ge 1$ for all $k \ge 1$.   ∎

Resuming the proof of theorem 10.25, we have seen via Lemmas 10.8 and 10.9 that $1 \le b_k \le [\sqrt{CD}]$ and $1 \le r_k \le 2[\sqrt{CD}]$ for all $k \ge 1$. therefore, if $m = 2[\sqrt{CD}]^2$, there exist at most m distinct ordered pairs (b_i, r_j). This implies that there exists t such that $1 \le t \le m$, $b_{t+1} = b_1$, and $r_{t+1} = r_1$. The least such t is called the *length of the period* of the continued fraction representation of $\sqrt{D/C}$. Now $x_{t+1} = (\sqrt{CD} + b_{t+1})/r_{t+1} = (\sqrt{CD} + b_1)/r_1 = x_1$, so $a_{t+1} = [x_{t+1}] = [x_1] = a_1$. By induction on k, we get $b_{t+k} = b_k$, $r_{t+k} = r_k$, and $a_{t+k} = a_k$ for all $k \ge 1$. (Also $x_{t+k} = x_k$ for all $k \ge 1$.) Therefore, $\sqrt{D/C} = [a_0, a_1, a_2, \cdots, a_t, a_1, a_2, \cdots, a_t, \cdots]$, which we write more conveniently as $\sqrt{D/C} = [a_0, \overline{a_1, a_2, \cdots, a_t}]$.   ∎

For example, we saw in Section 10.3 that $\sqrt{22} = [4, \overline{1, 2, 4, 2, 1, 8}]$. We could obtain this result without performing operations on irrational fractions, provided we make use of the inductive definitions of b_k, r_k, and a_k. Recall that, in general, $b_0 = 0$, $r_0 = C$, $a_0 = [\sqrt{D/C}]$, $b_k = a_{k-1}r_{k-1} - b_{k-1}$, $r_k = (CD - b_k^2)/r_{k-1}$, and finally, $a_k = [(\sqrt{CD} + b_k)/r_k]$. Using Theorem 1.15, we have $a_k = [([\sqrt{CD}] + b_k)/r_k]$. If we wish to obtain the continued fraction representation of $\sqrt{22}$, we note that $D = 22$ and $C = 1$ and construct the following table:

k	b_k	r_k	a_k
0	0	1	4
1	4	6	1
2	2	3	2
3	4	2	4
4	4	3	2
5	2	6	1
6	4	1	8
7	4	6	1

Now $b_7 = b_1 = 4$, $r_7 = r_1 = 6$, and $a_7 = a_1 = 1$. It is clear from the table that $t = 6$ is the least integer such that $b_{t+1} = b_1$, $r_{t+1} = r_1$, and $a_{t+1} = a_1$. Therefore, $\sqrt{22} = [4, \overline{1, 2, 4, 2, 1, 8}]$.

The reader may note that (1) the last term of the period, namely, 8, is double the initial nonperiodic term, 4, and (2) except for the last term, the period is symmetrical; that is, $a_{t-k} = a_k$ for all k such that $1 \le k \le t-1$. We shall see that these phenomena are not isolated, but occur in the general case.

In the several lemmas that follow, b_k, r_k, and a_k are as in the proof of Theorem 10.25.

Lemma 10.10 $b_j < b_k + r_k$ *for all* $j, k \ge 1$.

PROOF: Since $b_k = a_{k-1}r_{k-1} - b_{k-1}$, we have $b_{k-1} + b_k = a_{k-1}r_{k-1}$. Lemma 10.8 implies that $a_{k-1} \ge 1$, so $b_{k-1} + b_k > r_{k-1}$. Lemma 10.8 implies that $b_{k-1} < \sqrt{CD}$, so $\sqrt{CD} + b_k > b_{k-1} + b_k$, so $\sqrt{CD} + b_k > r_{k-1}$. However, $CD - b_k^2 = r_k r_{k-1}$, so $\sqrt{CD} - b_k < r_k$; thus $\sqrt{CD} < b_k + r_k$. Again, Lemma 10.8 implies that $b_j < \sqrt{CD}$, so $b_j < b_k + r_k$. ∎

Lemma 10.11 $b_t = b_1$ *and* $a_t = 2a_0$.

PROOF: $r_t = (CD - b_{t+1}^2)/r_{t+1} = (CD - b_1^2)/r_1 = r_0 = C$. Recall that $b_1 = Ca_0$. $b_t = a_t r_t - b_{t+1} = Ca_t - b_1 = Ca_t - Ca_0 = C(a_t - a_0)$. Lemma 10.10 implies that $b_1 < b_t + r_t$, that is, $b_1 < b_t + C$, so $b_1/C < b_t/C + 1$. Since b_1/C and b_t/C are integers, it follows that $b_1/C \le b_t/C$; thus, $b_1 \le b_t$. Now $\sqrt{CD} = b_1 + \sqrt{CD} - b_1 = b_1 + \sqrt{CD} - Ca_0 = b_1 + C(\sqrt{D/C} - [\sqrt{D/C}])$. Let $r = \sqrt{D/C}$, so $\sqrt{CD} = b_1 + C(r - [r])$. Lemma 10.8 implies that $b_t < \sqrt{CD}$, so $b_t < b_1 + C(r - [r])$. Therefore, $b_t/C < b_1/C + r - [r]$. Since b_1/C and b_t/C are integers and $0 < r - [r] < 1$, it follows that $b_t/C \le b_1/C$; hence $b_t \le b_1$. Therefore, $b_t = b_1$. Also, $a_t = (b_t + b_{t+1})/r_t = 2b_1/C = 2a_0$.   ∎

Lemma 10.12   *If* $1 \le k \le t - 1$, *then* $r_{t-k} = r_k$, $a_{t-k} = a_k$, *and* $b_{t-k} = b_{k+1}$.

PROOF: (Induction on k)   We have seen in the proof of Lemma 10.11 that $r_t = r_0$ and $b_t = b_1$. By induction hypothesis, we have $r_{t-k+1} = r_{k-1}$ and $b_{t-k+1} = b_k$. Now $r_{t-k} = (CD - b_{t-k+1}^2)/r_{t-k+1} = (CD - b_k^2)/r_{k-1} = r_k$. Also, $b_{t-k} + b_{t-k+1} = a_{t-k}r_{t-k}$, so $b_{t-k} + b_k = a_{t-k}r_{t-k}$. Moreover, $b_{k+1} + b_k = a_k r_k$, so $b_{k+1} + b_k = a_k r_{t-k}$. Subtracting, we get $b_{k+1} - b_{t-k} = r_{t-k}(a_k - a_{t-k})$, so $a_k - a_{t-k} = (b_{k+1} - b_{t-k})/r_{t-k}$. If $1 \le k \le t - 1$, then Lemma 10.10 implies that $(b_{k+1} - b_{t-k})/r_{t-k} < 1$. Therefore, $a_k - a_{t-k} \le 0$, so $a_k \le a_{t-k}$. If $1 \le k \le t - 1$, then also $1 \le t - k \le t - 1$, so we similarly obtain $a_{t-k} \le a_k$. Therefore, $a_{t-k} = a_k$, from which it follows that $b_{t-k} = b_{k+1}$.   ∎

We summarize our results in the following theorem:

Theorem 10.26   *If* C *and* D *are integers such that* $0 < C < D$ *and* $D/C \ne s^2$, *then the continued fraction representation for* $\sqrt{D/C}$ *is* $\sqrt{D/C} = [a_0, \overline{a_1, a_2, a_3, \cdots, a_3, a_2, a_1, 2a_0}]$

PROOF: This follows from Theorem 10.25 and Lemmas 10.11 and 10.12.

REMARKS: The first proof of Theorem 10.25 was published in 1770 by Lagrange (see Lagrange, 1770). The first proof of Theorem 10.26 was published in 1828 by Galois (see Galois, 1828). These distinguished mathematicians used more complex methods to obtain their results.

For example, let us develop the continued fraction expansion for $\sqrt{21/13}$. Here $C = 13$ and $D = 21$, so $CD = 273$. Following the notation of the proof of Theorem 10.25, we have $b_0 = 0$, $r_0 = C = 13$, and $a_0 = [\sqrt{21/13}] = 1$. For $k \geq 1$, we use the formulas

$$b_k = a_{k-1}r_{k-1} - b_{k-1}$$

$$r_k = \frac{CD - b_k^2}{r_{k-1}} = \frac{273 - b_k^2}{r_{k-1}}$$

$$a_k = \left[\frac{[\sqrt{CD}] + b_k}{r_k}\right] = \left[\frac{[\sqrt{273}] + b_k}{r_k}\right] = \left[\frac{16 + b_k}{r_k}\right]$$

The resulting data are listed in the following table:

k	b_k	r_k	a_k
0	0	13	1
1	13	8	3
2	11	19	1
3	8	11	2
4	14	7	4
5	14	11	2
6	8	19	1
7	11	8	3
8	13	13	2
9	13	8	3
$\vdots$			

Since $b_9 = b_1 = 13$, $r_9 = r_1 = 8$, and $a_9 = a_1 = 3$, we know that $t = 8$, so

$$\sqrt{21/13} = [1, \overline{3, 1, 2, 4, 2, 1, 3, 2}].$$

SECTION 10.5 EXERCISES

20. Prove that if $\sqrt{N} = [a, \overline{b, b, 2a}]$, then $(b^2 + 1) \mid (2ab + 1)$.

21. Find the infinite continued fraction representations for each of the following.

(a) $\sqrt{2}$

(b) $\sqrt{3}$

(c) $\sqrt{11}$

(d) $\sqrt{19}$

(e) $\sqrt{31}$

(f) $\sqrt{74}$

(g) $\sqrt{94}$

(h) $\sqrt{75}$

(i) $\sqrt{8/5}$

(j) $\sqrt{22/3}$

22. Prove that $\sqrt{m^2 + 1} = [m, \overline{2m}]$.

23. Prove that $\sqrt{m^2 - 1} = [m - 1, \overline{1, 2m - 2}]$.

24. Prove that $\sqrt{m^2 + 2} = [m, \overline{m, 2m}]$.

25. Prove that $\sqrt{m^2-2} = [m-1, \overline{1, m-2, 1,}$ $\overline{2m-2}]$ if $m \geq 3$.

26. Prove that $\sqrt{m^2+4} = [m, \overline{\frac{1}{2}m, 2m}]$ if m is even.

27. Prove that $\sqrt{m^2+4} = [m, \overline{\frac{1}{2}(m-1), 1, 1,}$ $\overline{\frac{1}{2}(m-1), 2m}]$ if m is odd and $m \geq 3$.

28. Prove that $\sqrt{m^2-4} = [m-1, \overline{1, \frac{1}{2}m-2,}$ $\overline{1, 2m-2}]$ if m is even and $m \geq 6$.

29. Prove that $\sqrt{m^2-4} = [m-1, \overline{1, \frac{1}{2}(m-3),}$ $\overline{2, \frac{1}{2}(m-3), 1, 2m-2}]$ if m is odd and $m \geq 5$.

30. Prove that $\sqrt{9m^2+3} = [3m, \overline{2m, 6m}]$.

31. Prove that $\sqrt{9m^2-3} = [3m-1, \overline{1, 2m-2,}$ $\overline{1, 6m-2}]$ if $m \geq 2$.

32. Prove that $\sqrt{9m^2+6} = [3m, \overline{m, 6m}]$.

33. Prove that $\sqrt{9m^2-6} = [3m-1, \overline{1, m-2,}$ $\overline{1, 6m-2}]$ if $m \geq 3$.

*34. Prove that if $\sqrt{N}$ has an infinite continued fraction representation whose period has odd length, then N is a sum of two squares.

SECTION 10.5 COMPUTER EXERCISE

35. Write a computer program to find the infinite continued fraction representation of $\sqrt{D/C}$, where C and D are integers such that $0 < C < D$ and $D/C \neq s^2$.

10.6   MORE ON PERIODIC CONTINUED FRACTIONS

The next chapter deals with nonlinear Diophantine equations, among them the *generalized Pell's equation:*

$$Cx^2 - Dy^2 = E$$

where C and D are integers such that $0 < C < D$ and $D/C \neq s^2$. We now present some additional properties of infinite periodic continued fractions that will be useful in obtaining the solutions of the generalized Pell's equation. First, some additional terminology is needed.

Definition 10.4

nth Complete Quotient of an Infinite Continued Fraction

Let $x_0 = [a_0, a_1, a_2, \cdots] = \lim_{t \to \infty} [a_0, a_1, a_2, \cdots, a_t]$. If $n \geq 1$, let $x_n = [a_n, a_{n+1}, a_{n+2}, \cdots] = \lim_{t \to \infty} [a_n, a_{n+1}, a_{n+2}, \cdots, a_{n+t}]$. We call x_n the *nth complete quotient of x_0.*

For example, if $x_0 = [0, 2, 4, 6, 8, 10, \cdots]$, then $x_3 = [6, 8, 10, 12, 14, 16, \cdots]$ and $x_5 = [10, 12, 14, 16, 18, 20, \cdots]$.

Actually, these nth complete quotients are not new; they are generated in the process of finding the terms a_n of the continued fraction representation of x_0, as the following theorem asserts.

Theorem 10.27 *Let the positive irrational number x_0 have the continued fraction representation $x_0 = [a_0, a_1, a_2, \cdots]$, where all a_i are integers, $a_0 \geq 0$, and $a_i > 0$ for all $i > 0$. Then the following are equivalent:*

(1) *x_n is defined inductively by $a_n = [x_n]$ for all $n \geq 0$, and $x_n = 1/(x_{n-1} - a_{n-1})$ for all $n \geq 1$.*
(2) *$x_n = [a_n, a_{n+1}, a_{n+2}, \cdots]$ for all $n \geq 0$.*

PROOF: (1) implies (2) (Induction on n) By hypothesis, $a_n + 1/x_{n+1} = x_n$. By induction hypothesis, $x_n = [a_n, a_{n+1}, a_{n+2}, \cdots] = \lim_{t \to \infty} [a_n, a_{n+1}, a_{n+2}, \cdots, a_{n+t}] = \lim_{t \to \infty} (a_n + 1/[a_{n+1}, a_{n+2}, \cdots, a_{n+t}]) = a_n + 1/\lim_{t \to \infty} [a_{n+1}, a_{n+2}, \cdots, a_{n+t}] = a_n + 1/[a_{n+1}, a_{n+2}, \cdots]$. Therefore, $x_{n+1} = [a_{n+1}, a_{n+2}, \cdots]$. In particular, since $x_0 = [a_0, a_1, a_2, \cdots]$ by hypothesis, we have $x_1 = [a_1, a_2, a_3, \cdots]$, so the proof is complete. (2) implies (1) If $n \geq 0$, then $x_n = [a_n, a_{n+1}, a_{n+2}, \cdots]$ and $x_{n+1} = [a_{n+1}, a_{n+2}, a_{n+3}, \cdots]$. Now $x_n = \lim_{t \to \infty} [a_n, a_{n+1}, \cdots, a_{n+t}] = \lim_{t \to \infty} (a_n + 1/[a_{n+1}, a_{n+2}, \cdots, a_{n+t}]) = a_n + 1/\lim_{t \to \infty} [a_{n+1}, a_{n+2}, \cdots, a_{n+t}] = a_n + 1/[a_{n+1}, a_{n+2}, \cdots] = a_n + 1/x_{n+1}$. Therefore, $x_{n+1} = 1/(x_n - a_n)$ for all $n \geq 0$ or $x_n = 1/(x_{n-1}, a_{n-1})$ for all $n \geq 1$. Since $n \geq 0$, $a_{n+1} > 1$ by hypothesis. Since $x_{n+1} = [a_{n+1}, a_{n+2}, \cdots]$, the same argument as in the proof of Theorem 10.11 implies that $x_{n+1} > 1$. Since $x_n = a_n + 1/x_{n+1}$, it follows that $a_n = [x_n]$. ∎

For example, referring again to the continued fraction expansion of $\sqrt{22}$ from Section 10.3, we have

$$x_0 = \sqrt{22} = [4, \overline{1, 2, 4, 2, 1, 8}]$$
$$x_1 = (\sqrt{22} + 4)/6 = [\overline{1, 2, 4, 2, 1, 8}]$$
$$x_2 = (\sqrt{22} + 2)/3 = [\overline{2, 4, 2, 1, 8, 1}]$$
$$x_3 = (\sqrt{22} + 4)/2 = [\overline{4, 2, 1, 8, 1, 2}]$$
$$x_4 = (\sqrt{22} + 4)/3 = [\overline{2, 1, 8, 1, 2, 4}]$$
$$x_5 = (\sqrt{22} + 2)/6 = [\overline{1, 8, 1, 2, 4, 2}]$$
$$x_6 = \sqrt{22} + 4 = [\overline{8, 1, 2, 4, 2, 1}]$$

We need not compute any further complete quotients here because $x_{6+k} = x_k$ for all $k \geq 1$.

Next, we show that an infinite continued fraction also can be expressed as the finite continued fraction whose last term is an nth complete quotient.

Theorem 10.28 *Let the positive irrational number x_0 have the continued fraction representation $x_0 = [a_0, a_1, a_2, \cdots]$, where the a_i are all integers, $a_0 \geq 0$, and $a_i > 0$ for all $i > 0$. Let $x_n = [a_n, a_{n+1}, a_{n+2}, \cdots]$, where $n \geq 1$. Then $x_0 = [a_0, a_1, a_2, \cdots, a_{n-1}, x_n]$.*

PROOF: (Induction on n)   By hypothesis and Theorem 10.27, $x_n = a_n + 1/x_{n+1}$ for all $n \geq 0$. In particular, $x_0 = a_0 + 1/x_1 = [a_0, x_1]$. By induction hypothesis, $x_0 = [a_0, a_1, a_2, \cdots, a_{n-1}, x_n]$. Therefore, $x_0 = [a_0, a_1, a_2, \cdots, a_{n-1}, a_n + 1/x_{n+1}]$. Now Theorem 10.1 implies that $x_0 = [a_0, a_1, a_2, \cdots, a_{n-1}, a_n, x_{n+1}]$. ■

For example, referring once again to the complete quotients associated with the continued fraction expansion of $\sqrt{22}$, we have

$$\sqrt{22} = x_0 = [a_0, x_1] = \left[4, \frac{\sqrt{22}+4}{6}\right]$$

$$\sqrt{22} = x_0 = [a_0, a_1, x_2] = \left[4, 1, \frac{\sqrt{22}+2}{3}\right]$$

$$\sqrt{22} = x_0 = [a_0, a_1, a_2, x_3] = \left[4, 1, 2, \frac{\sqrt{22}+4}{2}\right]$$

$$\sqrt{22} = x_0 = [a_0, a_1, a_2, a_3, x_4] = \left[4, 1, 2, 4, \frac{\sqrt{22}+4}{3}\right]$$

$$\sqrt{22} = x_0 = [a_0, a_1, a_2, a_3, a_4, x_5] = \left[4, 1, 2, 4, 2, \frac{\sqrt{22}+2}{6}\right]$$

$$\sqrt{22} = x_0 = [a_0, a_1, a_2, a_3, a_4, a_5, x_6] = [4, 1, 2, 4, 2, 1, \sqrt{22} + 4]$$

We can now derive the following useful identity.

Theorem 10.29 *Let C and D be integers such that $0 < C < D$ and $D/C \neq s^2$. Let $\sqrt{D/C} = [a_0, \overline{a_1, a_2, \cdots, a_t}]$ for some $t \geq 1$. If $k \geq 0$, let r_k be defined as in Theorem 10.25, and let $p_k/q_k = [a_0, a_1, a_2, \cdots, a_k]$. Then*

$$Cp_{k-1}^2 - Dq_{k-1}^2 = (-1)^k r_k$$

PROOF: By hypothesis and Theorem 10.25, we have $b_0 = 0$, $r_0 = C$, $x_k = (\sqrt{CD} + b_k)/r_k$, and $a_k = [x_k]$ for all $k \geq 0$ and $b_k = a_{k-1}r_{k-1} - b_{k-1}$ and $r_k = (CD - b_k^2)/r_{k-1}$ for all $k \geq 1$. Since $x_k = a_k + 1/x_{k+1}$ for all $k \geq 0$, Theorem 10.27 implies that $x_k = [a_k, a_{k+1}, a_{k+2}, \cdots]$ for all $k \geq 0$. Therefore, Theorem 10.28 implies that $x_0 = [a_0, a_1, a_2, \cdots, a_{k-1}, x_k]$. Now Theorem 10.5 implies that

$$x_0 = \frac{x_k p_{k-1} + p_{k-2}}{x_k q_{k-1} + q_{k-2}}$$

that is,

$$\sqrt{D/C} = \frac{(\sqrt{CD} + b_k)/r_k)p_{k-1} + p_{k-2}}{(\sqrt{CD} + b_k)/r_k)q_{k-1} + q_{k-2}} = \frac{(\sqrt{CD} + b_k)p_{k-1} + r_k p_{k-2}}{(\sqrt{CD} + b_k)q_{k-1} + r_k q_{k-2}}$$

so

$$Dq_{k-1} + (\sqrt{D/C})(b_k q_{k-1} + r_k q_{k-2}) = (\sqrt{CD})p_{k-1} + (b_k p_{k-1} + r_k p_{k-2})$$

Equating the irrational terms, we get

10.5 $b_k q_{k-1} + r_k q_{k-2} = Cp_{k-1}$

Equating the rational terms, we get

10.6 $b_k p_{k-1} + r_k p_{k-2} = Dq_{k-1}$

If we multiply equation 10.5 by p_{k-1}, equation 10.6 by q_{k-1}, and then subtract, we get

10.7 $Cp_{k-1}^2 - Dq_{k-1}^2 = r_k(p_{k-1}q_{k-2} - p_{k-2}q_{k-1})$
$$= (-1)^{k-2}r_k = (-1)^k r_k$$

by Theorem 10.6a.   ∎

For example, in the table on page 223 we computed the terms in the period of the continued fraction expansion for $\sqrt{21/13}$. Let us reproduce this data in the table below, but let us also compute the p_k and the q_k such that p_k/q_k is the kth convergent.

k	b_k	r_k	a_k	p_k	q_k
-2				0	1
-1				1	0
0	0	13	1	1	1
1	13	8	3	4	3
2	11	19	1	5	4
3	8	11	2	14	11
4	14	7	4	61	48
5	14	11	2	136	107
6	8	19	1	197	155
7	11	8	3	727	572
8	13	13	2	1651	1299
⋮					

Let us verify Theorem 10.29 using $C = 13$ and $D = 21$. We have

$$13p_{-1}^2 - 21q_{-1}^2 = 13(1)^2 - 21(0)^2 = 13 = r_0$$
$$13p_0^2 - 21q_0^2 = 13(1)^2 - 21(1)^2 = -8 = -r_1$$
$$13p_1^2 - 21q_1^2 = 13(4)^2 - 21(3)^2 = 19 = r_2$$
$$13p_2^2 - 21q_2^2 = 13(5)^2 - 21(4)^2 = -11 = -r_3$$

In the proof of Theorem 10.25, we saw that if x_k denotes the kth complete quotient in the continued fraction expansion of $\sqrt{D/C}$, where C and D are integers such that $0 < C < D$ and $D/C \neq s^2$, then $x_{t+k} = x_k$ for all $k \geq 1$, where t denotes the length of the period of the continued fraction expansion. We now prove that the complete quotients, which are periodic (starting with x_1), also have period of length t.

Theorem 10.30 *Let C and D be integers such that $0 < C < D$ and $D/C \neq s^2$. Let $\sqrt{D/C}$ have the continued fraction representation $\sqrt{D/C} = [a_0, \overline{a_1, a_2, \cdots, a_t}]$. Let x_k be the kth complete quotient, that is, $x_k = (\sqrt{CD} + b_k)/r_k$, for $k \geq 0$, where b_k and r_k are defined as in Theorem 10.25. Then the x_k are periodic with period of length t; that is, $x_{t+k} = x_k$ for all $k \geq 1$, and t is the least such integer.*

PROOF: We have seen in the proof of Theorem 10.25 that $x_{t+k} = x_k$ for all $k \geq 1$. Therefore, it suffices to show that if $0 < j < k$ and $x_j = x_k$, then $k > t$. Suppose that $0 < j < k$ and $x_j = x_k$. Then $(\sqrt{CD} + b_j)/r_j = (\sqrt{CD} + b_k)/r_k$, so $r_k\sqrt{CD} + b_jr_k = r_j\sqrt{CD} + b_kr_j$. This implies that $(r_k - r_j)\sqrt{CD} = b_k r_j - b_j r_k$. Since the left side of this equation is an integer, we must have $r_j - r_k = 0 = b_k r_j - b_j r_k$, hence also $b_j = b_k$, since $r_j \neq 0$. Therefore, by definition of t, we have $k - j \geq t$, so $k > t$.   ∎

As an example, the reader may again look at the complete quotients $x_1, x_2, x_3, x_4, x_5,$ and x_6 associated with continued fraction expansion of $\sqrt{22}$ and verify that they are all distinct, but that $x_7 = x_1, x_8 = x_2$, etc. We can now say more about the complete quotients.

Theorem 10.31 *If x_k and t are defined as in Theorem* 10.30, $1 \leq j \leq t$, *and $j < k$, then $x_k = x_j$ if and only if $k = j + th$ for some $h \geq 1$.*

PROOF: (Sufficiency)   By Theorem 10.25 and the definition of t, we have $b_{t+j} = b_j$ and $r_{t+j} = r_j$ for all $j \geq 1$. By induction on h, we can show that $b_{th+j} = b_j$ and $r_{th+j} = r_j$ for all $h, j \geq 1$. Since $x_k = (\sqrt{CD} + b_k)/r_k$, it follows that $x_{th+j} = x_j$ for all $h, j \geq 1$.   (Necessity)   Let $k = th + r$, where $0 \leq r \leq t$. By hypothesis, $x_k = x_j$; that is, $x_{th+r} = x_j$. Theorem 10.30 implies that $x_{th+r} = x_r$, so $x_r = x_j$. If $r = 0$, then $x_{th} = x_0$, so $b_{th} = b_0$. This is impossible because $b_{th} = b_t \neq 0$ and $b_0 = 0$. Therefore, $1 \leq r \leq t$. Since $1 \leq r, j \leq t$, Theorem 10.30 implies that $r = j$, so $k = th + j$ for some $h \geq 1$.   ∎

Referring to the continued fraction expansion of $\sqrt{D/C}$, where C and D are as in Theorem 10.30, we have seen that r_k is periodic; that is, $r_{t+k} = r_k$ for all $k \geq 1$. Since $r_t = r_0 = C$, we can say that $r_{t+k} = r_k$ for all $k \geq 0$. We now show that if C is square-free, then r_k assumes the value C only at the end of each period.

Theorem 10.32 *Let $C, D, t, b_k, r_k,$ and a_k be defined as in Theorem* 10.30. *If C is square-free and $r_k = C$, where $k > 1$, then $k = th$ for some $h \geq 1$.*

PROOF: Since $r_k r_{k-1} = CD - b_k^2$ and $r_k = C$ by hypothesis, we have $C \mid b_k^2$. Since C is square-free by hypothesis, it follows that $C \mid b_k$, so b_k/C is an integer. Now $a_k = [x_k] = [(\sqrt{CD} + b_k)/r_k] = [\sqrt{D/C} + b_k/C] = [\sqrt{D/C}] + b_k/C = (b_1/C) + (b_k/C) = (b_1 + b_k)/C$, so $Ca_k - b_k = b_1$. However, $b_{k+1} = a_k r_k - b_k = Ca_k - b_k$, so $b_{k+1} = b_1$. Also, $r_{k+1} = (CD - b_{k+1}^2)/r_k = (CD - b_1^2)/c = (CD - b_1^2)/r_0 = r_1$. Therefore, $x_{k+1} = x_1$. Now Theorem 10.31 implies that $k + 1 = th + 1$ for some $h \geq 1$; that is, $k = th$ for some $h \geq 1$.   ∎

For example, if we refer to the table on page 225, which lists the r_k associated with continued fraction expansion of $\sqrt{21/13}$, we may note that $t = 8$ and $r_0 = r_8 = 13$, but $r_k \neq 13$ if $0 < k < 8$.

By virtue of Theorem 10.29, if C, D, p_k, q_k, and r_k are as in Theorem 10.29, it is not surprising that the quantity $Cp_{k-1}^2 - Dq_{k-1}^2$ is periodic. If T is the length of the period of $Cp_{k-1}^2 - Dq_{k-1}^2$, the following theorem relates T to t, the period of the continued fraction expansion of $\sqrt{D/C}$.

Theorem 10.33 *If C and D are integers such that $0 < C < D$, $k \geq 0$, and p_k/q_k is the kth convergent in the continued fraction expansion of $\sqrt{D/C}$, which has a period of length t, the $Cp_{k-1}^2 - Dq_{k-1}^2$ is periodic with period of length T, where*

$$T = \begin{cases} t & \text{if } t \text{ is even} \\ 2t & \text{if } t \text{ is odd} \end{cases}$$

PROOF: By virtue of Theorem 10.29, it suffices to investigate the behavior of $(-1)^k r_k$. This quantity is periodic with period of length at most $2t$, since if $k = 2t + j$, where $0 \leq j < 2t$, then $(-1)^k r_k = (-1)^{2t+j} r_{2t+j} = (-1)^j r_j$. If T is the length of the period of $(1)^k r_k$, we have $T \leq 2t$. Since $(-1)^T r_T = (-1)^0 r_0 = C$, we must have $2 \mid T$. Since $r_T = C$, Theorem 10.32 implies that $t \mid T$. Therefore, LCM$[2, t] \mid T$. If t is odd, then LCM$[2, t] = 2t$, so $2t \mid T$. However, recall that $T \leq 2t$, so $T = 2t$. If t is even and j is any nonnegative integer, then $(-1)^{t+j} r_{t+j} = (-1)^j r_j$, so $T \leq t$. However, $t \mid T$, so $t \leq T$. Therefore, $T = t$.   ∎

For example, let us develop the continued fraction expansion for $\sqrt{41}$, listing the relevant data in the following table:

k	b_k	r_k	a_k	p_k	q_k	$p_{k-1}^2 - 41q_{k-1}^2$
-2				0	1	
-1				1	0	
0	0	1	6	6	1	1
1	6	5	2	13	2	-5
2	4	5	2	32	5	5
3	6	1	12	397	62	-1
4	6	5	2	826	129	5
5	4	5	2	2049	320	-5
6	6	1	12	25414	3969	1

We have $\sqrt{41} = [6, \overline{2, 2, 12}]$ and $t = 3$, which is odd, so $T = 6$.

We have devoted some attention to the infinite continued fraction expansion of $\sqrt{D/C}$, where C and D are integers such that $0 < C < D$ and $D/C \neq s^2$. We have seen that these continued fractions are periodic (with initial nonperiodic term). More generally, if x is a positive irrational number, it can be shown that $x = [b_0, b_1, b_2, \cdots, b_{m-1}, \overline{a_1, a_2, \cdots, a_t}]$ if and only if x is a quadratic irrational number; that is, the continued fraction expansion of x is eventually periodic if and only if x satisfies a quadratic equation with integer coefficients. We conclude this chapter by introducing a generalization of periodic continued fractions.

Definition 10.5   ***Arithmetic Progression of Degree m***
If $m \geq 0$, let $g(x)$ be a function such that (1) $g(n)$ is a natural number for all $n \geq 0$ and (2) $g(x) = f(x)/m!$, where $f(x)$ is a polynomial of degree m with integer coefficients. Then the sequence $\{g(0), g(1), g(2), \cdots\}$ is an *arithmetic progression of order m*.

For example, if $g(x) = [(x+1)(x+2)(x+3)]/6$, we obtain the sequence $\{1, 4, 10, 20, 35, 56, \cdots\}$, an arithmetic progression of order 3. If $g(x)$ is constant, then we get a constant sequence, which is an arithmetic progression of order 0.

Now suppose that we have k functions $g_1(x), g_2(x), \cdots, g_k(x)$, all of which satisfy the conditions of Definition 10.5.

Definition 10.6   ***Hurwitz-Type Continued Fraction***
If $x = [a_0, a_1, a_2, \cdots, a_{b-1}, g_1(0), g_2(0), \cdots, g_k(0), g_1(1), g_2(1), \cdots, g_k(1), g_1(2), g_2(2), \cdots, g_k(2), \cdots]$, then we say x is a *Hurwitz-type continued fraction*. We denote x more conveniently as $x = [a_0, a_1, a_2, \cdots, a_{b-1}, \overline{g_1(n), g_2(n), \cdots, g_k(n)}]_{n=0}^{\infty}$. The functions $g_i(x)$ are assumed to satisfy the conditions of Definition 10.5.

If each $g_i(x)$ is constant, then x is simply a periodic continued fraction. The following results, which we present without proof, are given in terms of Hurwitz-type continued fractions.

Theorem 10.34 *If n is a natural number, then*

$$\coth \frac{1}{n} = \frac{e^{2/n}+1}{e^{2/n}-1} = [n, 3n, 5n, 7n, 9n, \cdots] = [\overline{(2m+1)n}]_{m=0}^{\infty}$$

In particular,

$$\coth 1 = \frac{e^2+1}{e^2-1} = [1, 3, 5, 7, 9, \cdots] = [\overline{2m+1}]_{m=0}^{\infty}$$

Theorem 10.35 $e = [2, \overline{1, 2 + 2n, 1}]_{n=0}^{\infty} = [2, 1, 2, 1, 1, 4, 1, 1, 6, 1, 1, 8, 1, \cdots]$

Theorem 10.36 *If* $m > 1$, *then*

$$e^{1/m} = [\overline{1, (1 + 2n)m - 1, 1}]_{n=0}^{\infty}$$
$$= [1, m - 1, 1, 3m - 1, 1, 1, 5m - 1, 1, 1, 7m - 1, 1, \cdots]$$

For the proofs of these last three theorems, we refer the interested reader to Perron (1913).

SECTION 10.6 EXERCISES

36. Let x be a positive irrational number whose infinite continued fraction representation is a purely periodic; that is, $x = [\overline{a_0, a_1, a_2, \cdots, a_{m-1}}]$ for some $m \geq 1$. Prove that x is a quadratic irrational number.

37. Given the periodic infinite continued fraction representation of each of the following irrationals, find the corresponding closed-form representation. [For example, we have seen that if $x = [\overline{1}]$, then $x = \frac{1}{2}(1 + \sqrt{5})$.]

    (a) $[\overline{1, 2}]$ (b) $[\overline{2, 1}]$

    (c) $[1, \overline{1, 2}]$ (d) $[\overline{1, 2, 3}]$

    (e) $[\overline{2, 3, 1}]$ (f) $[\overline{3, 1, 2}]$

    (g) $[1, 2, \overline{3, 4}]$ (h) $[3, 5, 7, 9, 11, \cdots]$

    (i) $[5, 7, 9, 11, 13, \cdots]$ (j) $[1, 1, 1, 5, 1, 1, 1, 9, 1, 1, 1, 13, \cdots]$

38. Let y be a positive irrational number whose infinite continued fraction representation is periodic; that is, $y = [b_0, b_1, b_2, \cdots, b_{n-1}, \overline{a_0, a_1, a_2, \cdots, a_{m-1}}]$ for some $m, n \geq 1$. Prove that y is a quadratic irrational number.

SECTION 10.6 COMPUTER EXERCISE

39. Using Theorem 10.35, write a computer program to approximate e to 100 decimal places.

CHAPTER ELEVEN

NONLINEAR DIOPHANTINE EQUATIONS

11.1 INTRODUCTION

A *Diophantine equation* is an equation in two or more integer unknowns. Such equations are named after Diophantus of Alexandria, who lived about 250 A.D. In Section 4.5, we learned how to solve linear Diophantine equations in two unknowns, that is equations $ax + by = c$, where a, b, and c are given integers and x and y are unknown. In Section 4.9, we learned how to find all solutions of the Pythagorean equation

$$x^2 + y^2 = z^2$$

In this chapter we will consider several additional types of nonlinear Diophantine equations, that is, Diophantine equations in which each unknown term has degree at least 2.

11.2 FERMAT'S LAST THEOREM

In a letter to Mersenne that he wrote in 1638, Fermat claimed that the equation

11.1 $$x^n + y^n = z^n$$

has no solution in positive integers for any $n \geq 3$. Fermat did not provide a proof of this assertion, which is now known as *Fermat's last theorem*. (Fermat claimed that he had found a proof while reading Bachet's *Diophantus* but had lacked sufficient space to write the proof in the margin of the book.)

A more appropriate name for Fermat's last theorem is *Fermat's conjecture*. The case $n = 2^k$ can be excluded, as a result of Exercise 66 in Chapter 4. Therefore, let $n = mp$, where p is an odd prime. If equation 11.1 holds, then we have

11.2 $$(x^m)^p + (y^m)^p = (z^m)^p$$

Therefore, we need only consider the case where $n = p$, an odd prime.

To prove (or disprove) Fermat's conjecture is one of the most famous unsolved problems in mathematics. Fermat's conjecture is currently known to be true for all primes p such that $3 \leq p \leq 10^6$. Before discussing Fermat's conjecture further, however, we must define some additional concepts.

Definition 11.1   ***Bernoulli Numbers***

Let $\frac{t}{e^t - 1} = 1 + \sum_{n=1}^{\infty} \frac{B_n t^n}{n!}$. B_n is called the nth *Bernoulli number*.

The first few Bernoulli numbers are $B_1 = \frac{-1}{2}$, $B_2 = \frac{1}{6}$, $B_3 = 0$, $B_4 = \frac{-1}{30}$, $B_5 = 0$, and $B_6 = \frac{1}{42}$.

It is easily seen that $B_n = 0$ for all odd n such that $n \geq 3$. In addition, one can show that the Bernoulli numbers satisfy the inductive relation

11.3 $$B_{2n} = \frac{1}{2} - \frac{1}{2n+1}\left(1 + \sum_{k=1}^{n-1}\binom{2n+1}{2k}B_{2k}\right) \quad \text{if } n \geq 2$$

Another property of Bernoulli numbers is given by the following theorem.

Theorem 11.1   *(von Staudt)*

Let $B_{2n} = N_{2n}/D_{2n}$, where $(N_{2n}, D_{2n}) = 1$. Then D_{2n} is the product of all primes p such that $(p - 1) \mid 2n$.

PROOF: Omitted, but see Hardy and Wright (1979).

For example, $(p - 1) \mid 6$ if and only if $p - 1 = 1, 2,$ or 6; that is, $p = 2, 3,$ or 7. Therefore, $D_6 = 2 \cdot 3 \cdot 7 = 42$.

The concept of a *regular* prime, which has an important application to Fermat's conjecture, may be defined in terms of Bernoulli numbers.

Definition 11.2   *Regular Prime*

If p is prime and $p \geq 5$, we say that p is *regular* if $p \nmid B_{2k}$ for all k such that $2 \leq 2k \leq p - 3$.

Note that to say $p \nmid B_{2k}$ means $p \nmid N_{2k}$, where N_{2k} is the numerator of B_{2k}.

For example, the prime 5 is regular because $B_2 = \frac{1}{6}$ and $5 \nmid 1$. Also 7 is regular because $B_4 = \frac{-1}{30}$ and $7 \nmid \pm 1$, so $7 \nmid B_2$ and $7 \nmid B_4$.

In 1850, Kummer achieved a significant result by proving that Fermat's conjecture is valid for all exponents that are regular primes. Not all primes are regular, however. It is not even known whether there exist infinitely many regular primes. [See Ribenboim (1979).]

The smallest irregular prime is 37. (One can verify that $37 \mid B_{32}$.) There are two additional irregular primes below 100, namely, 59 and 67. If one uses Definition 11.2 in order to determine whether the prime p is regular or irregular, then one must compute all the B_{2k} such that $2 \leq 2k \leq p - 3$. This is a rather tedious undertaking, since the numerators of the Bernoulli numbers grow rapidly. For example, $B_{12} = 236{,}364{,}091/2730$. It is known that there exist infinitely many irregular primes. It has been conjectured that as x tends to infinity, the proportion of primes below x that are regular approaches $e^{-1/2} = 0.61$. [See Ribenboim (1979).]

On two occasions during the nineteenth century, the Academie des Sciences de Paris offered a prize of 3000 francs and a gold medal to the first French mathematician to prove Fermat's conjecture, but to no avail. In 1908, the Konigliche Gesellschaft der Wissenschaften offered the Wolfskehl Prize of 100,000 marks for the first proof of Fermat's conjecture to be published by September 13, 2007. (The current value of the Wolfskehl prize is about $6500.)

SECTION 11.2 EXERCISES

1. Prove that $B_n = 0$ if n is odd and $n \geq 3$.

2. Use equation 11.3 to compute B_{2n} for $n = 4,$ 5, 6, and 7.

3. Using the results of Exercise 2, verify that each of the primes 11, 13, and 17 is regular.

SECTION 11.2 COMPUTER EXERCISE

4. Write a computer program to compute B_{2n}
for $n \geq 2$ using equation 11.3.

11.3 PELL'S EQUATION: $x^2 - Dy^2 = 1$

John Pell (1611–1685) was an English mathematician who during the 1640s taught mathematics at the universities of Amsterdam and Breda in Holland. In a letter to Goldbach dated August 10, 1732, Euler attributed the equation

11.4 $$x^2 - Dy^2 = 1$$

to Pell. The name has stuck, although most mathematicians consider this attribution to be erroneous.

Note that equation 11.4 is really a one-parameter family of equations, the parameter being D. Without loss of generality, assume that x and y are both nonnegative. It is easily seen that for any D, $x = 1$, $y = 0$ is a solution of equation 11.4, which we call the *trivial solution*. Furthermore, if D is a square, then the trivial solution is the *unique solution*. We therefore assume henceforth that D is a positive, nonsquare integer. We shall see that if $x = a$, $y = b$ is a solution of equation 11.4, then a/b is an approximation of $\sqrt{D}$.

Throughout the ages, many mathematicians have taken an interest in Pell's equation. Baudhayana (India, fourth century B.C.) approximated $\sqrt{2}$ with 17/12 and 577/408. These ratios correspond to solutions of Pell's equation with $D = 2$. Archimedes (third century B.C.) approximated $\sqrt{3}$ by 1351/780, which corresponds to a solution of equation 11.4 with $D = 3$. Pell's equation is discussed at length in both Brahmagupta's *Brahma-Sputa-Siddhanta* (seventh century A.D.) and Bhaskara's *Bija-Ganita* (twelfth century A.D.).

In more recent times, Fermat, that inveterate correspondent, sparked an interest in Pell's equation among European mathematicians. In a letter of February, 1657 to Frenicle de Bessy, Fermat claimed that equation 11.4 has infinitely many integer solutions. He challenged some English mathematicians to provide the proof.

Later that same year, 1657, Lord William Brouncker found a constructive method to find infinitely many solutions of equation 11.4. His method, which depends on the continued fraction representation of $\sqrt{D}$, was published by his colleague, John Wallis. This method was subsequently rediscovered by Euler. Lagrange noted that the earlier efforts of

Wallis and Euler were incomplete. In 1766 Lagrange managed to fill the gaps by providing a completely rigorous proof of Fermat's assertion that Pell's equation has infinitely many solutions. (For further details, see Lagrange, 1770.)

We shall see that if C and D are square-free integers such that $0 < C < D$ and if E is an integer such that $|E| < \sqrt{CD}$, then all solutions, if any, of the generalized Pell's equation

11.5 $$Cx^2 - Dy^2 = E$$

can be obtained from the convergents in the infinite continued fraction representation of $\sqrt{D/C}$.

Theorem 11.2   *Let C and D be square-free integers such that $0 < C < D$, let E be an integer such that $|E| < \sqrt{CD}$, let r_k be defined as in Theorem 10.25, and let T be the length of the period of $(-1)^k r_k$. Then (a) equation 11.5 has solutions if and only if there exists k such that $0 \leq k < T$ and $E = (-1)^k r_k$, and (b) for each such E and each such k, all solutions of equation 11.5 are given by*

$$x_{n,k} = p_{k-1+nT} \quad and \quad y_{n,k} = q_{k-1+nT}$$

where $n = 0, 1, 2, 3, \cdots$ and p_j/q_j is the jth convergent in the continued fraction representation of $\sqrt{D/C}$.

PROOF OF (a):   (Sufficiency)   If $E = (-1)^k r_k$ for some k such that $0 \leq k < T$, then Theorem 10.29 implies that $x = p_{k-1}$ and $y = q_{k-1}$ is a solution of equation 11.5.   (Necessity)   If equation 11.5 holds, then $|Cx^2 - Dy^2| = |E|$. Now $|E| < \sqrt{CD}$ by hypothesis, so $|Cx^2 - Dy^2| < \sqrt{CD}$; thus

$$\left| x^2 - \frac{D}{C}y^2 \right| < \sqrt{\frac{D}{C}}$$

Let $r = \sqrt{D/C}$, so $|x^2 - r^2 y^2| < r$. This implies that $|(x/y) - r| \, |x + yr| < r/y$, so $|(x/y) - r| < r/y(ry+x)$ or $|(x/y) - r| < 1/y[y + (x/r)]$. By Theorem 10.24, in order to prove that x/y is a convergent to $\sqrt{D/C}$, it suffices to prove that $y + (x/r) > 2y - 1$; that is, $x/r > y - 1$. This inequality is trivially true if $y \leq 1$. By hypothesis, we have

$$\frac{Cx^2}{D} = y^2 + \frac{E}{D} > y^2 - \frac{\sqrt{CD}}{D}$$

that is, $(x/r)^2 > y^2 - (1/r)$. Since $0 < C < D$, we have $r > 1$, so $y^2 - (1/r) > y^2 - 1 > (y - 1)^2$. Therefore, $(x/r)^2 > (y - 1)^2$, so $x/r > y - 1$, and we are done. Therefore, there exists k such that $x = p_{k-1}$ and $y = q_{k-1}$. Equation 11.5 and Theorem 10.29 imply that $E = (-1)^k r_k$ for some k. Since $Cp_{k-1}^2 - Dq_{k-1}^2$ is periodic with period T, according to Theorem 10.33, we may further specify that $0 \le k < T$. We call this solution a *fundamental solution*.   ∎

PROOF OF (b):   In the proof of (a) we saw that if equation 11.5 holds, then $x = p_{j-1}$ and $y = q_{j-1}$ for some j. Let $k = j - T[j/T]$. Then $0 \le k < T$ and $Cp_{k-1}^2 - Dq_{k-1}^2 = (-1)^k r_k = (-1)^j r_j = E$ by hypothesis and by definition of T, so $x = p_{k-1}, y = q_{k-1}$ is a fundamental solution of equation 11.5 associated with the solution $x = p_{j-1}$ and $y = q_{j-1}$. By virtue of Theorems 10.29 and 10.33, each fundamental solution, which we now denote $x_{0,k} = p_{k-1}, y_{0,k} = q_{k-1}$, generates an infinite family of solutions $x_{n,k} = p_{k-1+nT}$ and $y_{n,k} = q_{k-1+nT}$, where $n = 0, 1, 2, 3, \cdots$. Also, there are no additional solutions of equation 11.5.   ∎

For example, consider the equation

11.6 $$2x^2 - 5y^2 = -3$$

Since $|-3| = 3 < \sqrt{10}$, Theorem 11.5 is applicable. We therefore develop the continued fraction expansion of $\sqrt{5/2}$, following the notation of Theorem 10.25. We list the results in Table 11.1. Note that $t = 3$ and $T =$

Table 11.1.

k	b_k	r_k	a_k	p_k	q_k	$(-1)^k r_k$
-2				0	1	
-1				1	0	
0	0	2	1	1	1	2
1	2	3	1	2	1	-3
2	1	3	1	3	2	3
3	2	2	2	8	5	-2
4	2	3	1	11	7	3
5	1	3	1	19	12	-3
6	2	2	2	49	31	2
7	2	3	1	68	43	-3
8	1	3	1	117	74	3
9	2	2	2	302	191	-2
10	2	3	1	419	265	3
11	1	3	1	721	456	-3

6. We obtain two fundamental solutions of equation 11.6; namely, for $k = 1$, we get $x_{0,1} = p_0 = 1$ and $y_{0,1} = q_0 = 1$; for $k = 5$ we get $x_{0,5} = p_4 = 11$ and $y_{0,5} = q_4 = 7$. Therefore, all solutions of equation 11.6 are given by the two families: (i) $x_{n,1} = p_{0+6n}, y_{n,1} = q_{0+6n}$ and (ii) $x_{n,5} = p_{5+6n}$, $y_{n,5} = q_{5+6n}$, where $n = 0, 1, 2, 3, \cdots$.

In particular, we may use Theorem 11.2 to obtain all solutions of Pell's equation, that is, equation 11.4.

Theorem 11.3   *Let D be a positive nonsquare integer. Let t be the length of the period of the continued fraction expansion of $\sqrt{D}$. Then Pell's equation (11.4) has infinitely many solutions, all given as follows: (i) if t is even, then $x_n = p_{nt-1}, y_n = q_{nt-1}$ for $n = 0, 1, 2, 3, \cdots$, and (ii) if t is odd, then $x_n = p_{2nt-1}, y_n = q_{2nt-1}$ for $n = 0, 1, 2, 3, \cdots$.*

PROOF: We apply Theorem 11.2 with $C = E = 1$. Therefore, if T is the length of the period of $(-1)^k r_k$, there exists k such that $0 \leq k < T$ and $(-1)^k r_k = 1$. This implies that $r_k = 1 = C$ and $2 \mid k$. In general, if $(-1)^j r_j = 1$, then Theorem 10.32 implies that $t \mid j$. Therefore, $t \mid k$. Since also $2 \mid k$, we have LCM$[2, t] \mid k$. If t is odd, then LCM$[2, t] = 2t = T$, so $T \mid k$. This implies that $k = 0$. If t is even, then Theorem 10.33 implies that $T = t$, so again $T \mid k$; hence $k = 0$. Since we obtain a single family of solutions, following Theorem 11.2, we may denote these solutions as $x_n = p_{nT-1}, y_n = q_{nT-1}$, where $n = 0, 1, 2, 3, \cdots$. The conclusion now follows by appeal to Theorem 10.33.   ∎

For example, suppose we wish to solve the Pell's equation

11.7 $$x^2 - 7y^2 = 1$$

Now $\sqrt{7} = [2, \overline{1, 1, 1, 4}]$, so $t = T = 4$, and our solutions are $x = p_{4n-1}$ and $y = q_{4n-1}$, where $n \geq 0$. Let us compute some convergents to $\sqrt{7}$ using the recursion relations of Theorem 10.5. We list our results in Table 11.2.

From Table 10.3, we see that the first few solutions of equation 11.7 are

$$x_0 = p_{-1} = 1 \qquad y_0 = q_{-1} = 0$$
$$x_1 = p_3 = 8 \qquad y_1 = q_3 = 3$$
$$x_2 = p_7 = 127 \qquad y_3 = q_7 = 48$$
$$x_3 = p_{11} = 2024 \qquad y_3 = q_{11} = 765$$

As a second example, let us solve the Pell's equation

11.8 $$x^2 - 41y^2 = 1$$

Now $\sqrt{41} = [6, \overline{2, 2, 12}]$, so $t = 3$, $T = 6$, and our solutions are $x_n = p_{6n-1}$ and $y_n = q_{6n-1}$, where $n \geq 0$. Let us compute some convergents to $\sqrt{41}$. We list our results in Table 11.3.

From Table 11.3, we see that the first few solutions of equation 11.8 are

$$\begin{array}{ll} x_0 = p_{-1} = 1 & y_0 = q_{-1} = 0 \\ x_1 = p_5 = 2049 & y_1 = q_5 = 320 \\ x_2 = p_{11} = 8,396,801 & y_2 = q_{11} = 1,311,360 \end{array}$$

The equation

11.9 $$x^2 - Dy^2 = -1$$

is sometimes called the *associated Pell's equation*. The following theorem, whose proof is left as an exercise, concerns the solutions of equation 11.9.

Theorem 11.4   *Let D be a positive nonsquare integer. Let t be the length of the period of the continued fraction expansion of $\sqrt{D}$. If t is even, then equation 11.9 has no solutions. If t is odd, then equation 11.9 has infinitely many solutions, all given by $x_n = p_{nt-1}$, $y_n = q_{nt-1}$, where $n = 1, 3, 5, \cdots$.*

PROOF: Exercise.

Table 11.2.

k	a_k	p_k	q_k
-2		0	1
-1		1	0
0	2	2	1
1	1	3	1
2	1	5	2
3	1	8	3
4	4	37	14
5	1	45	17
6	1	82	31
7	1	127	48
8	4	590	223
9	1	717	271
10	1	1307	494
11	1	2024	765
12	4	9403	3554

Table 11.3.

k	a_k	p_k	q_k
-2		0	1
-1		1	0
0	6	6	1
1	2	13	2
2	2	32	5
3	12	397	62
4	2	826	129
5	2	2049	320
6	12	25414	3969
7	2	52877	8258
8	2	131168	20485
9	12	1626893	254078
10	2	3384954	528641
11	2	8396801	1311360

For example, the equation

11.10 $$x^2 - 7y^2 = -1$$

has no solutions, since $\sqrt{7} = [2, \overline{1, 1, 1, 4}]$, so $t = 4$. [We also could reach this conclusion by considering quadratic residues (mod 7).]

Again, the equation

11.11 $$x^2 - 41y^2 = -1$$

has solutions, since $\sqrt{41} = [6, \overline{2, 2, 12}]$, so $t = 3$. The first two such solutions are

$$x_1 = p_2 = 32 \qquad y_1 = q_2 = 5$$
$$x_2 = p_8 = 131,168 \qquad y_2 = q_8 = 20,485$$

Next, we define the fundamental solution of Pell's equation, and we develop a way to obtain all solutions from the fundamental solution. This method will make use of two-by-two matrices and will use continued fractions to a lesser extent.

Definition 11.3   Let T be the length of the period of $(-1)^k r_k$ in the continued fraction expansion of $\sqrt{D}$, where D is a positive nonsquare integer. Then

$$x_1 = p_{T-1} \qquad \text{and} \qquad y_1 = q_{T-1}$$

is called the *fundamental solution* of equation 11.4.

For example, we have just see that the fundamental solution of equation 11.7 is

$$x_1 = p_3 = 8 \qquad \text{and} \qquad y_1 = q_3 = 3$$

Also, the fundamental solution of equation 11.8 is

$$x_1 = p_5 = 2049 \qquad \text{and} \qquad y_1 = q_5 = 320$$

REMARK: The solution $x_0 = p_{-1} = 1, y_0 = q_{-1} = 0$ is called the *trivial solution* of any equation 11.4.

In order to show how all solutions of equation 11.4 can be obtained from the fundamental solution, we must make a brief excursion into algebraic number theory.

Definition 11.4   If D is a positive, nonsquare integer, let $S(D)$ be the set of all real numbers $a + b\sqrt{D}$, where a and b are integers.

For example, $S(7) = \{a + b\sqrt{7} : a, b \text{ are in } Z\}$. In general, the set $S(D)$ is closed under multiplication; that is,

Theorem 11.5   *If z_1 and z_2 belong to $S(D)$, then so does $z_1 z_2$.*

PROOF: Let $z_1 = a_1 + b_1\sqrt{D}$ and $z_2 = a_2 + b_2\sqrt{D}$. Then

$$z_1 z_1 = (a_1 + b_1\sqrt{D})(a_2 + b_2\sqrt{D})$$
$$= (a_1 a_2 + Db_1 b_2) + (a_1 b_2 + a_2 b_1)\sqrt{D}$$

so $z_1 z_2$ belongs to $S(D)$.   ∎

For example, if $z_1 = 2 + 3\sqrt{7}$ and $z_2 = 3 + 5\sqrt{7}$, then $z_1 z_2 = 111 + 19\sqrt{7}$. If z belongs to $S(D)$, it is convenient to assign to z an integer that we call its *norm* which we define as follows:

Definition 11.5   *Norm*
If $z = a + b\sqrt{D}$, let $N(z) = a^2 - Db^2$. The integer $N(z)$ is called the *norm* of z.

For example, $N(2 + 3\sqrt{7}) = 2^2 - 7(3^2) = -59$, and $N(3 + 5\sqrt{7}) = 3^2 - 7(5^2) = -166$. The norm of z is a completely multiplicative function of z; that is:

Theorem 11.6   *If z_1 and z_2 belongs to $S(D)$, then $N(z_1 z_2) = N(z_1)N(z_2)$.*

PROOF: $N(z_1)N(z_2) = (a_1^2 - Db_1^2)(a_2^2 - Db_2^2) = a_1^2 a_2^2 - Da_1^2 b_2^2 - Da_2^2 b_1^2 + D^2 b_1^2 b_2^2 = (a_1^2 a_2^2 + D^2 b_1^2 b_2^2 + 2Da_1 a_2 b_1 b_2) - D(a_1^2 b_2^2 + a_2^2 b_1^2 + a_1 a_2 b_1 b_2) = (a_1 a_2 + Db_1 b_2)^2 - D(a_1 b_2 + a_2 b_1)^2 = N(z_1 z_2)$.   ∎

For example, we have seen that if $z_1 = 2 + 3\sqrt{7}$ and $z_2 = 3 + 5\sqrt{7}$, then $N(z_1) = -59$ and $N(z_2) = -166$; also, $z_1 z_2 = 111 + 19\sqrt{7}$. Now $N(z_1 z_2) = 111^2 - 7(19)^2 = 9794 = (-59)(-166) = N(z_1)N(z_2)$.

In order to establish the connection between Pell's equation and the set $S(D)$, we need to define what are known as *units*.

Definition 11.6   *Unit*

Let $u = a + b\sqrt{D}$ belong to $S(D)$. We say that u is a *unit* if $N(u) = 1$, that is, $a^2 - Db^2 = 1$.

For example, $8 + 3\sqrt{7}$ is a unit in $S(7)$, since $N(8 + 3\sqrt{7}) = 8^2 - 7(3^2) = 1$. Note that $u = a + b\sqrt{D}$ is a unit in $S(D)$ if and only if $x = a$ and $y = b$ is a solution of the Pell's equation $x^2 - Dy^2 = 1$. We now develop some properties of units in $S(D)$. Let $u^{-1} = 1/u$.

Theorem 11.7   *If u is a unit in $S(D)$, then so is u^{-1}.*

PROOF: First, we must verify that u^{-1} belongs to $S(D)$. If $u = a + b\sqrt{D}$, then

$$u^{-1} = \frac{1}{u} = \frac{1}{a + b\sqrt{D}} = \frac{a - b\sqrt{D}}{a^2 - Db^2} = \frac{a - b\sqrt{D}}{1} = a - b\sqrt{D}$$

so u^{-1} belongs to $S(D)$. Now $N(u^{-1}) = N(u^{-1})1 = N(u^{-1})N(u) = N(u^{-1}u) = N(1) = 1$, so u^{-1} is a unit in $S(D)$.   ∎

For example, if $u = 8 + 3\sqrt{7}$ in $S(7)$, then $u^{-1} = 8 - 3\sqrt{7}$. We have just seen that the (multiplicative) inverse of a unit is a unit. We are about to see that the product of two units is also a unit.

Theorem 11.8   *If u_1 and u_2 are both units in $S(D)$, then so is u_1u_2.*

PROOF: By hypothesis and Theorem 11.5, u_1u_2 belongs to $S(D)$. By Theorem 11.6, $N(u_1u_2) = N(u_1)N(u_2) = 1^2 = 1$, so u_1u_2 is a unit in $S(D)$.   ∎

For example, $u_1 = 3 + 2\sqrt{2}$ and $u_2 = 17 + 12\sqrt{2}$ are both units in $S(2)$, since $N(u_1) = 3^2 - 2(2^2) = 1$ and $N(u_2) = 17^2 - 2(12^2) = 1$. Now $u_1u_2 = 99 + 70\sqrt{2}$, which is also a unit in $S(2)$, since $N(u_1u_2) = 99^2 - 2(70^2) = 1$.

If $u = r + s\sqrt{D}$ is a unit in $S(D)$, we say that r is the *rational part* of u and s is the *irrational part* of u. The units of $S(D)$ are real numbers, and therefore, distinct units may be compared in size. If we look at the units whose rational and irrational parts are both nonnegative, we can say more.

Theorem 11.9   *Let $u_1 = a_1 + b_1\sqrt{D}$ and $u_2 = a_2 + b_2\sqrt{D}$ be units in $S(D)$ such that $a_1, a_2, b_1,$ and b_2 are all nonnegative. Then the following statements are equivalent: (i) $a_1 < a_2$, (ii) $b_1 < b_2$, and (iii) $u_1 < u_2$.*

PROOF: Consider also the statements (iv) $a_1 = a_2$, (v) $b_1 = b_2$, and (vi) $u_1 = u_2$. Clearly, if (i) and (ii) both hold, then so does (iii). Furthermore, if (iv) and (v) hold, then so does (vi). Also, if (vi) holds, then so do (iv) and (v). Therefore, it suffices to show that each of (i) and (ii) implies the other. By hypothesis, $a_1^2 - Db_1^2 = N(u_1) = 1 = N(u_2) = a_2^2 - Db_2^2$, so $a_1^2 + Db_2^2 = a_2^2 + Db_1^2$. Now $a_1 < a_2$ iff $a_1^2 < a_2^2$ iff $Db_2^2 > Db_1^2$ iff $Db_1^2 < Db_2^2$ iff $b_1^2 < b_2^2$ iff $b_1 < b_2$.   ∎

For example, $u_1 = 3 + 2\sqrt{2}$ and $u_2 = 17 + 12\sqrt{2}$ are units of $S(2)$ whose rational and irrational parts are all nonnegative. Note that $3 < 17$, $2 < 12$, and $u_1 < u_2$.

We are about to see that all units in $S(D)$ with nonnegative rational and irrational parts are powers of a single generating unit. First, however, we need the following lemma.

Lemma 11.1   *If $u = a + b\sqrt{D}$ is a unit in $S(D)$ such that $u > 1$, then $a \geq 2$ and $b \geq 1$.*

PROOF: By hypothesis, $(a - b\sqrt{D})(a + b\sqrt{D}) = a^2 - Db^2 = N(u) = 1$ and $a + b\sqrt{D} = u > 1$, so $0 < a - b\sqrt{D} < 1$. Therefore, $a - 1 < b\sqrt{D} < a$. If $a \leq 0$, then $b < 0$, so $u = a + b\sqrt{D} < 0$, contrary to hypothesis. If $a = 1$, then $1 - Db^2 = 1$, so $b = 0$. However, then $u = 1$, contrary to hypothesis. Therefore, $a \geq 2$. This implies that $1 < b\sqrt{D}$, so $b \geq 1$.   ∎

Theorem 11.10   *Let U be the set of units u of $S(D)$ such that $u > 1$. Then there exists θ in U such that for all v in U there exists $k \geq 1$ such that $v = \theta^k$.*

PROOF: By Theorem 11.7, if t is the length of the period of the continued fraction expansion of $\sqrt{D}$, then $u = p_{t-1} + q_{t-1}\sqrt{D}$ is a unit and $u > 1$, so U is nonempty. Let A be the set of all natural numbers r such that $r + s\sqrt{D} = u$ for some u in U. Since U is nonempty, it follows from Lemma 11.1 that A is nonempty. By the well-ordering principle, A has a least element a. If $\theta = a + b\sqrt{D}$ is the corresponding element of U, Theorem 11.9 implies that θ is the least element of U. Now let v belong to U. Since $v > 1$ and $\theta > 1$, there exists $k \geq 1$ such that $\theta^k \leq v < \theta^{k-1}$. If $\theta^k < v$, then $1 < \theta^{-k}v < \theta$. Let $w = \theta^{-k}v$. Theorems 11.7 and 11.8 imply that w belongs to U. However, $w < \theta$, which contradicts the definition of θ. Therefore, $v = \theta^k$.   ∎

For example, we saw earlier that $u_1 = 3 + 2\sqrt{2}$ and $u_2 = 17 + 12\sqrt{2}$ are units in $S(D)$. One can show that if $a + b\sqrt{D}$ is a unit in $S(D)$, then $a \not\equiv D \pmod 2$. If U is the set of units u in $S(2)$ such that $u > 1$ and $\theta = a + b\sqrt{2}$ generates U, then a is odd. Therefore, Lemma 11.1 implies that $a \geq 3$. Therefore, Theorem 11.9 implies that $\theta = u_1$. One can verify that $u_2 = \theta^2$.

The following theorem enables us to obtain all solutions of Pell's equation (11.4) from the fundamental solution.

Theorem 11.11 *If D is a positive, nonsquare integer, and if the fundamental solution of*

11.4 $x^2 - Dy^2 = 1$

is $x = x_1$ and $y = y_1$, then all nontrivial solutions are given by $x = x_n$ and $y = y_n$, where $x_n + y_n\sqrt{D} = (x_1 + y_1\sqrt{D})^n$.

PROOF: $x_n^2 - Dy_n^2 = N(x_n + y_n\sqrt{D}) = N[(x_1 + y_1\sqrt{D})^n] = [N(x_1 + y_1\sqrt{D})]^n = 1^n = 1$, so $x = x_n, y = y_n$ is a solution of equation 11.4. If $x = a, y = b$ is a solution of equation 11.4 such that $a \geq 0$ and $b \geq 0$, then $u = a + b\sqrt{D}$ is a unit in $S(D)$ and $u > 1$. Therefore, Theorem 11.10 implies that $a + b\sqrt{D} = (x_1 + y_1\sqrt{D})^n = x_n + y_n\sqrt{D}$ for some $n \geq 1$.   ∎

For example, we have seen that in $S(2), x_1 = 3$ and $y_1 = 2$. Now $x_2 + y_2\sqrt{2} = (x_1 + y_1\sqrt{2})^2 = (3 + 2\sqrt{2})^2 = 17 + 12\sqrt{2}$, so $x_2 = 17$ and $y_2 = 12$. Likewise, $x_3 + y_3\sqrt{2} = (x_1 + y_1\sqrt{2})^3 = (3 + 2\sqrt{2})^3 = 99 + 70\sqrt{2}$, so $x_3 = 99$ and $y_3 = 70$.

In practice, using Theorem 11.11 to compute the solutions of equation 11.4 would be laborious. The following theorem, which makes use of two-by-two matrices, offers a shortcut, though.

Theorem 11.12 *Let D be a positive, nonsquare integer. Let $x_0 = 1, y_0 = 0$ be the trivial solution of equation 11.4. Then all solutions of equation 11.4 in nonnegative integers may be expressed by each of the two following forms:*

(a) $\begin{bmatrix} x_n \\ y_n \end{bmatrix} = \begin{bmatrix} x_1 & Dy_1 \\ y_1 & x_1 \end{bmatrix} \begin{bmatrix} x_{n-1} \\ y_{n-1} \end{bmatrix}$ *where $n \geq 1$*

(b) $$\begin{bmatrix} x_n \\ y_n \end{bmatrix} = \begin{bmatrix} x_1 & Dy_1 \\ y_1 & x_1 \end{bmatrix}^n \begin{bmatrix} 1 \\ 0 \end{bmatrix}$$ *where* $n \geq 0$

PROOF: Since (b) follows from (a), it suffices to prove (a). Theorem 11.11 implies that $x_n + y_n\sqrt{D} = (x_1 + y_1\sqrt{D})^n = (x_1 + y_1\sqrt{D})(x_1 + y_1\sqrt{D})^{n-1} = (x_1 + y_1\sqrt{D})(x_{n-1} + y_{n-1}\sqrt{D}) = x_1x_{n-1} + Dy_1y_{n-1} + (x_1y_{n-1} + x_{n-1}y_1)\sqrt{D}$. Therefore, $x_n = x_1x_{n-1} + Dy_1y_{n-1}$ and $y_n = x_1y_{n-1} + x_{n-1}y_1$, which implies (a).   ∎

For example, if $D = 7$, we saw previously that the fundamental solution of equation 11.4 is $x_1 = 8, y_1 = 3$. Using Theorem 11.12a, we have

$$\begin{bmatrix} x_2 \\ y_2 \end{bmatrix} = \begin{bmatrix} x_1 & Dy_1 \\ y_1 & x_1 \end{bmatrix}\begin{bmatrix} x_1 \\ y_1 \end{bmatrix} = \begin{bmatrix} 8 & 21 \\ 3 & 7 \end{bmatrix}\begin{bmatrix} 8 \\ 3 \end{bmatrix} = \begin{bmatrix} 127 \\ 48 \end{bmatrix}$$

$$\begin{bmatrix} x_3 \\ y_3 \end{bmatrix} = \begin{bmatrix} x_1 & Dy_1 \\ y_1 & x_1 \end{bmatrix}\begin{bmatrix} x_2 \\ y_2 \end{bmatrix} = \begin{bmatrix} 8 & 21 \\ 3 & 7 \end{bmatrix}\begin{bmatrix} 127 \\ 48 \end{bmatrix} = \begin{bmatrix} 2024 \\ 765 \end{bmatrix}$$

In general, to solve a Pell's equation, one must first use the continued fraction expansion of $\sqrt{D}$ to obtain the fundamental solution and then obtain all other solutions via Theorem 11.12.

For example, in order to solve the Pell's equation

11.12 $x^2 - 6y^2 = 1$

we begin by obtaining the continued fraction expansion of $\sqrt{6}$. Following the notation of Theorems 10.25 and 10.5, we obtain data that we list in Table 11.4. We see that $\sqrt{6} = [2, \overline{2, 4}]$, so $t = 2$. Now the fundamental solution of equation 11.12 is given by $x_1 = p_{t-1} = p_1 = 5, y_1 = q_{t-1} = q_1 = 2$. Therefore, by Theorem 11.12, all nonnegative solutions of equation 11.12 are given by

$$\begin{bmatrix} x_n \\ y_n \end{bmatrix} = \begin{bmatrix} 5 & 12 \\ 2 & 5 \end{bmatrix}^n \begin{bmatrix} 1 \\ 0 \end{bmatrix}$$ where $n \geq 0$

Table 11.4.

k	b_k	r_k	a_k	p_k	q_k
-2				0	1
-1				1	0
0	0	1	2	2	1
1	2	2	2	5	2
2	2	1	4	22	9
3	2	2	2	49	20

SECTION 11.3 EXERCISES

5. For each of the following values of D, find the first three nontrivial solutions of the corresponding Pell's equations.

   (a) $D = 5$ (b) $D = 10$

   (c) $D = 13$ (d) $D = 39$

6. Prove Theorem 11.4.

7. Prove that if $x = x_1, y = y_1$ is the fundamental solution of the associated Pell's equation

11.9 $x^2 - Dy^2 = -1$

   then all solutions of equation 11.9 are given by $x = x_n, y = y_n$, where

$$\begin{bmatrix} x_n \\ y_n \end{bmatrix} = \begin{bmatrix} x_1 & Dy_1 \\ y_1 & x_1 \end{bmatrix}^{2n-1} \begin{bmatrix} 1 \\ 0 \end{bmatrix} \quad \text{with } n = 1, 2, 3, \cdots$$

8. Does the equation $x^2 - 15y^2 = -1$ have any solution? Explain using (a) continued fractions and (b) quadratic residues.

9. Prove that the equation $x^4 - 2y^2 = 1$ has only the trivial solution $x = 1, y = 0$ using (a) continued fractions and (b) other considerations.

10. Let t be the length of the period of the continued fraction expansion of $\sqrt{D}$, where D is a positive nonsquare integer. Prove that if D has a prime factor p such that $p \equiv 3 \pmod 4$, then $2 \mid t$.

11. Prove that there are infinitely many n such that each of $n, n + 1$, and $n + 2$ is a sum of two squares.

12. Find the first four solutions of the generalized Pell's equation $3x^2 - 5y^2 = -2$.

SECTION 11.3 COMPUTER EXERCISE

13. Write a computer program to find the first 20 solutions of the equation $x^2 - 11y^2 = 1$ using Theorem 11.12a. (The fundamental solution is $x_1 = 10, y_1 = 3$.)

11.4 MORDELL'S EQUATION: $x^3 = y^2 + k$

In 1621 the French mathematician Bachet stated that if $k = 2$, then the equation

11.13 $x^3 = y^2 + k$

has the unique solution $x = 3, y = 5$. In 1657 Fermat stated that if $k = 4$, then equation 11.13 has only the two solutions: $x = y = 2$ and $x = 5, y = 11$. Two centuries elapsed before it was possible to verify the validity of these assertions. In 1738 Euler proved that if $k = -1$, then the unique solution of equation 11.13 is $x = 2, y = 3$. In 1869 Lebesgue proved that if $k = -7$, then equation 11.13 has no solution. We give this proof in Theorem 11.13 below. In 1930 the Norwegian mathematician Nagell proved that if $k = -17$, then equation 11.13 has eight solutions. Louis J. Mordell (1888–1972) was Sadlerian Professor of Mathematics at the University of Cambridge in England. He wrote extensively on Diophantine equations in general, and on equation 11.13 in particular. His name is now associated with equation 11.13, where k represents a given integer.

Note that Mordell's equation is inhomogeneous; that is, the variable terms that appear in equation 11.13 do not have the same degree. As a consequence, Mordell's equation is more difficult to solve for most values of k than the other Diophantine equations considered earlier in this chapter. Usually, the solution requires the techniques of algebraic number theory, which are not treated in this text. The cases $k = -7$ and $k = 16$, however, can be handled by elementary methods.

Theorem 11.13 *(Lebesgue)*
The equation

11.14 $$x^3 = y^2 - 7$$

has no solution in integers.

PROOF: Assume the contrary. If $2 \mid x$, then $y^2 \equiv 7 \pmod 8$, an impossibility. If $x \equiv 3 \pmod 4$, then $x^3 \equiv 3 \pmod 4$, so $y^2 \equiv 2 \pmod 4$, an impossibility. Therefore, $x \equiv 1 \pmod 4$, so $x + 2 \equiv 3 \pmod 4$. Now equation 11.14 implies that $x^3 + 8 = y^2 + 1$, so $(x + 2)(x^2 - 2x + 4) = y^2 + 1$. Since $x + 2 \equiv 3 \pmod 4$, there exists a prime p such that $p \equiv 3 \pmod 4$ and $p \mid (x + 2)$. However, then $p \mid (y^2 + 1)$, which contradicts Theorem 4.12. ∎

Theorem 11.16 *The equation*

11.15 $$x^3 = y^2 + 16$$

has no solution in integers.

PROOF: Assume the contrary. Then x and y have the same parity. If x and y are both even, then $x = 2u$ and $y = 2v$, so equation 11.15 implies that $8u^3 = 4v^2 + 16$. Therefore, $8 \mid 4v^2$, so $v = 2t$; hence $8u^3 = 16t^2 + 16$. Therefore, $16 \mid 8u^3$, so $u = 2w$; hence $64w^3 = 16t^2 + 16$. However, this implies that $4w^3 = t^2 + 1$, so $t^2 \equiv -1$ (mod 4), an impossibility. If x and y are both odd, then $x^2 \equiv y^2 \equiv 1$ (mod 8), so equation 11.15 implies that $x \equiv 1$ (mod 8). However, equation 11.15 implies $y^2 + 8 = x^3 - 8 = (x-2)(x^2 + 2x + 4)$. If p is a prime such that $p \mid (x-2)$, then $p \mid (y^2 + 8)$, so $1 = (-8/p) = (-2/p)$. This implies that $p \equiv 1$ or 3 (mod 8). Since $x - 2$ is the product of such primes, it follows that $x - 2 \equiv 1$ or 3 (mod 8); hence $x \equiv 3$ or 5 (mod 8), an impossibility. If $x - 2$ has no prime divisor, then $x = 3$, which is also impossible.

■

SECTION 11.4 EXERCISES

Prove that each of the two following Mordell's equations has no solution.

14. $x^3 = y^2 + 80$

15. $x^3 = y^2 - 215$

CHAPTER 12

COMPUTATIONAL NUMBER THEORY

12.1 INTRODUCTION

Computational number theory is the newest branch of number theory. It has developed rapidly in recent years, as high-speed computers have become increasingly available to scientists. Computational number theory is *not* pure mathematics, although it makes ample use thereof. It is a blend of mathematics with computer science, with some of the flavor of engineering thrown in.

An *algorithm* is a specific procedure that accepts a given input and produces a desired output. Computational number theory takes a hard look at the algorithms of elementary number theory. Some of these algorithms require too much computer time or computer memory for a given sufficiently large input. Such algorithms are called *computationally infeasible* and are therefore rejected. An example of a computationally infeasible algorithm is trial division as a test of primality for large integers. Computational number theory seeks ever more efficient algorithms, that is, algorithms that require less time or space than those currently in use. Computational number theory is closely related to that branch of computer science known as *complexity theory*.

It is worth mentioning that computational number theory is an *experimental* science with a very tolerant viewpoint, such as one finds in engineering. That is to say, an algorithm may be considered useful that works for some, but not all, inputs of a given size. Furthermore, many of the algorithms of computational number theory are *nondeterministic*; that is, they require the use of random (actually pseudorandom) numbers. The emphasis is on "what works." It is not unusual in the literature of computational number theory to find the phrase "the cost of factoring an integer." Computational number theory is also subject to transient influences, such as the cost and availability of certain computer hardware items.

Computational number theory has its theorems, but it is largely concerned with algorithms. Just as today's computers have rendered obsolete those of yesterday and will surely themselves be rendered obsolete by those of tomorrow, it is reasonable to expect that many of the algorithms yet to come will make museum pieces of those presently in use.

The achievements of computational number theory include fast primality tests for large integers and less rapid, but often workable methods for the factorization of large composite integers. An important application of computational number theory has been to cryptology, which is discussed in Chapter 13. We begin the study of computational number theory by introducing the concept of *computational feasibility*.

Definition 12.1   An algorithm is said to be *computationally feasible* for a given input if it can be executed in a reasonable amount of time and requires no more than a reasonable amount of computer memory.

"A reasonable amount of time" means seconds, minutes, hours, days, weeks, or even months, but *not* years or centuries. An algorithm which, for a given input, requires an excessive amount of time or space is called *computationally infeasible*. Some algorithms are computationally feasible for small inputs but computationally infeasible for large inputs. This is the case for trial division as a test of primality. Other algorithms, such as Euclid's algorithm for finding the greatest common divisor of two natural numbers, are computationally feasible for all inputs.

At the end of Chapter 2 we presented the binary number system. This is the system by which number are represented in computers. Recall that each natural number has a binary representation that consists of a string of 0s and 1s. For example, $34_{10} = 100010_2$. The 0s and 1s that appear are called *binary digits*, or *bits*. We define a *bit operation* as an addition, subtraction, or multiplication of two 1-bit integers, the division of a 2-bit integer by a 1-bit integer, or a one-place shift of a binary integer. Computer time is usually estimated in terms of the number of bit operations needed to execute an algorithm. In arriving at such an estimate, it is customary to use "big-oh" notation, whose definition follows.

Definition 12.2   Let $f(x)$ and $g(x)$ be real-valued functions defined on the positive reals. We say "$f(x)$ is big oh of $g(x)$" and write $f(x) = O(g(x))$ if there exists a positive constant k such that $|f(x)| \leq kg(x)$ for all sufficiently large x.

For example $x + \sin x = O(x)$, since $|x + \sin x| \leq |x| + |\sin x| \leq x + 1 \leq 2x$ if $x \geq 1$. Generally, if $f(x) = O(x^n)$ for some n, we seek the smallest n for which this is true. Note that if $f(x) = O(1)$, then $f(x)$ is bounded. We leave as exercises the proofs of the two following theorems.

Theorem 12.1   *If $f(x) = O(g(x))$ and c is an arbitrary constant, then $cf(x) = O(g(x))$.*

PROOF: Exercise

For example, let $f(x) = x^2 + x$ and $g(x) = x^2$. Then $f(x) = O(g(x))$, since $x^2 + x \leq 2x^2$ if $x \geq 1$. Let $c = 2$. Now $2f(x) = 2(x^2 + x) \leq 4x^2$ if $x \geq 1$, so $2f(x) = O(x^2)$.

Theorem 12.2   *Let $f_1 = O(g_1(x))$ and $f_2(x) = O(g_2(x))$. Then (a) $f_1(x) + f_2(x) = O(g_1(x) + g_2(x))$ and (b) $f_1(x)f_2(x) = O(g_1(x)g_2(x))$.*

PROOF: Exercise.

For example, $\sqrt{x+1} = O(x^{1/2})$ and $\sqrt{x^2-x+1} = O(x)$, so $\sqrt{x^3+1} = \sqrt{(x+1)(x^2-x+1)} = O(x^{3/2})$.

As a consequence, it can be shown that (1) adding or subtracting two n-bit integers requires $O(n)$ bit operations and (2) a conventional multiplication of an m-bit integer by an n-bit integer requires $O(mn)$ bit operations. In particular, the conventional multiplication of two n-bit integers requires $O(n^2)$ bit operations. [By means of a technique known as the *fast Fourier transform*, the multiplication of two n-bit integers can be performed in $O(n \log n)$ bit operations.] To conventionally compute the integer part of the quotient of a $2n$-bit integer by an n-bit integer requires $O(n^2)$ bit operations.

Next, we analyze Euclid's algorithm for finding the greatest common divisor of two given natural numbers.

Theorem 12.3   *If a and b are integers such that $a > b > 0$, then the number of bit operations needed to compute (a, b) using Euclid's algorithm is $O((\log a)^3)$.*

PROOF: If Euclid's algorithm requires n iterations to compute (a, b), then Lame's theorem (Theorem 9.2) implies that $n < 1 + \log_\alpha b$, where $\alpha = \frac{1}{2}(1 + \sqrt{5})$, so $n = O(\log b)$; hence $n = O(\log a)$. Each iteration is a division that requires $O((\log a)^2)$ bit operations. Therefore, Theorem 12.2b implies that the total number of bit operations needed is $O((\log a)^3)$.   ∎

For example, suppose that a and b are 100-digit integers, while c and d are 200-digit integers. Theorem 12.3 implies that computing (c, d) will take about eight times longer than computing (a, b).

Definition 12.3   An algorithm that accepts n as input is said to be *polynomial time* if it requires $O(n^c)$ bit operations for some $c > 0$.

For example, the preceding discussion shows that Euclid's algorithm is polynomial time, in log n, with $c = 3$.

As a contrast, suppose that n is a given integer such that $(6, n) = 1$ and we wish to determine whether n is prime or composite. Trial division consists in dividing n by every integer d such that $(6, d) = 1$ and $5 \leq d \leq \sqrt{n}$. In other words, we are interested in potential divisors d of the form $6k \pm 1$. If n/d is an integer for such a d, then n is composite and d is a factor of n. If no quotient n/d is an integer for $5 \leq d \leq \sqrt{n}$, then n is prime.

It would be sufficient, in theory, to consider only those potential divisors d which are prime. However, this would entail either (1) testing each d for primality or (2) storing all the primes up to $\sqrt{n}$. Such an endeavor would use more computer time or space than simply trial division by all integers $6k \pm 1$ not exceeding $\sqrt{n}$, even though many of these divisions are redundant.

In trial division, the number of divisors is $O(\sqrt{n})$, and each division requires $O((\log_2 n)^2)$ bit operations, so the total number of bit operations needed to test n for primality (or to factor n) is $O(\sqrt{n}(\log_2 n)^2)$. Therefore, trial division is *not* a polynomial time algorithm and is computationally infeasible for large n. Nevertheless, in several current factorization techniques, trial division is still used to find *small* factors of a large n, that is, factors d such that $d < (\log_2 n)^2$.

In the work ahead, we will discuss computationally feasible tests of compositeness and primality, as well as methods of factorization of composite integers. How significant is this enterprise? Let us quote Gauss: "The problem of distinguishing prime numbers from composite numbers and resolving the latter into their prime factors is known to be one of the most important and useful in arithmetic . . . . the dignity of the science itself seems to require that every possible means be explored for the solution of a problem so elegant and so celebrated." (*Disquisitiones Arithmeticae*, Art. 329)

12.2A TESTS FOR COMPOSITENESS AND PRIMALITY: PSEUDOPRIMES AND CARMICHAEL NUMBERS

Given a large odd integer n, we wish to determine whether n is prime or composite. Recall that the prime number theorem implies that the proportion of primes below x is approximately $1/\log x$. Therefore, if n is chosen at random, n is far more likely to be composite than to be prime.

If b is an integer such that $(b, n) = 1$ and $b^{n-1} \not\equiv 1 \pmod{n}$, then Fermat's little theorem (Theorem 4.4) implies that n is composite. For example, if $n = 221$, let $b = 2$. Now $2^{220} \equiv 16 \not\equiv 1 \pmod{221}$, so we may conclude that 221 is composite. Unfortunately, this test for compositeness does not always work, since it may occur that $(b, n) = 1$ and $b^{n-1} \equiv 1 \pmod{n}$ even though n is composite. For example, $341 = 11 \cdot 31$ is composite and $(2, 341) = 1$, yet $2^{340} \equiv 1 \pmod{341}$. This leads us to the following definition.

Definition 12.4 n is a *pseudoprime to base b* if n is composite, $(b, n) = 1$, and $b^{n-1} \equiv 1 \pmod{n}$. If so, we write n is PSP(b).

The preceding example shows that 341 is a pseudoprime to base 2; the reader may verify that so are 561 and 645. Pseudoprimes to base 2 are rare. Of all the odd integers below 10^{10}, only 14,884 of them are pseudoprimes to base 2, whereas 455,051,511 of them are prime. Nevertheless, as we shall see shortly, there are infinitely many pseudoprimes to base 2. First, we need the following lemma.

Lemma 12.1 If $b \neq 1$ and $m \mid n$, then $(b^m - 1) \mid (b^n - 1)$.

PROOF: By hypothesis, we have $n = km$, so $b^n - 1 = b^{km} - 1 = (b^m)^k - 1$. Let $a = b^m$. Now $a^k - 1 = (a - 1)(a^{k-1} + a^{k-2} + \cdots + a^2 + a + 1)$, so $(a - 1) \mid (a^k - 1)$; hence $(b^m - 1) \mid (b^n - 1)$.

Theorem 12.4   *There are infinitely many pseudoprimes to base* 2.

PROOF: Let n be a PSP(2); that is, n is an odd composite number such that $2^{n-1} \equiv 1 \pmod{n}$. If $m = 2^n - 1$, then m is certainly odd; Lemma 12.1 implies that m is composite. Now $2^n \equiv 2 \pmod{n}$, so $m - 1 = 2^n - 2 = kn$ for some k. Now Lemma 12.1 implies that $m \mid (2^{kn} - 1)$; that is, $2^{kn} \equiv 1 \pmod{m}$. Therefore, $2^{m-1} \equiv 1 \pmod{m}$, so m is also a PSP(2). Since $n_1 = 341$ is a PSP(2), we can generate an infinite sequence of PSP(2)'s $\{n_1, n_2, n_3, \cdots\}$ via the formula

$$n_k = 2^{n_{k-1}} - 1 \qquad (k \geq 2)$$ ■

REMARK: $n_2 = 2^{341} - 1$ is an integer with over 100 decimal digits!

If n is pseudoprime to base b, we may yet unmask its composite nature. Very few composite numbers are pseudoprime to several different bases simultaneously. We may therefore proceed as follows. Choose integers $a_1, a_2, \cdots, a_m$ such that $(a_i, n) = 1$ for each index i. Compute $a_i^{n-1} \pmod{n}$ for $i = 1, 2, 3, \cdots, m$. If for some index i we get $a_i^{n-1} \not\equiv 1 \pmod{n}$, then n is composite.

For example, $(2, 341) = 1$ and $2^{340} \equiv 1 \pmod{341}$, but $(3, 341) = 1$ and $3^{340} \equiv 56 \not\equiv 1 \pmod{341}$, so 341 is composite.

This test for compositeness has its limitations, however, since even if $b_{n-1} \equiv 1 \pmod{n}$ for *all* b such that $(b, n) = 1$, n may still be composite.

Definition 12.5 n is a *Carmichael number* if n is composite and $b^{n-1} \equiv 1 \pmod{n}$ for all b such that $(b, n) = 1$.

The smallest Carmichael number is $561 = 3 \cdot 11 \cdot 17$. Let us verify that 561 is indeed a Carmichael number. Suppose that $(b, 561) = 1$. Therefore, $(b, 3) = (b, 11) = (b, 17) = 1$. Fermat's little theorem implies that $b^2 \equiv 1 \pmod{31}$, so $b^{560} \equiv (b^2)^{280} \equiv 1^{280} \equiv 1 \pmod{3}$. Similarly, $b^{10} \equiv 1 \pmod{11}$, so $b^{560} = (b^{10})^{56} \equiv 1^{56} \equiv 1 \pmod{11}$. Finally,

$b^{16} \equiv 1 \pmod{17}$, so $b^{560} = (b^{16})^{35} \equiv 1^{35} \equiv 1 \pmod{17}$. Therefore, $b^{560} \equiv 1 \pmod{\text{LCM}[3, 11, 17]}$; that is, $b^{560} \equiv 1 \pmod{561}$.

The several theorems that follow give a characterization of Carmichael numbers.

Theorem 12.5   *If $n = \prod_{i=1}^{r} p_i$, where $r \geq 2$ and the p_i are distinct odd primes such that $(p_i - 1) \mid (n - 1)$ for each i, then n is a Carmichael number.*

PROOF: By hypothesis, n is composite. If $(b, n) = 1$, then $(b, p_i) = 1$ for all i, so $b^{p_i - 1} \equiv 1 \pmod{p_i}$. By hypothesis, $n - 1 = k_i(p_i - 1)$, so $b^{n-1} = b^{k_i(p_i - 1)} \equiv (b^{p_i - 1})^{k_i} \equiv 1^{k_i} \equiv 1 \pmod{p_i}$. Therefore, $b^{n-1} \equiv 1 \pmod{\prod_{i=1}^{r} p_i}$; that is, $b^{n-1} \equiv 1 \pmod{n}$, so n is a Carmichael number.   ■

We will soon see that every Carmichael number is of the type mentioned in Theorem 12.5, but first we need the following lemma.

Lemma 12.2   *If p is an odd prime factor of n and $b \geq 1$, then there exists an integer r such that $(r, n) = 1$ and r is a primitive root $\pmod{p^b}$.*

PROOF: Let g be a primitive root $\pmod{p^b}$ such that $0 < g < p^b$. If $(g, n) = 1$, then $r = g$ and we are done. Otherwise, let $t_j = g + jp^b$, where $j = 1, 2, 3, \cdots$. Since $(g, p) = 1$, Dirichlet's theorem (Theorem 3.12) implies that there are infinitely many j such that t_j is prime. Pick the least j such that t_j is prime and t_j exceeds all the prime factors of n. Then $(t_j, n) = 1$. Also, t_j is a primitive root $\pmod{p^b}$, so let $r = t_j$.   ■

For example, let $n = 900 = 2^2 3^2 5^2$, $p = 3$, and $b = 2$. Now $g = 2$ is a primitive root $\pmod{3^2}$, but $(g, n) = (2, 900) = 2 \neq 1$. However, $t_1 = g + p^2 = 2 + 3^2 = 11$, which is prime, and $11 > 5$, so $r = 11$.

Theorem 12.6   *Let n be the product of distinct odd primes p_i. If n is a Carmichael number, then $(p_i - 1) \mid (n - 1)$ for each index i.*

PROOF: For each index i, Lemma 11.2 implies that there exists an integer r_i such that $(r_i, n) = 1$ and r_i is a primitive root $\pmod{p_i}$. By hypothesis, $r_i^{n-1} \equiv 1 \pmod{n}$; hence $r_i^{n-1} \equiv 1 \pmod{p_i}$. Now Theorem 6.1 implies that $(p_i - 1) \mid (n - 1)$.   ■

Theorem 12.7   *Every Carmichael number is square-free.*

PROOF: Suppose that n is Carmichael number that is not square-free, so $n = p^b m$, where p is prime, $b \geq 2$, and $p \nmid m$. If $p = 2$, let q be a prime such that $q > m$ and $q \equiv 3 \pmod{2^b}$. (Dirichlet's theorem guarantees the existence of such a q.) Therefore $(q, n) = 1$, so hypothesis implies that $q^{n-1} \equiv 1 \pmod{n}$, from which it follows that $q^{2^b m - 1} \equiv 1 \pmod{2^b}$. Since q is odd, Euler's theorem implies that $q^{2^{b-1}} \equiv 1 \pmod{2^b}$. Since $2^{b-1} \mid 2^b m$, it follows that $q^{2^b m} \equiv 1 \pmod{2^b}$, so $q \equiv 1 \pmod{2^b}$; that is, $3 \equiv 1 \pmod{2^b}$, an impossibility.

If p is odd, then Lemma 12.2 implies that there exists r such that $(r, n) = 1$ and r is a primitive root $\pmod{p^2}$. Let $n = sp^2$. Then hypothesis implies that $r^{sp^2 - 1} \equiv 1 \pmod{n}$, so $r^{sp^2 - 1} \equiv 1 \pmod{p^2}$. Now Theorem 6.1 implies that $\phi(p^2) \mid (sp^2 - 1)$; that is, $p(p - 1) \mid (sp^2 - 1)$, an impossibility.   ∎

Theorem 12.8   *Every Carmichael number is odd.*

PROOF: Exercise

Theorem 12.9   *Every Carmichael number has at least three distinct prime factors.*

PROOF: Exercise

We can now state the following:

Theorem 12.10   *Every Carmichael number n is a product of three or more distinct odd prime factors. Furthermore, if p is an odd prime, then $p \mid n$ if and only if $(p - 1) \mid (n - 1)$.*

PROOF: This follows from Theorems 12.5 through 12.9.   ∎

For example, 561 is a Carmichael number because $561 = 3 \cdot 11 \cdot 17$, and 560 is divisible by each of 2, 10, and 16. Also, 8911 is a Carmichael number because $8911 = 7 \cdot 19 \cdot 67$, and 8910 is divisible by each of 6, 18, and 66.

REMARKS: Among all the integers below 10^{10}, only 1547 are Carmichael numbers. It is suspected, however, that there exist infinitely many Carmichael numbers.

SECTION 12.2A EXERCISES

1. Prove Theorem 12.1.

2. Prove Theorem 12.2.

3. If $f(x)$ is a polynomial of degree n, prove that $f(x)$ is $O(x^n)$.

4. Prove that $\pi(x)$, the prime counting function, is $O(x/\log x)$.

5. Use Fermat's little theorem to prove that 645 is composite.

6. Prove that if each of $6m + 1$, $12m + 1$, and $18m + 1$ is prime, then their product is a Carmichael number.

7. Prove that if p and q are primes such that $3 < p < q$ and $3pq$ is a Carmichael number, then $p = 11$ and $q = 17$.

8. Prove Theorem 12.8.

9. Prove Theorem 12.9.

10. Find all Carmichael numbers, if any, of the form $5pq$, where p and q are primes such that $5 < p < q$.

SECTION 12.2A COMPUTER EXERCISES

11. Write a computer program to find the smallest pseudoprime to base 3 that is not a Carmichael number.

12. Write a computer program that, given an odd prime p, finds all Carmichael numbers, if any, of the form pqr, where q and r are primes such that $p < q < r$.

13. Write a computer program to find all Carmichael numbers below 10^5 that have four distinct prime factors.

12.2B MILLER'S TEST AND STRONG PSEUDOPRIMES

Let n be a large odd integer. We have seen that we may use Fermat's little theorem to prove that n is composite, unless n is a pseudoprime or a Carmichael number. Miller's test, which we describe below, was devised to cope with the obstacle posed by pseudoprimes and Carmichael numbers.

Suppose that x is an integer such that $x^2 \equiv 1 \pmod{n}$. If n is prime, then Theorem 4.5 implies that $x \equiv \pm 1 \pmod{n}$. Therefore, if $x^2 \equiv 1 \pmod{n}$ and $x \not\equiv \pm 1 \pmod{n}$, we may conclude that n is composite.

In particular, if b is an integer such that $(b, n) = 1$ and $b^{n-1} \equiv 1 \pmod{n}$ but $b^{\frac{1}{2}(n-1)} \not\equiv \pm 1 \pmod{n}$, then n is composite.

For example, let $n = 561$ and $b = 5$. now $(5, 561) = 1$ and $5^{560} \equiv 1 \pmod{561}$, but $5^{280} \equiv 67 \not\equiv \pm 1 \pmod{561}$, so 561 is composite.

More generally, let n be odd, and let b be an integer such that $(b, n) = 1$ and $b^{n-1} \equiv 1 \pmod{n}$. Let $n - 1 = 2^k m$, where $k \geq 1$ and m is odd. Let j be the least nonnegative integer such that $b^{2^j} \equiv 1 \pmod{n}$. (By hypothesis, j exists and $j \leq k$.) If n is prime and $j > 0$, then Theorem 4.5 implies that $b^{2^{j-1}} \equiv -1 \pmod{n}$. Therefore, if $j > 0$ and $b^{2^{j-1}} \not\equiv -1 \pmod{n}$, we know that n is composite. These considerations lead to what is known as *Miller's test*.

Definition 12.6　　***Miller's Test***

Let $n - 1 = 2^k m$, where $k \geq 1$ and m is odd. Let b be an integer such that $(b, n) = 1$. We say that ***n passes Miller's test to base b*** if either

(i)
$$b^m \equiv \pm 1 \pmod{n}$$

or

(ii)
$$b^{2^j m} \equiv -1 \pmod{n}$$

for some j such that $1 \leq j \leq k - 1$.

The following theorem helps to establish the utility of Miller's test.

Theorem 12.11　*If p is an odd prime and $(b, p) = 1$, then p passes Miller's test to base b.*

PROOF:　Let $p - 1 = 2^k m$, where $k \geq 1$ and m is odd. Let $(b, p) = 1$. It suffices to show that either (i) $b^m \equiv 1 \pmod{p}$ or (ii) $b^{2^j m} \equiv -1 \pmod{p}$ for some j such that $0 \leq j \leq k - 1$. Let h belong to $b \pmod{p}$. By Theorems 4.4 and 6.1, we have $h \mid (p - 1)$. If h is odd, then $h \mid m$, so $m = ht$ for some t. Now $b^m = b^{ht} = (b^h)^t \equiv 1^t \equiv 1 \pmod{p}$. If h is even, let $h = 2^{j+1} r$, where $j \geq 0$ and r is odd. Since $h \mid (p - 1)$, we have $2^{j+1} r \mid 2^k m$, which implies that $j \leq k - 1$ and $r \mid m$. Since $b^{2^{j+1} r} \equiv 1 \pmod{p}$, it follows from Theorem 4.5 that $b^{2^j r} \equiv \pm 1 \pmod{p}$. Since $2^j r = \frac{1}{2} h < h$, it follows from the definition of h that $b^{2^j r} \not\equiv 1 \pmod{p}$. Therefore, $b^{2^j r} \equiv -1 \pmod{p}$. Now $m = rt$, with t odd, so $b^{2^j m} = (b^{2^j r})^t \equiv (-1)^t \equiv -1 \pmod{\text{p}}$.　∎

As a consequence of Theorem 12.11, if $(b, n) = 1$ and n fails Miller's test to base b, then we know that n is composite.

For example, let $n = 6601$. Then $n - 1 = 6600 = 2^3 \cdot 825$. Now $2^{825} \equiv 2738 \pmod{6601}$; $2^{1650} \equiv 4509 \pmod{6601}$ and $2^{3300} \equiv 1 \pmod{6601}$. Therefore, 6601 fails Miller's test to base 2, so 6601 is composite.

If n passes Miller's test to base b, then we suspect that n is prime. We cannot be certain of this, since some composite numbers pass Miller's test, for example, $2047 - 1 - 2 \cdot 1023$, and $2^{1023} \equiv 1 \pmod{2047}$. Therefore, 2047 passes Miller's test to base 2, yet $2047 = 23 \cdot 89$, so 2047 is composite. This leads to the following definition.

Definition 12.7   If n is composite, yet n passes Miller's test to base b, then we say n is a ***strong pseudoprime to base b***, and we write n is SPSP(b).

The example immediately preceding Definition 12.7 shows that 2047 is a strong pseudoprime to base 2. Although strong pseudoprimes are rare, nevertheless we shall demonstrate that there are infinitely many strong pseudoprimes to base 2.

Theorem 12.12   *There are infinitely many strong pseudoprimes for base* 2.

PROOF: Since there are infinitely many PSP(2)'s, it suffices to show that if n is a PSP(2) and $m = 2^n - 1$, then m is a SPSP(2). By hypothesis, n is odd and composite; consequently, so is m. By hypothesis, $2^{n-1} \equiv 1 \pmod{n}$, so $2^{n-1} - 1 = kn$ for some odd k. Now $m - 1 = 2^n - 2 = 2(2^{n-1} - 1) = 2kn$. Since $m = 2^n - 1$, we have $2^n \equiv 1 \pmod{m}$. Therefore, $2^{\frac{1}{2}(m-1)} = 2^{kn} = (2^n)^k \equiv 1^k \equiv 1 \pmod{m}$. Therefore, m is a SPSP(2). ■

A Carmichael number may be considered as a universal pseudoprime. There is no analogy for strong pseudoprimes. It can be shown (see Rosen) that if n is odd and composite, then n passes Miller's test for at most $\frac{1}{4}(n - 1)$ bases b with $0 < b < n$. As a result, if n passes Miller's test for more than $\frac{1}{4}(n - 1)$ different such bases, then n must be prime. Such a primality test would be computationally infeasible, even worse than trial division. Still, something may be salvaged here, providing that one is willing to sacrifice certainty for computational feasibility.

Theorem 12.13   *(Rabin's Probabilistic Primality Test)*
Let n be an odd positive integer. Pick k distinct bases b such that $0 < b < n$ and $(b, n) = 1$. Perform Miller's test for each base. If n is composite, then the probability than n passes all k tests is less than $(\frac{1}{4})^k$.

PROOF: If n is composite and p is the probability that n passes Miller's test for base b, where b is a randomly chosen integer such that $0 < b < n$ and $(b, n) = 1$, then $p \leq (n - 1)/4n < \frac{1}{4}$. If the k different bases are chosen independently, then the probability that n passes all k tests is $p^k < (\frac{1}{4})^k$.

For example, given n, let us pick 30 distinct integers b at random such that $0 < b < n$ and $(b, n) = 1$. Let us perform Miller's test for each of these 30 bases. If n is composite, then the probability that n passes all 30 tests is less than 10^{-18}.

If we suspect that n is prime, and if we are able to obtain at least a partial factorization of $n - 1$, then we may be able to prove that n is prime by means of Pocklington's theorem, which follows.

Theorem 12.14   *(Pocklington's Theorem)*
Let $s \mid (n - 1)$. Let p be a prime divisor of n. If there exists c such that $c^{n-1} \equiv 1 \pmod{n}$ and $(c^{(n-1)/q} - 1, n) = 1$ for all primes q such that $q \mid s$, then $p \equiv 1 \pmod{s}$. In particular, if $s \geq \sqrt{n}$, then n is prime.

PROOF: Let p be a prime divisor of n. Let $b \equiv c^{(n-1)/s} \pmod{n}$. Then $b^s \equiv c^{n-1} \equiv 1 \pmod{n}$, so $b^s \equiv 1 \pmod{p}$. Let b belong to $h \pmod{p}$. Then $h \mid s$ and $h \mid (p - 1)$. We will show that, in fact, $h = s$. Now $b^{s/q} \equiv c^{(n-1)/q} \pmod{p}$. By hypothesis $(c^{(n-1)/q} - 1, n) = 1$, so $(c^{(n-1)/q} - 1, p) = 1$. Therefore, $c^{(n-1)/q} \not\equiv 1 \pmod{p}$, so $b^{s/q} \not\equiv 1 \pmod{p}$. Therefore, $h \nmid s/q$ for all primes q such that $q \mid s$. This implies that $h = s$, so $s \mid (p - 1)$; that is, $p \equiv 1 \pmod{s}$. If also $s \geq \sqrt{n}$, then $p > \sqrt{n}$, so $p = n$.   ■

For example, let $n = 3851$. Then $n - 1 = 3850 = 2 \cdot 5^2 \cdot 7 \cdot 11$. Let $c = 2$. Now $2^{3850} \equiv 1 \pmod{3851}$. Let $s = 77 = 7 \cdot 11 > \sqrt{3851}$. Now $2^{3850/7} = 2^{550} \equiv 1192 \pmod{3851}$, so $(2^{550} - 1, 3851) = (1191, 3851) = 1$. Also, $2^{3850/11} = 2^{350} \equiv 9 \pmod{3851}$, so $(2^{350} - 1, 3851) = (8, 3851) = 1$. Therefore, 3851 is prime.

To prove that n is prime using Pocklington's theorem, we may proceed as follows: Randomly select integers c such that $0 < c < n$ and $(c, n) = 1$ until one is found that satisfies the hypothesis of Pocklington's theorem. If such a c is found, then n is prime. If numerous attempts fail to find such a c, then we suspect that n is composite.

Note that this method requires that we obtain at least a partial factorization of $n - 1$.

SECTION 12.2B EXERCISES

14. Prove that if n is a SPSP(b), then n is a PSP(b).

15. Verify that none of the following is a SPSP(2).

   (a) 561   (b) 1105

   (c) 2465   (d) 1729

   (e) 8911

16. Determine whether 15,841 is a strong pseudoprime (a) for base 2 and (b) for base 3.

17. Use Pocklington's theorem to verify the primality of each of the following.

   (a) 997   (b) 5003

   (c) 10,007   (d) 10,009

   (e) 99,991   (f) 100,003

   (g) 166,667   (h) 1,000,003

18. Prove that if p is a prime such that $2^p - 1$ is composite, then $2^p - 1$ is a SPSP(2).

SECTION 12.2B COMPUTER EXERCISE

19. Write a computer program to test the primality of $\frac{1}{2}(3^{16} + 1)$ using Pocklington's theorem.

12.3A FACTORING: FERMAT'S METHOD AND THE CONTINUED FRACTION METHOD

Let n be an odd composite positive integer. We know that n is a square if and only if $n = [n^{1/2}]^2$. If n is not a square, then $n = pq$, where $p > q > 1$. We seek to factor n, that is, to determine the values p and q such that $n = pq$. The case of greatest interest is where p and q are primes. We shall consider several methods of factorization, namely, trial division, Fermat's method, the continued fraction method, the quadratic sieve method, and the Pollard $p - 1$ method.

Trial Division

As we saw earlier (see p. 254), the number of bit operations needed to factor n using trial division is $O(\sqrt{n} (\log_2 n)^2)$. Therefore, trial division is computationally infeasible for large n. Nevertheless, trial division is a practical method for obtaining prime factors of n that are less than a suitable bound, such as $(\log n)^2$.

Fermat's Method

Let $n = pq$, where p and q are odd primes such that $p > q$. Then also $n = x^2 - y^2$, where $x = \frac{1}{2}(p + q)$ and $y = \frac{1}{2}(p - q)$. To factor n, we seek a positive integer y such that $n + y^2$ is a square. We therefore compute $n + y^2$ for $y = 1, 2, 3$, etc. We succeed when we try $y = \frac{1}{2}(p - q)$, but not sooner. Then $p = x + y$ and $q = x - y$.

For example, let us use Fermat's method to factor 377.

$$377 + 1^2 = 378 \neq x^2$$
$$377 + 2^2 = 381 \neq x^2$$
$$377 + 3^2 = 386 \neq x^2$$
$$377 + 4^2 = 393 \neq x^2$$
$$377 + 5^2 = 402 \neq x^2$$
$$377 + 6^2 = 413 \neq x^2$$
$$377 + 7^2 = 426 \neq x^2$$
$$377 + 8^2 = 441 = 21^2$$

Therefore, $x = 21$ and $y = 8$, so $p = 21 + 8 = 29$, and $q = 21 - 8 = 13$. Our result is $377 = 29 \cdot 13$.

Note that the number of iterations needed to factor by Fermat's method is $\frac{1}{2}(p - q)$, which is $O(p)$, thus greater than $O(\sqrt{n})$. Therefore, unless $p - q$ is small, Fermat's method is computationally infeasible.

Continued Fraction Method

Fermat's method of factoring an odd integer n consists of finding integers x and y such that $n = x^2 - y^2$. Instead, let us try to find integers x and y such that $n \mid (x^2 - y^2)$, that is, $x^2 \equiv y^2 \pmod{n}$, or $x^2 - y^2 = mn$ for some positive integer m. Without loss of generality, we may specify that $0 < y < x < n$. Once we obtain such a pair of integers x and y, we compute $d = (x - y, n)$. If d is a nontrivial divisor of n, that is, $1 < d < n$, then $d = p$ or q, and $n/d = q$ or p. We have therefore succeeded in factoring n.

Let p_k/q_k denote the kth convergent in the continued fraction expansion of $\sqrt{n}$. Recall that in order to obtain p_k and q_k, we need to compute b_k, r_k, and a_k (see Theorem 10.25). Furthermore, recall the identity

$$p_{k-1}^2 - nq_{k-1}^2 = (-1)^k r_k$$

If we can find an *even* k such that $r_k = y^2$, then we have

$$p_{k-1}^2 - nq_{k-1}^2 = y^2$$

or $n \mid (p_{k-1}^2 - y^2)$. We then compute $d = (p_{k-1} - y, n)$, hoping to find a nontrivial factor of n. [It suffices to compute the $p_k \pmod{n}$.]

For example, let us factor 481 by the continued fraction method. We list our results in computing the continued fraction expansion of $\sqrt{481}$ as follows:

k	b_k	r_k	a_k	$p_k \pmod{481}$
-2				0
-1				1
0	0	1	21	21
1	21	40	1	22
2	19	3	13	307
3	20	27	1	329
4	7	16		

Since $r_4 = 16 = 4^2$, w compute $d = (p_3 - 4, n) = (329 - 4, 481) = 13$. Therefore, $n/d = 481/13 = 37$. Our result is $481 = 13 \cdot 37$.

It may occur that $r_k = y^2$ for even k, yet $d = (p_{k-1} - y, n)$ is a trivial factor of n. For example, if $p_{k-1} \equiv y \pmod{n}$, then $d = (0, n) = n$. Should this situation arise, one option is to proceed further with the continued expansion of n, hoping to find a larger even k such that $r_k = y^2$ and $p_{k-1} \not\equiv \pm y \pmod{n}$.

For example, let us factor 161 by the continued fraction method. Listing our results in tabular form, we have

k	b_k	r_k	a_k	$p_k \pmod{161}$
-2				0
-1				1
0	0	1	12	12
1	12	17	1	13
2	5	8	2	38
3	11	5	4	4
4	9	16	1	42
5	7	7	2	88
6	7	16		

Now $r_4 = 16 = 4^2 = p_3^2$, so $(p_3 - 4, 161) = (4 - 4, 161) = (0, 161) = 161$. If we continue, we find that $r_6 = 4^2$ and $(p_5 - 4, 161) = (88 - 4, 161) = (84, 161) = 7$. Therefore, $161 = 7 \cdot 23$.

If numerous iterations in the continued fraction expansion of $\sqrt{n}$ fail to yield nontrivial factors of n, it may pay to start over by computing the continued fraction expansion of $\sqrt{jn}$, where $j = 2, 6, 30$, etc. In general, j is the product of the first few primes.) This variation of the continued fraction method corresponds to finding integers x and y such that $x^2 \equiv y^2 \pmod{jn}$ and then computing $(x - y, jn)$, hoping to obtain a factor of n.

For example, let us try to factor 66043 by the continued fraction method. We have

k	b_k	r_k	a_k	p_k (mod 66043)	Comments
-2				0	
-1				1	
0	0	1	256	256	
1	256	507	1	257	
2	251	6	84	21,844	
3	253	339	1	22,101	
4	86	173	1	43,945	
5	87	338	1	3	
6	251	9	56	44,113	$r_6 = 9 = p_5^2$
7	253	226	2	22,186	
8	199	117	3	44,628	
9	152	367	1	771	
10	215	54	8	50,796	
11	217	351	1	51,567	
12	134	137	2	21,844	
13	140	339	1	7,368	
14	199	78	5	58,684	
15	191	379	1	9	
16	188	81	5	58,729	$r_{16} = 81 = p_{15}^2$
17	217	234	2	51,424	
18	251	13	39	16,932	
19	256	39	13	7,368	
20	251	78	6	61,140	
21	217	243	1	2,465	
22	26	269	1	63,605	
23	243	26	19	22,186	
24	251	117	4	20,263	
25	217	162	2	62,712	

26	107	337	1	16,932	
27	230	39	12	1,724	
28	238	241	2	20,380	
29	244	27	18	38,349	
30	242	277	1	58,729	
31	35	234	1	31,035	
32	199	113	4	50,783	
33	253	18	28	13	
34	251	169			$r_{34} = 169 = p_{33}^2$

Our efforts so far have been fruitless. Let us shift to the continued fraction expansion of $\sqrt{2 \cdot 66043} = \sqrt{132086}$. We have

k	b_k	r_k	a_k	p_k (mod 132,086)
-2				0
-1				1
0	0	1	363	363
1	363	317	2	727
2	271	185	3	2,544
3	284	278	2	5,815
4	272	209	3	19,989
5	355	29	24	89,293
6	341	545	1	109,282
7	204	166	3	20,881
8	294	275	2	18,958
9	256	242	2	58,797
10	228	331	1	77,755
11	103	367	1	4,466
12	264	170	3	91,153
13	246	421	1	95,619
14	175	241	2	18,219
15	307	157	4	36,409
16	321	185	3	127,446
17	234	418	1	31,769
18	184	235	2	58,898
19	286	214	3	76,377
20	356	25		

Now $(p_{19} - 5, 132086) = (76,377 - 5, 132086) = (76372, 132086) = 313$, so $66043 = 313 \cdot 211$.

SECTION 12.3A EXERCISES

20. Use Fermat's method to factor each of the following.

    (a) 299   (b) 629

    (c) 5141   (d) 7493

    (e) 16,549

In each of Exercises 21 through 26, factor the given integer using the continued fraction method.

21. 10,511

22. 17,711

23. 32,639

24. 121,393

25. 141,191

26. 196,423

SECTION 12.3A COMPUTER EXERCISE

27. Write a computer program to factor composite integers using the continued fraction method.

12.3B   FACTORING: THE QUADRATIC SIEVE METHOD

As in the continued fraction method, given an odd composite integer n, we seek positive integers x and y such that

12.1
$$x^2 \equiv y^2 \pmod{n}$$

Once we find x and y, we compute $d = (x - y, n)$. If d is nontrivial, that is, if $1 < d < n$, then we have found a factor of n, namely, d. In the quadratic sieve method, congruence 12.1 is obtained by multiplying together several preliminary congruences, as we shall explain.

We begin by choosing a factor base F whose k elements are 2 and the first $k - 1$ primes p such that the Legendre symbol $(n/p) = 1$. It has been suggested that the optimal k for given n is approximately $\sqrt{L(n)}$, where $L(n) = \exp(\sqrt{\log n \log \log n})$. Here $\exp(t) = e^t$. Now define the polynomial $g(x) = x^2 - n$. Therefore, $x^2 \equiv g(x) \pmod{n}$. Let $x_i = i + [\sqrt{n}]$, where $i = 1, 2, 3$, etc. We are interested in finding values x_i such that all prime factors of $g(x_i)$ belong to the factor base F. For each of these x_i, we obtain a congruence of type

12.2 $$x_i^2 \equiv y_i \pmod{n}$$

where

12.3 $$y_i = \prod_{j=1}^{k} p_{ij}^{a_{ij}},$$

each prime factor p_{ij} is in F, and each exponent $a_{ij} \geq 0$. Suppose we find r such congruences, where

12.4 $$\prod_{i=1}^{r} y_i = y^2$$

Then, letting $x = \prod_{i=1}^{r} x_i$, we obtain the desired congruence (12.1). Note that for a particular value of i, $g(x_i)$ may well fail to factor completely in F. If so, we discard this value.

Each congruence of type 12.2 gives rise to a k-dimensional vector:

12.5 $$v_i = \langle v_{i1}, v_{i2}, v_{i3}, \cdots, v_{ik} \rangle$$

where each component v_{ij} is the least positive residue of the exponent $a_{ij} \pmod 2$. Therefore, $v_{ij} = 0$ or 1 for all i and j.

Suppose we generate m congruences of type 12.2, where $m > k$. Then certainly $m - k$ dependency relations exist among the corresponding m vectors of type 12.5 [Vector addition is performed (mod 2).] A typical dependency relation might be $v_2 + v_3 + v_7 = 0$, where 0 denotes the k-dimensional zero vector. For each subset of dependent vectors, multiplying together the corresponding congruences of type 12.2 leads to the desired congruence of type 12.1.

The congruence of type 12.2 that we have obtained may yield a trivial d, that is, $d = 1$ or n. It can be shown that the probability of such an event is at most $\frac{1}{2}$. If this occurs, then we must generate an additional congruence of type 12.1 via other dependency relations among our vectors. If we let $m = k + 10$, then the probability that all 10 congruences of type 12.1 that we obtain will yield a trivial d is at most 2^{-10}. In other words, we are better than 99.9 percent certain of obtaining a nontrivial factor of n.

The following remarks are pertinent:

(1) We exclude from the factor base all odd primes p such that $(n/p) = -1$, because for such primes, $x_i^2 \not\equiv n \pmod{p}$, so $p \nmid g(x_i)$.

(2) A dependency relation among the vectors is certain to arise once we have at least $k + 1$ vectors but may arise among fewer vectors.

(3) It is useful to note that if $p \mid g(x_i)$, then $p \mid g(x_i + jp)$ for all integers j.

For example, let us use the quadratic sieve method to factor $n = 28,421$. Note that $[\sqrt{n}] = 168$. The criteria given above indicate that the factor base F should contain 11 primes: 2 and the first 10 odd primes p such that $(28,421/p) = 1$. This yields

$$F = \{2, 5, 7, 13, 19, 23, 29, 31, 41, 43, 53\}$$

In Table 12.1 below, for each index i such that $1 \le i \le 18$, we list $i, x_i = i + 168, g(x_i) = x_i^2 - 28,421$, its factorization over F (where possible), and the corresponding vector.

Table 12.1.

i	x_i	$x_i^2 - 28421$	Vector
1	169	$140 = 2^2 \cdot 5 \cdot 7$	$\langle 0, 1, 1, 0, 0, 0, 0, 0, 0, 0, 0 \rangle = v_1$
2	170	479	None
3	171	$820 = 2^2 \cdot 5 \cdot 41$	$\langle 0, 1, 0, 0, 0, 0, 0, 0, 1, 0, 0 \rangle = v_2$
4	172	1163	None
5	173	$1508 = 2^2 \cdot 13 \cdot 29$	$\langle 0, 0, 0, 1, 0, 0, 1, 0, 0, 0, 0 \rangle = v_3$
6	174	$1855 = 5 \cdot 7 \cdot 53$	$\langle 0, 1, 1, 0, 0, 0, 0, 0, 0, 0, 1 \rangle = v_4$
7	175	$2204 = 2^2 \cdot 19 \cdot 29$	$\langle 0, 0, 0, 0, 1, 0, 1, 0, 0, 0, 0 \rangle = v_5$
8	176	$2555 = 5 \cdot 7 \cdot 73$	None
9	177	$2908 = 2^2 \cdot 727$	None
10	178	$3263 = 13 \cdot 251$	None
11	179	$3620 = 2^2 \cdot 5 \cdot 181$	None
12	180	$3979 = 23 \cdot 173$	None
13	181	$4340 = 2^2 \cdot 5 \cdot 7 \cdot 31$	$\langle 0, 1, 1, 0, 0, 0, 0, 1, 0, 0, 0 \rangle = v_6$
14	182	4703	None
15	183	$5068 = 2^2 \cdot 7 \cdot 181$	None
16	184	$5435 = 5 \cdot 1087$	None
17	185	$5804 = 2^2 \cdot 1451$	None
18	186	$6175 = 5^2 \cdot 13 \cdot 19$	$\langle 0, 0, 0, 1, 1, 0, 0, 0, 0, 0, 0 \rangle = v_7$

At this point, although we have generated only 7 (not 12) vectors, we may notice the dependency relation: $v_3 + v_5 + v_7 = 0$. The corresponding congruences of type 12.2 are

$$173^2 \equiv 2^2 \cdot 13 \cdot 29 \pmod{n}$$
$$175^2 \equiv 2^2 \cdot 19 \cdot 29 \pmod{n}$$
$$186^2 \equiv 5^2 \cdot 13 \cdot 19 \pmod{n}$$

Multiplying them together, we get

$$(173 \cdot 175 \cdot 186)^2 \equiv (2^2 \cdot 5 \cdot 13 \cdot 19 \cdot 29)^2 \pmod{n}$$

that is,

$$3792^2 \equiv 1155^2 \pmod{n}$$

Finally, $d = (3792 - 1155, n) = (2637, 28421) = 293$, and $28421 = 293 \cdot 97$.

SECTION 12.3B EXERCISES

Determine which of the following integers are composite. Factor each composite integer using the quadratic sieve method.

28. 30,941

29. 31,621

30. 64,079

31. 838,861

32. 1,149,847

33. 1,016,801

34. 1,346,269

12.3C   FACTORING: THE POLLARD $P - 1$ METHOD

Let p be a prime factor of the odd composite integer N. let Q be an integer such that $(p - 1) \mid Q$. Then Fermat's little theorem implies that $p \mid (2^Q - 1)$. Let $d = (2^Q - 1, N)$. Therefore, $p \mid d$. If also $N \nmid (2^Q - 1)$, then $1 < d < N$, so we have obtained a nontrivial factor of N.

We can obtain such a Q as follows: If k is a natural number, let $M(k)$ be the least common multiple of all integers from 1 to k; that is, $M(k) = [1, 2, 3, \cdots, k]$. For sufficiently large k, such as $k = p - 1$, we know that $(p - 1) \mid M(k)$, so we can let $Q = M(k)$. In order for this method of factorization to work efficiently, it is necessary that the least k such that $(p - 1) \mid M(k)$ be small. This, in turn, requires that all prime factors of

$p - 1$ be small. Incidentally, if k is not a prime or a power of a prime, then $M(k) = M(k - 1)$. Therefore, it suffices to generate $M(k)$ only for primes and their powers.

We begin by listing all primes and prime powers not exceeding a preassigned bound B. In Table 12.2, for $B = 16$, we list the following: in the top row, the natural numbers from 1 to 10; in the middle row, the first 10 primes or prime powers; and in the bottom row, the corresponding primes, denoted p_n.

Table 12.2.

n	1	2	3	4	5	6	7	8	9	10
p_k	2	3	4	5	7	8	9	11	13	16
p_n	2	3	2	5	7	2	3	11	13	2

Given N, we start by letting $b_1 = 2$. For $n \geq 2$, we let $b_n \equiv b_{n-1}^{p_{n-1}}$ (mod N). Then we compute $d_n = (b_n - 1, N)$. If $1 < d_n < N$, then we have found a nontrivial factor of N. Otherwise, we compute b_{n+1}, d_{n+1}, etc.

For example, let us factor $N = 10,001$ by the $p - 1$ method. We list the results of our computations in Table 12.3. Therefore, $73 \mid 10,001$; indeed, $10,001 = 73 \cdot 137$.

Let N have the prime factor p. If q is a prime factor of $p - 1$, suppose that $q^k \mid (p - 1)$ but $q^{k+1} \nmid (p - 1)$. If for each such q and each such k it is true that $q^k \leq B$, then the $p - 1$ method will succeed in identifying p as a factor of N. In factor, if q^k is the largest prime power factor of $p - 1$, and if q^k corresponds to p_n in Table 12.2 (extended if necessary), then at most $n + 1$ iterations are needed to obtain the factor p.

In the preceding example, $73 - 1 = 72 = 2^3 3^2$. The largest prime power factor of 72 is $3^2 = 9$, which corresponds to $n = 7$ in Table 12.2. Therefore, eight iterations were needed to obtain the factor 73. Note that at most 11 iterations would be necessary to obtain any prime factor p such that $p - 1 = 2^a 3^b 5^c 7^d 11^e 13^f$, where $1 \leq a \leq 4, 0 \leq b \leq 2$, and each of c, d, e, and f is 0 or 1.

Additional methods of factorization exist, which we will mention briefly. These include *Pollard's rho method, the p + 1 method, the elliptic curve method, and the number field sieve.* By virtue of the number field sieve, it was recently possible to factor the ninth Fermat number: $f_9 = 2^{512} + 1$. This 155-digit integer turned out to be the product of three primes, having 7, 49, and 99 digits, respectively.

Table 12.3.

n	p_n	b_n (mod N)	$b_n - 1$ (mod N)	d_n
1	2	2	1	1
2	3	4	3	1
3	2	64	63	1
4	5	4096	4095	1
5	7	8532	8531	1
6	2	5466	5465	1
7	3	4169	4168	1
8	11	3578	3577	73

All methods of factorization currently in use have in common that the factorization of N requires approximately $\exp(\sqrt{\log N \log \log N})$ bit operations. This means that it takes 3.9 hours to factor a 50-digit integer, 104 days for a 75-digit integer, and 74 years for a 200-digit integer. It is not known whether factoring large integers is intrinsically time-consuming. The apparent difficulty in factoring large integers is put to good use in public-key cryptosystems, which are discussed in the concluding chapter of this book.

SECTION 12.3C EXERCISES

35. Extend Table 12.2 to accommodate the bound $B = 30$.

Use the $p - 1$ method to factor each of the following:

36. 30,607

37. 76,201

38. 145,921

39. 175,717

40. 876,437

CHAPTER 13

CRYPTOLOGY

13.1 INTRODUCTION

The ability to transmit confidential information securely is of great importance in diplomatic, military, and commercial affairs. The success of a given enterprise may well require that a transmitted message remain secret to all but the intended recipient. The process of encoding a message in order to protect its secrecy is known as *enciphering*, *encryption*, or *cryptography*. Having received an encoded message, the intended recipient uses a key to decode it. The key must, of course, itself be kept secret.

If an encoded message is sent by an adversary or competitor, it is desirable to be able to decode the message so as to render it comprehensible. The latter process is called *deciphering*, *decryption*, or *cryptanalysis*. The science of *cryptology*, which includes the complementary operations of cryptography and cryptanalysis, relies heavily on the use of numbers and, therefore, on number theory.

The military use of cryptology dates back at least as far as Julius Caesar (100–44 B.C.), to whom are attributed the "Caesar Ciphers," which are discussed below. During the 1580s, France and Spain were at odds. Francois Vieta (1540-1603), a Frenchman, succeeded in deciphering the Spanish code, which employed more than 400 characters. As a result, the French were able for some time to thwart the designs of their Spanish adversaries. Vieta, although a lawyer by profession, was an outstanding mathematician of his time.

At the same time, in England, Mary, Queen of Scots, plotted to overthrow Queen Elizabeth I. Mary's enciphered messages were intercepted and deciphered by the British government. As a consequence, Mary was beheaded in 1587.

In the 1640s, during the English Civil War, the mathematician John Wallis (1616–1703) deciphered encoded messages for the Parliamentarians. In recognition of his service and his abilities, he was named Savilian Professor of Geometry at Oxford in 1649.

In August of 1914, at the outbreak of World War I, the failure of two Russian armies to encipher their communications led to a humiliating defeat at the hands of the Germans at Tannenberg. In 1917 British cryptologists deciphered the Zimmermann telegram, in which the German foreign minister proposed an alliance with Mexico against America. Soon after this was made public, America, which until then had remained neutral, entered the war against Germany.

Before America entered World War II, in 1941, American cryptologists had broken the Japanese code. As a result, it was usually possible for America to monitor Japanese naval movements. The Japanese fleet that attacked Pearl Harbor, commanded by Admiral Isoroku Yamamoto, maintained total radio silence in order to achieve surprise. Just a few months after the catastrophe of Pearl Harbor, the numerically inferior American navy was able to defeat the Japanese in the Battle of the Coral Sea and in the Battle of Midway Island, thanks to continued cryptological access to Japanese communications. A planned Japanese invasion of Australia was thereby averted.

In April of 1943, American forces learned of a planned inspection visit by Admiral Yamamoto to some Japanese-held islands in the South Pacific. Yamamoto's plane, although escorted by six fighters, was intercepted by American aircraft and shot down. It has been estimated that because of American use of cryptology, World War II was shortened by at least a year.

Before presenting the details of cryptology, we introduce some notation. The term *literal plaintext* will denote the original message, expressed in the English language. The term *numerical plaintext* will denote a numerical equivalent of the literal plaintext, obtained as follows: Let f be the bijective function whose domain is the alphabet in its usual order and whose range is the set of integers from 00 to 25. This is defined by Table 13.1.

For example, the literal plaintext MATH corresponds to the numerical plaintext 12001907; also, the numerical plaintext 180402201781924 corresponds to the literal plaintext SECURITY. Note that an m-letter literal plaintext can be converted via Table 13.1 to a $2m$-digit numerical plaintext, and vice versa.

Table 13.1. Converting Letters to Numbers

x	A	B	C	D	E	F	G	H	I	J	K	L	M
$f(x)$	00	01	02	03	04	05	06	07	08	09	10	11	12
x	N	O	P	Q	R	S	T	U	V	W	X	Y	Z
$f(x)$	13	14	15	16	17	18	19	20	21	22	23	24	25

Similarly, the term *literal ciphertext* denotes an encoded message in literal form, whereas the term *numerical ciphertext* denotes an encoded message in numerical form.

13.2 CHARACTER CIPHERS

In a *substitution*, or *character*, *cipher*, each letter in the literal plaintext is transformed to another letter in the literal ciphertext. One type of character cipher is the *linear cipher*. For example, if n is a two-digit unit of numerical plaintext, we might encipher via the key

$$E(n) \equiv an + b \quad (\text{mod } 26)$$

where $1 \leq a \leq 25$, $(a, 26) = 1$, $0 \leq b \leq 25$, and $0 \leq E(n) \leq 25$. A linear cipher with $a = 1$ is known as a Caesar cipher.

Suppose our literal plaintext is the proverb SPEAK SOFTLY BUT CARRY A BIG STICK, rewritten in 4-letter blocks as SPEA KSOF TYLB UTCA RRYA BIGS TICK. The corresponding numerical plaintext is 18140400 10181405 19112401 20190200 17172400 01080618 19080210. Suppose that we encipher using a linear cipher with $a = 7$ and $b = 3$, so that the enciphering key is $E(n) \equiv 7n + 3 \quad (\text{mod } 26)$. Enciphering 18, we have $E(18) \equiv 7(18) + 3 \equiv 129 \equiv 25 \quad (\text{mod } 26)$. Continuing in like fashion, we obtain the numerical ciphertext

$$25230503 \ 21252132 \ 06021510 \ 13061703 \ 18181503$$
$$10071925 \ 06071721$$

which corresponds to the literal ciphertext

ZXFD VZXM GCPK NGRD SSPD KHTZ GHRV

Decryption is achieved by converting the literal ciphertext back to numerical ciphertext and then applying the inverse transformation

$$D(n) \equiv 15n + 7 \quad (\text{mod } 26)$$

For example, to decipher Z, we apply the decryption key to its numerical equivalent, 25:

$$D(25) \equiv 15(25) + 7 \equiv 382 \equiv 18 \quad (\text{mod } 26)$$

Therefore, Z deciphers as the literal equivalent of 18, namely, S.

The decryption key for a linear cipher has the form

$$D(n) \equiv cn + d \quad (\text{mod } 26)$$

where $1 \le c \le 25$, $(c, 26) = 1$, and $0 \le d \le 25$. There are 12 possible values of c and 26 possible values of d; therefore, there are $12 \cdot 26 = 312$ possible decryption keys. Using a computer, one could try all these possibilities until the right one appeared. This method of decryption is called *exhaustive cryptanalysis*.

Another method of decryption is *frequency analysis*. It is claimed that the ranking of the letters of the alphabet by relative frequency in the English language is given by Table 13.2.

Table 13.2. Letters of English Language in Order of Relative Frequency

E T A O I N S R H D L C U M F P G W Y B V K X J Q Z

Assume that the following message has been encoded using a linear cipher: DWALQ OWLYE RTQYF YRUYE LQEIW WALDW LYDEV NYLQJ LWAIN VYWLV NWRVN SQOWR VSRYR VERYK REVSW ROWRO YSFYT SRHSJ YLVUE RTTYT SOEVY TVWVN YBLWB WQSVS WRVNE VEHHM YRELY OLYEV YTYGA EH. We now list the letters of the ciphertext in descending order of frequency:

Y	19	O	6	U	2
V	15	Q	6	G	1
W	15	A	4	K	1
R	14	H	4	M	1
E	13	D	3	C	0
L	12	B	2	P	0
S	9	F	2	X	0
T	7	I	2	Z	0
N	6	J	2		

Apparently, Y deciphers as E, and V or W deciphers as T. Assume that the former is true. We can now find the values of c and d in the decryption key by passing to numerical equivalents. We have $D(24) \equiv 4 \pmod{26}$ and $D(21) \equiv 19 \pmod{26}$. This yields the system of simultaneous linear congruences

$$24c + d \equiv 4 \pmod{26}$$
$$21c + d \equiv 19 \pmod{26}$$

To solve the system, we first subtract, obtaining $3c \equiv -15 \pmod{26}$. Therefore, $c \equiv -5 \equiv 21 \pmod{26}$; also, $d \equiv 4 - 24c \equiv 4 - 24(-5) \equiv 124 \equiv 20 \pmod{26}$. We have succeeded in obtaining the decryption key: $D(n) \equiv 21n + 20 \pmod{26}$. The ciphertext may not be transformed to the following plaintext message: FOURS COREA NDSEV ENYEA RSAGO OURFO REFAT HERSB ROUGH TFORT HONTH ISCON TINEN TANEW NATIO NCONC EIVED INLIB ERTYA NDDED ICATE DTOTH EPROP OSITI ONTHA TALLM ENARE CREAT EDEQU AL. Finally, converting from 5-letter blocks to ordinary English, we have the first sentence of Lincoln's Gettysburg Address.

More generally, a character cipher might consist of a permutation of the alphabet such as given by the following encryption key:

Plaintext: A B C D E F G H I J K L M N O P Q R S T U V W X Y Z
Ciphertext: T H E Q U I C K B R O W N F X J M P S V L A Z Y D G

There are 26!, or about $4 \cdot 10^{26}$, possible decryption keys for a character cipher, so exhaustive cryptanalysis is no longer computationally feasible. Frequency analysis, however, provides an effective means of decrypting character ciphers.

13.3 BLOCK CIPHERS

In a block or polygraph cipher, every n-letter block of literal plaintext is transformed to an n-letter block of literal ciphertext. For example, a *digraph*, or *two-character block cipher*, might be encoded with the following encryption key: $E(P_1 P_2) = C_1 C_2$, where $C_1 \equiv 3P_1 + 4P_2 \pmod{26}$ and $C_2 \equiv 5P_1 + 7P_2 \pmod{26}$. Here P_1 and P_2 denote the numerical equivalents of plaintext letters, while C_1 and C_2 denote the numerical equivalents of ciphertext letters. The plaintext message SPEAK SOFTL YBUTC ARRYA BIGST ICK would be transformed to the ciphertext KNMUY UKBXQ YXGZG PPWUQ MFMAL UNZ. Our encryption key may be represented conveniently in matrix form:

$$\begin{bmatrix} C_1 \\ C_2 \end{bmatrix} \equiv \begin{bmatrix} 3 & 4 \\ 5 & 7 \end{bmatrix} \begin{bmatrix} P_1 \\ P_2 \end{bmatrix} \pmod{26}$$

The matrix that appears, which we designate M, is called the *encryption matrix*. In order to decipher, we need M^{-1}, the inverse of M (mod 26). It can be shown that M^{-1} always exists, provided that $(D, 26) = 1$, where D is the determinant of M. In the preceding case,

$$M = \begin{pmatrix} 3 & 4 \\ 5 & 7 \end{pmatrix} \quad \text{and} \quad M^{-1} = \begin{pmatrix} 7 & -4 \\ -5 & 3 \end{pmatrix} \equiv \begin{pmatrix} 7 & 22 \\ 21 & 3 \end{pmatrix} \quad (\text{mod } 26)$$

In general, let P represent the column vector whose components are P_1 and P_2, while C represents the column vector whose components are C_1 and C_2. Then encryption is performed via the matrix congruence $C \equiv MP$ (mod 26), while decryption is performed via $P \equiv M^{-1}C$ (mod 26). To decipher a digraph, one needs to know the four elements of M^{-1}. One might compute the frequencies of the various digraphs that appear in the ciphertext, place these digraphs in descending order of frequency, and then make use of a digraph table comparable to Table 13.2. In general, the most frequently occurring digraph is TH.

The procedure desired above can be generalized to the n-block cipher (also known as the *Hill cipher*). Again, encryption is performed via the matrix congruence $C \equiv MP$ (mod 26). Here M denotes an n by n matrix with integer entries such that $(\det M, 26) = 1$, while P and C denote n-dimensional column vectors corresponding to blocks of numerical plaintext and ciphertext, respectively. For n sufficiently large, decryption by frequency analysis is no longer feasible.

Although computers play a large role in modern cryptology, this chapter contains no exercises specifically designated as computer exercises. The exercises in this chapter have been written so as to be capable of solution either with or without the use of computers, according to the reader's preference.

SECTION 13.3 EXERCISES

1. The ciphertext NDJGC JBQTG XHJE has been enciphered with a Caesar cipher. Decipher it by exhaustive cryptanalysis.

2. The ciphertext ZKJPN KYGPD AXKWP has been enciphered with a Caesar cipher. Decipher it by exhaustive cryptanalysis.

3. Encipher the plaintext HIT THE ROAD JACK with the liner cipher $E(n) \equiv 5n + 8$ (mod 26).

4. The ciphertext VKYAQ VAKEC has been enciphered with the linear cipher $E(n) \equiv 17n + 10$ (mod 26). Decipher it.

5. Encipher the plaintext WHAT A SURPRISE three times in succession with the linear cipher $E(n) \equiv 3n + 2$ (mod 26).

6. The ciphertext RLOI NRLO PZHP HOIN TVNN YFUO INHZ PNYZ NHLY GYVR SNAO INFA XPHO INTV NNYF URLO INRL OPZH

has been enciphered with a linear cipher. Decipher it using frequency analysis.

7. The ciphertext TGVE DQVA KQGT HLGV AQTY QGDQ VRQV RDQQ IDQG VLRO MAYA GTM has been enciphered with a linear cipher. Decipher it using frequency analysis.

8. If any linear cipher is applied to a given plaintext 312 times in succession, what is the result? Why is this so?

9. A linear cipher $E(n) \equiv an + b \pmod{26}$ leaves N unchanged. What value or values must b have?

10. If a linear cipher interchanges S and E, what does it do to X?

11. Encipher the plaintext PRACTICE MAKES PERFECT using the block cipher $C_1 \equiv 9P_1 +$ $5P_2 \pmod{26}$ and $C_2 \equiv 3P_1 + 4P_2 \pmod{26}$.

12. The ciphertext TX NT QM TW HT YB SK ER AT UT QE has been enciphered with the block cipher $C_1 \equiv 5P_1 + 2P_2 \pmod{26}$ and $C_2 \equiv 7P_1 + 3P_2 \pmod{26}$. Decipher it.

13. It is known that the two most frequently occurring digraphs in the English language are TH and HE. In a ciphertext, the two most frequently occurring digraphs are XY and BT. Assuming that the plaintext was enciphered with a block cipher so that $P_1 \equiv aC_1 + bC_2 \pmod{26}$ and $P_2 \equiv cC_1 + dC_2 \pmod{26}$, determine a, b, c, and d.

13.4 ONE-TIME PAD

Suppose that we wish to encipher an n-character plaintext message. The corresponding numerical plaintext may be considered as a vector $P = \langle P_1, P_2, P_3, \cdots, P_n \rangle$, where each component P_i is a two-digit integer with $0 \leq P_i \leq 25$. Let the encryption vector $K = \langle K_1, K_2, K_3, \cdots, K_n \rangle$, where the components K_i satisfy the same conditions as the P_i. The corresponding numerical ciphertext is a vector $C = \langle C_1, C_2, C_3, \cdots, C_n \rangle$, obtained via $C_i \equiv P_i + K_i \pmod{26}$ for all i.

For example, suppose our literal plaintext message is

$$S \quad W \quad I \quad M \quad M \quad I \quad N \quad G \quad I \quad S \quad F \quad U \quad N$$

Then

$$P = \langle\ 18,\ 22,\ 08,\ 12,\ 12,\ 08,\ 13,\ 06,\ 08,\ 18,\ 05,\ 20,\ 13,\ \rangle$$

Let

$$K = \langle\ 19,\ 04,\ 03,\ 04,\ 06,\ 07,\ 04,\ 00,\ 09,\ 10,\ 02,\ 09,\ 21,\ \rangle$$

Then

$$C = \langle\ 11,\ 00,\ 11,\ 16,\ 18,\ 15,\ 17,\ 06,\ 17,\ 02,\ 07,\ 03,\ 08,\ \rangle$$

This corresponds to the literal ciphertext

L   A   L   Q   S   P   R   G   R   C   H   D   I

On the other hand, suppose our literal plaintext message is

G   O   T   Y   O   U   R   N   U   M   B   E   R

Then

$$P = \langle\ 06,\ 14,\ 19,\ 24,\ 14,\ 20,\ 17,\ 13,\ 20,\ 12,\ 01,\ 04,\ 17,\ \rangle$$

Let

$$K = \langle\ 05,\ 12,\ 18,\ 18,\ 04,\ 21,\ 00,\ 19,\ 23,\ 16,\ 06,\ 25,\ 08,\ \rangle$$

Then

$$C = \langle\ 11,\ 00,\ 11,\ 16,\ 18,\ 15,\ 17,\ 06,\ 17,\ 02,\ 07,\ 03,\ 08,\ \rangle$$

so that again the literal ciphertext is

L   A   L   Q   S   P   R   G   R   C   H   D   I

An unintended recipient without knowledge of the encryption vector *K* has no way of obtaining the plaintext from the cipher text, *provided that K is used only once*. In general, there are 26^n possible values of *K*, so exhaustive cryptanalysis is not feasible. This cryptosystem, known as the *one-time pad*, was invented by Joseph O. Mauborgne of the U.S. Army Signal Corps during World War I. It is an unbreakable code and has been used in the "hot line" from Washington to Moscow.

Nevertheless, the one-time pad has its disadvantages. The key, which must be sent in advance to the intended recipient, must be changed very frequently. Furthermore, since the key is as long as the message, this code is not practical for the transmission of lengthy messages.

13.5   EXPONENTIAL CIPHERS

Let a literal plaintext message be split into *m*-letter blocks, that are converted to $2m$-digit integer blocks of numerical plaintext via Table 13.1. (If necessary, extra letters can e added to the literal plaintext so that the total number of letters is a multiple of *m*.) Choose a prime *p* such that $\frac{25}{99}(10^{2m} - 1) < p < 10^{2m}$ and $\frac{1}{2}(p - 1)$ is prime. Choose an exponent *j* such that $2 \leq j \leq p - 2$ and $(j, p - 1) = 1$. Now transform each $2m$-digit block *P* of numerical plaintext into a block *C* of numerical ciphertext via the congruence

$$C \equiv P^j \quad (\text{mod } p)$$

where $0 < C < p$. The bounds on the prime p ensure that distinct blocks of numerical plaintext are transformed to distinct blocks of numerical ciphertext.

For example, suppose our plaintext message is

BEWARE THE IDES OF MARCH

and we encipher it using an exponential cipher with $m = 2, p = 9887,$ $j = 3$. Then our numerical plaintext is

0104 2200 1704 1907 0408 0304 1814 0512 0017 0207

We encipher via the congruence $C \equiv P^3 \quad (\text{mod } 9887)$. This procedure yields the ciphertext

7633 7497 0367 1798 3509 5497 3312 1703 4913 1104

In order to decipher a block C of ciphertext, we need to know k, the multiplicative inverse of $j \quad (\text{mod } p - 1)$. That is, k is the unique integer such that $2 \leq k \leq p - 2$ and $jk \equiv 1 \quad (\text{mod } p - 1)$. Then $C^k \equiv (P^j)^k \equiv P^{jk} \equiv P \quad (\text{mod } p)$, so we decipher by raising each block of ciphertext to the kth power $(\text{mod } p)$. In the preceding example, $3k \equiv 1 \quad (\text{mod } 9886)$ implies that $k \equiv 6591 \quad (\text{mod } 9886)$, so we decipher via

$$P \equiv C^{6591} \quad (\text{mod } 9886)$$

Even if one block of numerical plaintext P is known, decryption requires solving the congruence $C \equiv P^j \quad (\text{mod } p)$ for j. We call j the logarithm of C to the base $P \quad (\text{mod } p)$. The best algorithms for finding logarithms $(\text{mod } p)$ require approximately $\exp(\sqrt{\log p \log \log p})$ bit operations unless all factors of $p - 1$ are small. We avoid this exception by insisting that $\frac{1}{2}(p - 1)$ be prime. As in the case of factoring a large integer, the number of bit operations required for large p renders the algorithm computationally infeasible.

SECTION 13.5 EXERCISES

14. What one-time pad will encipher the plaintext KEEPTHEFAITH as PLANFARAHEAD?

15. What one-time pad will encipher the plaintext WARTOENDWARS as PLANFARAHEAD?

16. Use an exponential cipher with $m = 2, p = 9973,$ and $j = 5$ to encipher the plaintext GIVE ME LIBERTY OR GIVE ME DEATH.

17. Use an exponential cipher with $m = 2, p = 9949$, and $j = 7$ to encipher the plaintext DAMN THE TORPEDOES FULL SPEED AHEAD.

18. The following ciphertext has been enciphered with an exponential cipher with $m = 2, p = 9967$, and $j = 5$: 9219 2474 7105 2804 6178 7554 2790 4198 1189 9219 3441 2416 9565. Decipher it.

19. The ciphertext 28 08 15 20 31 14 03 21 05 19 has been enciphered with an exponential cipher with $m = 1$ and $p = 37$. IT is also known that the ciphertext 20 corresponds to the plaintext 19. Decipher it.

13.6 PUBLIC-KEY CRYPTOGRAPHY

Under the various cryptosystems described heretofore, the problem of key management becomes unwieldy if secure communications are required in a network of many users. Each pair of participants needs an enciphering key that is kept secret from all the others. The keys themselves must be transmitted through a secure channel, and they must be changed frequently.

The key-management problem is effectively solved by a new type of cipher system known as the *public-key cryptosystem*, which is a variation of the exponential cipher.

Let $n = pq$, where p and q are distinct, large (over 100 digits), suitably chosen primes. Let j be an integer such that $2 < j < \phi(n)$ and $(j, \phi(n)) = 1$. Let numerical plaintext be split into blocks P of equal length. We obtain a corresponding block C of numerical ciphertext via $C = E(P) \equiv P^j \pmod{n}$, where $0 < C < n$.

Let k be the multiplicative inverse of $j \pmod{\phi(n)}$, that is, $2 < k < \phi(n)$ and $jk \equiv 1 \pmod{\phi(n)}$. We decipher via $P = D(C) \equiv C^k \pmod{n}$. This works because $C^k \equiv (P^j)^k \equiv P^{jk} \equiv P \pmod{n}$.

This enciphering key, namely, the values of n and j, is made *public* and is listed in a directory, while k, p and q are kept secret. An unintended listener needs the value of k in order to decipher a ciphertext message. However, this requires knowing that $\phi(n) = (p - 1)(q - 1)$. Since $p + q = n - \phi(n) + 1$ and $p - q = \sqrt{(p+q)^2 - 4n}$, this is equivalent to knowing p and q. Therefore, in order to decipher, the unintended listener must factor n. For n sufficiently large, this is computationally infeasible.

How does the intended recipient know that a message received originated from an authorized source? This problem is also disposed of in public-key cryptography by what are known as *signatures*. Let participant

Alice have encryption key E_A and decryption key D_A. Similarly, let participant Bob have encryption key E_B and decryption key D_B. Let M be a brief plaintext message that identifies Alice. At the end of an enciphered message from Alice to Bob, Alice transmits $E_B(D_A(M))$. Bob deciphers this message via $E_A(D_B(E_B(D_A(M)))) = E_A(D_A(M)) = M$. Only Bob can decipher $E_B(D_A(M))$, since only Bob possesses D_B. Only Alice can transmit $E_B(D_A(M))$ to Bob, since only Alice possesses D_A. Therefore, Bob can be sure that the author of the message was indeed Alice.

Public-key cryptosystems were first proposed by W. Diffie and R. Hellman in 1975. The public-key cipher described above is known as the *RSA cryptosystem*. The initials are those of its inventors, Ronald Rivest, Adi Shamir, and Leonard Adleman, who published their results in 1977. Since that time, the RSA cryptosystem, whose security is based on the computational infeasibility of factoring large integers, has been under intense scrutiny by numerous investigators. It has survived all attacks and is still in use.

SECTION 13.6 EXERCISES

20. If $n = pq = 19,939$ and $\phi(n) = 19,656$, find p and q.

21. If $n = pq = 63,083$ and $\phi(n) = 62,568$, find p and q.

22. Encipher the plaintext message ENDINSIGHT using RSA with $n = 8633$ and $j = 5$.

23. The ciphertext 7482 6330 4952 4707 0705 has been enciphered using RSA with $n = 8051$ and $j = 5$. Decipher it.

24. Suppose that Alice's RSA enciphering key is $n = 7663$ and $j = 5$, while Bob's RSA enciphering key is $n = 8881$ and $j = 3$. (a) Encipher Alice's signature to Bob, namely, I AM ALICE. (b) Encipher Bob's signature to Alice, namely, BOBS BIG BOY.

APPENDIX A

SOME OPEN QUESTIONS IN ELEMENTARY NUMBER THEORY

1. Is there any n other than $n = 2, 8$ such that 2^n is the sum of distinct powers of 3?

2. Are there infinitely many n such that $105 \mid \binom{2n}{n}$?

3. Are there infinitely many pairs of twin primes; that is, are there infinitely many n such that $2n - 1$ and $2n + 1$ are both prime?

4. Are there infinitely many primes of the form $n^2 + 1$?

5. Are there infinitely many primes of the form $n^2 + n + 1$?

6. Are there infinitely many primes p such that $2p + 1$ is prime?

7. Are there infinitely many prime Fibonacci numbers?

8. Are there infinitely many Mersenne primes; that is, are there infinitely many primes p such that $2^p - 1$ is prime?

9. Is there any $n \geq 5$ such that $2^{2^n} + 1$ (the corresponding Fermat number) is prime?

10. Is it true that for every $n \geq 4$ there exist primes p and q such that $p + q = 2n$ (Goldbach's conjecture)?

11. Is it true for all n that there is a prime p such that $n^2 < p < (n + 1)^2$?

12. Is it true that for every odd prime p the equation $x^p + y^p = z^p$ has no solution in natural numbers (Fermat's conjecture)?

13. Does there exist an odd perfect number, that is, an odd natural number n such that $\sigma(n) = 2n$?

14. Does there exist n such that the equation $\emptyset(x) = n$ has a unique solution?

15. Are there infinitely many primes p such that 2 is a primitive root (mod p)?

16. In the infinite continued fraction representation of $\sqrt[3]{2}$, is it true that the partial quotients are unbounded?

17. Are there infinitely many Carmichael numbers?

APPENDIX B

ANSWERS TO ODD-NUMBERED EXERCISES

Chapter 1

3. See Table 9.1.

17. n = 70

19. F_{n+1}/F_n approaches $\frac{1}{2}(1 + \sqrt{5})$.

21. The sum approaches 2.

23. $2^n(n!)$

25. (a) 10   (c) 126   (e) 66

33. $\frac{n+k}{n}\binom{n}{k} = \binom{n}{k} + \binom{n-1}{k-1}$ if $k \geq 1$

39. F_{2n}

45. F_n

47. F_{n+1}

Chapter 2

7. 20: 1, 2, 4, 5, 10, 20; 25: 1, 5, 25; 30: 1, 2, 3, 5, 6, 10, 15, 20

9. (a) 2   (c) 5   (e) 1   (g) 1   (i) 9

11. (a) 13/20   (c) 7/19   (e) 5/6

23. $(F_m, F_n) - F_{(m,n)} = 0$ for all m, n

25. (a) $[44, 102] = 2244$ (c) $[75, 95] = 1425$ (e) $[29, 123] = 3567$ (g) $[27, 32] = 864$ (i) $[126, 621] = 8694$

31. $(30, 42, 70) = 2$, $[30, 42, 70] = 210$

33. $(296, 444, 555) = 37$, $[296, 444, 555] = 440$

35. $(20, 30\ 60) = 10$, $[20, 30\ 60] = 60$

37. $(90, 120, 135) = 15$, $[90, 120, 135] = 1080$

39. $(45, 81, 87) = 3$, $[45, 81, 87] = 11{,}745$

45. $r_2 = (1 + q_1q_2)a - q_2b; r_3 = -(q_1 + q_3 + q_1q_2q_3)a + (1 + q_2q_3)b$

53. (a) $13_{10} = 1101_2$ (c) $127_{10} = 1111111_2$ (e) $1001_{10} = 1111101001_2$

Chapter 3

5. $10^2 + 1 = 101$

7. 3, 7, 13, 31, 43

11. Estimate: $4.343 \cdot 10^8$; relative error = 4.6 percent

17. 223, 227, 229, 239, 241

19. (a) $13 \cdot 23$ (c) $3 \cdot 7 \cdot 37$ (e) $5 \cdot 29$ (g) $2^3 \cdot 3^2$ (i) $5 \cdot 7 \cdot 11 \cdot 13$ (k) $2^7 \cdot 7$ (m) $2^3 \cdot 13$

35. 249

Chapter 4

3. (b) and (d) are equivalence relations.

5. 26

9. (a) $x = 1$ (c) $x = 3, 4$

21. (a) $x \equiv 9$ (mod 28)   (d) $x \equiv 17$ (mod 280)   (g) $x \equiv 94$ (mod 140)

23. 17, 10, or 3 chickens; 1, 5, or 9 geese

29. $x \equiv 35, 83$ (mod 121)

31. No solutions

33. $x \equiv 3, 6, 7$ (mod 8)

35. $x \equiv 7$ (mod 81)

37. $x \equiv 24$ (mod 49)

43. (b) $1 \cdot 1 \equiv 2 \cdot 6 \equiv 3 \cdot 4 \equiv 5 \cdot 9 \equiv 7 \cdot 8 \equiv 10 \cdot 10 \equiv 1$ (mod 11)

63. 12, 16, 20; 20, 21, 29; 20, 48, 52

71. x 2 3 4   5   6   7
y 3 40 63 312 323 1200
z 5 41 65 313 325 1201

Chapter 5

1. n 40 41 42 43 44 45 46 47 48 49 50
$\tau(n)$ 8 2 8 2 6 6 4 2 10 3 6
$\sigma(n)$ 90 42 96 44 84 78 72 48 124 57 93

13. 28

25. 482

33. $b(p) = -(p + 1); b(p^2) = p; b(p^k) = 0$ for $k \geq 3$

39. (a) 1, 2   (c) 5, 8, 10, 12   (e) 15, 16, 20, 24, 30

43. $\frac{1}{2}m\phi(m)$

Chapter 6

1. t 1 2 3 4 5 6 7 8 9 10 11 12
$o_{13}(t)$ 1 12 3 6 4 12 12 4 3 6 12 2

3. $2^2 \equiv 4$ (mod 19); $2^3 \equiv 8$ (mod 19); $2^6 \equiv 7$ (mod 19); $2^9 \equiv -1$ (mod 19). The primitive roots (mod 19) are 2, 3, 10, 13, 14, and 15.

5. (a) 2, 3   (c) 11, 13   (e) 17, 31   (g) 29, 37, 43

11. No

13. $p \equiv 5$ (mod 6)

19. (a) 1, 3, 9   (c) 1, 7, 8, 11, 12, 18

21. 3, 5, 12, 18, 19, 20, 26, 28, 29, 30, 33, 34

35. $x \equiv 4$ (mod 11)

37. No solution

39. $x \equiv 6$ (mod 17)

41. No solution

Chapter 7

1. $x = 4y + 3; x^2 \equiv 10$ (mod 11)

3. $x = 12y + 1; x^2 \equiv 7$ (mod 17)

5. $x = 16y + 7; x^2 \equiv 18$ (mod 23)

13. (a) $\frac{1}{2}(11 - 1) = 5; 5, 10, 15, 20, 25; 5, \underline{10}, 4, \underline{9}, 3$, so $(5/11) = (-1)^2 = 1$ (c) $\frac{1}{2}(17 - 1) = 8; 6, 12, 18, 24, 30, 36, 42, 48; 6, \underline{12}, 1, 7, \underline{13}, 2, 8, \underline{14}$, so $(6/17) = (-1)^3 = -1$

15. (a) $\left(\frac{3}{17}\right) = \left(\frac{17}{3}\right) = \left(\frac{2}{3}\right) = -1$ (c) $\left(\frac{105}{113}\right) = \left(\frac{3}{113}\right)\left(\frac{5}{113}\right)\left(\frac{7}{113}\right) = \left(\frac{113}{3}\right)\left(\frac{113}{5}\right)\left(\frac{113}{7}\right) = \left(\frac{2}{3}\right)\left(\frac{3}{5}\right)\left(\frac{1}{7}\right) = (-1)\left(\frac{5}{3}\right)1 = -\left(\frac{2}{3}\right) = -(-1) = 1$ (e) $\left(\frac{11}{37}\right) = \left(\frac{37}{11}\right) = \left(\frac{4}{11}\right) = 1$ (g) 1   (i) -1

23. $x \equiv \pm 10$ (mod 37)

25. No solution

31. $x \equiv \pm 105$ (mod 577)

35. $x \equiv \pm 85$ (mod 361)

37. $x \equiv \pm 623$ (mod 4913)

39. $x \equiv \pm 17, \pm 61$ (mod 143)

41. $x \equiv \pm 45, \pm 110$ (mod 403)

45. $\left(\frac{5}{77}\right) = -1$; no solution

47. (a) 1   (c) -1 (e) 1

Chapter 8

1. (a) $233 = 13^2 + 8^2$ (c) $613 = 18^2 + 17^2$ (e) $1409 = 28^2 + 25^2$

3. (a) $65 = 8^2 + 1^2 = 7^2 + 4^2$ (c) $221 = 14^2 + 5^2 = 11^2 + 10^2$

7. $72, 73, 74$

13. $111, 112$

17. $63 = 7^2 + 3^2 + 2^2 + 1^2 = 6^2 + 5^2 + 1^2 + 1^2 = 6^2 + 3^2 + 3^2 + 3^2 = 5^2 + 5^2 + 3^2 + 2^2$
    $71 = 7^2 + 3^2 + 3^2 + 2^2 = 6^2 + 5^2 + 3^2 + 1^2$
    $95 = 9^2 + 3^2 + 2^2 + 1^2 = 7^2 + 6^2 + 3^2 + 1^2 = 6^2 + 5^2 + 5^2 + 3^2$

Chapter 9

57. $u_n = [2^n - (-1)^n]/3$

Chapter 10

1. (a) $100/37 = [2, 1, 2, 2, 1, 3]$ (c) $21/13 = [1, 1, 1, 1, 1, 2]$ (e) $1000/301 = [3, 3, 9, 1, 2, 3]$

11. $\sqrt{41} = [6, 2, 2, 12, 2, 2, 12, \cdots]$; $397/62$

15. $\sqrt{6} = [2, \overline{2, 4}]$; $n = 4$

21. (a) $[1, \overline{2}]$ (c) $[3, \overline{3, 6}]$ (e) $[5, \overline{1, 1, 3, 5, 3, 1, 1, 10}]$ (g) $[9, \overline{1, 2, 3, 1, 1, 5, 1, 8, 1, 5, 1, 1, 3, 2, 1, 18}]$ (i) $[1, \overline{3, 1, 3, 2, 3, 1, 3, 2}]$

37. (a) $\frac{1}{2}(1 + \sqrt{3})$ (c) $\sqrt{3}$ (e) $\frac{1}{4}(3 + \sqrt{37})$ (g) $1 + \frac{1}{4}\sqrt{3}$ (i) $2/(e^2 - 7)$

Chapter 11

5. (a) $(x_1, y_1) = (9, 4)$, $(x_2, y_2) = (161, 72)$, $(x_3, y_3) = (2889, 1292)$ (c) $(x_1, y_1) = (649, 180)$, $(x_2, y_2) = (842{,}401, 233{,}640)$, $(x_3, y_3) = (1{,}093{,}435{,}849, 313{,}264{,}540)$

Chapter 12

5. $7^{644} \equiv 436 \not\equiv 1 \pmod{645}$

15. (a) $2^{140} \equiv 67 \pmod{561}$, $2^{280} \equiv 1 \pmod{561}$ (c) $2^{308} \equiv 1886 \pmod{2465}$, $2^{616} \equiv 1 \pmod{2465}$ (e) $2^{4455} \equiv 6434 \pmod{8911}$, $2^{8910} \equiv 1 \pmod{8911}$

17. (a) $996 = 2^2 \cdot 3 \cdot 83$, $2^{996} \equiv 1 \pmod{997}$, $(2^{12} - 1, 997) = 1$ (c) $10006 = 2 \cdot 5003$, $2^{10{,}006} \equiv 1 \pmod{10{,}007}$, $(2^2 - 1, 10{,}007) = 1$ (e) $99{,}990 = 2 \cdot 3^2 \cdot 5 \cdot 11 \cdot 101$, $2^{99{,}990} \equiv 1 \pmod{99{,}991}$, $(2^{9090} - 1, 99{,}991) = (2^{990} - 1, 99{,}991) = 1$

21. $10{,}511 = 23 \cdot 457$

23. $43{,}691$ is prime.

25. $174{,}763$ is prime.

29. $31{,}621 = 103 \cdot 307$

31. $131{,}071$ is prime.

33. $1{,}149{,}851 = 59 \cdot 19{,}489$

37. $76{,}201 = 181 \cdot 421$

39. $175{,}717 = 199 \cdot 883$

Chapter 13

1. YOUR NUMBER IS UP

3. RVZZR CPAHW BHSG

5. WHAT A SURPRISE

7. NATURE TIME AND PATIENCE ARE THE THREE GREAT PHYSICIANS

9. 0 OR 13

11. MJKI DLMW EKGU DKRC NFJE

13. $a = 7, b = 14, c = 15, d = 11$

15. 19, 11, 09, 20, 17, 22, 04, 23, 11, 04, 09, 11

17. 7048 1518 8150 5578 4880 7201 4031 8375 4241 8039 8991 5538 7048 2874 2187

19. DONT GIVE UP

21. $p = 317, q = 199$

23. HALLELUJAH

APPENDIX C

TABLES

Table 1. Primes below 10,000

2	73	179	283	419	547	661	811	947	1087
3	79	181	293	421	557	673	821	953	1091
5	83	191	307	431	563	677	823	967	1093
7	89	193	311	433	569	683	827	971	1097
11	97	197	313	439	571	691	829	977	1103
13	101	199	317	443	577	701	839	983	1109
17	103	211	331	449	587	709	853	991	1117
19	107	223	337	457	593	719	857	997	1123
23	109	227	347	461	599	727	859	1009	1129
29	113	229	349	463	601	733	863	1013	1151
31	127	233	353	467	607	739	877	1019	1153
37	131	239	359	479	613	743	881	1021	1163
41	137	241	367	487	617	751	883	1031	1171
43	139	251	373	491	619	757	887	1033	1181
47	149	257	379	499	631	761	907	1039	1187
53	151	263	383	503	641	769	911	1049	1193
59	157	269	389	509	643	773	919	1051	1201
61	163	271	397	521	647	787	929	1061	1213
67	167	277	401	523	653	797	937	1063	1217
71	173	281	409	541	659	809	941	1069	1223

1229	1453	1663	1901	2131	2371	2621	2833	3083	3343
1231	1459	1667	1907	2137	2377	2633	2837	3089	3347
1237	1471	1669	1913	2141	2381	2647	2843	3109	3359
1249	1481	1693	1931	2143	2383	2657	2851	3119	3361
1259	1483	1697	1933	2153	2389	2659	2857	3121	3371
1277	1487	1699	1949	2161	2393	2663	2861	3137	3373
1279	1489	1709	1951	2179	2399	2671	2879	3163	3389
1283	1493	1721	1973	2203	2411	2677	2887	3167	3391
1289	1499	1723	1979	2207	2417	2683	2897	3169	3407
1291	1511	1733	1987	2213	2423	2687	2903	3181	3413
1297	1523	1741	1993	2221	2437	2689	2909	3187	3433
1301	1531	1747	1997	2237	2441	2693	2917	3191	3449
1303	1543	1753	1999	2239	2447	2699	2927	3203	3457
1307	1549	1759	2003	2243	2459	2707	2939	3209	3461
1319	1553	1777	2011	2251	2467	2711	2953	3217	3463
1321	1559	1783	2017	2267	2473	2713	2957	3221	3467
1327	1567	1787	2027	2269	2477	2719	2963	3229	3469
1361	1571	1789	2029	2273	2503	2729	2969	3251	3491
1367	1579	1801	2039	2281	2521	2731	2971	3253	3499
1373	1583	1811	2053	2287	2531	2741	2999	3257	3511
1381	1597	1823	2063	2293	2539	2749	3001	3259	3517
1399	1601	1831	2069	2297	2543	2753	3011	3271	3527
1409	1607	1847	2081	2309	2549	2767	3019	3299	3529
1423	1609	1861	2083	2311	2551	2777	3023	3301	3533
1427	1613	1867	2087	2333	2557	2789	3037	3307	3539
1429	1619	1871	2089	2339	2579	2791	3041	3313	3541
1433	1621	1873	2099	2341	2591	2797	3049	3319	3547
1439	1627	1877	2111	2347	2593	2801	3061	3323	3557
1447	1637	1879	2113	2351	2609	2803	3067	3329	3559
1451	1657	1889	2129	2357	2617	2819	3079	3331	3571

3581	3823	4073	4327	4591	4861	5099	5393	5641	5861
3583	3833	4079	4337	4597	4871	5101	5399	5647	5867
3593	3847	4091	4339	4603	4877	5107	5407	5651	5869
3607	3851	4093	4349	4621	4889	5113	5413	5653	5879
3613	3853	4099	4357	4637	4903	5119	5417	5657	5881
3617	3863	4111	4363	4639	4909	5147	5419	5659	5897
3623	3877	4127	4373	4643	4919	5153	5431	5669	5903
3631	3881	4129	4391	4649	4931	5167	5437	5683	5923
3637	3889	4133	4397	4651	4933	5171	5441	5689	5927
3643	3907	4139	4409	4657	4937	5179	5443	5693	5939
3659	3911	4153	4421	4663	4943	5189	5449	5701	5953
3671	3917	4157	4423	4673	4951	5197	5471	5711	5981
3673	3919	4159	4441	4679	4957	5209	5477	5717	5987
3677	3923	4177	4447	4691	4967	5227	5479	5737	6007
3691	3929	4201	4451	4703	4969	5231	5483	5741	6011
3697	3931	4211	4457	4721	4973	5233	5501	5743	6029
3701	3943	4217	4463	4723	4987	5237	5503	5749	6037
3709	3947	4219	4481	4729	4993	5261	5507	5779	6043
3719	3967	4229	4483	4733	4999	5273	5519	5783	6047
3727	3989	4231	4493	4751	5003	5279	5521	5791	6053
3733	4001	4241	4507	4759	5009	5281	5527	5801	6067
3739	4003	4243	4513	4783	5011	5297	5531	5807	6073
3761	4007	4253	4517	4787	5021	5303	5557	5813	6079
3767	4013	4259	4519	4789	5023	5309	5563	5821	6089
3769	4019	4261	4523	4793	5039	5323	5569	5827	6091
3779	4021	4271	4547	4799	5051	5333	5573	5839	6101
3793	4027	4273	4549	4801	5059	5347	5581	5843	6113
3797	4049	4283	4561	4813	5077	5351	5591	5849	6121
3803	4051	4289	4567	4817	5081	5381	5623	5851	6131
3821	4057	4297	4583	4831	5087	5387	5639	5857	6133

6143	6373	6679	6947	7211	7507	7727	8039	8293	8599
6151	6379	6689	6949	7213	7517	7741	8053	8297	8609
6163	6389	6691	6959	7219	7523	7753	8059	8311	8623
6173	6397	6701	6961	7229	7529	7757	8069	8317	8627
6197	6421	6703	6967	7237	7537	7759	8081	8329	8629
6199	6427	6709	6971	7243	7541	7789	8087	8353	8641
6203	6449	6719	6977	7247	7547	7793	8089	8363	8647
6211	6451	6733	6983	7253	7549	7817	8093	8369	8663
6217	6469	6737	6991	7283	7559	7823	8101	8377	8669
6221	6473	6761	6997	7297	7561	7829	8111	8387	8677
6229	6481	6763	7001	7307	7573	7841	8117	8389	8681
6247	6491	6779	7013	7309	7577	7853	8123	8419	8689
6257	6521	6781	7019	7321	7583	7867	8147	8423	8693
6263	6529	6791	7027	7331	7589	7873	8161	8429	8699
6269	6547	6793	7039	7333	7591	7877	8167	8431	8707
6271	6551	6803	7043	7349	7603	7879	8171	8443	8713
6277	6553	6823	7057	7351	7607	7883	8179	8447	8719
6287	6563	6827	7069	7369	7621	7901	8191	8461	8731
6299	6569	6829	7079	7393	7639	7907	8209	8467	8737
6301	6571	6833	7103	7411	7643	7919	8219	8501	8741
6311	6577	6841	7109	7417	7649	7927	8221	8513	8747
6317	6581	6857	7121	7433	7669	7933	8231	8521	8753
6323	6599	6863	7127	7451	7673	7937	8233	8527	8761
6329	6607	6869	7129	7457	7681	7949	8237	8537	8779
6337	6619	6871	7151	7459	7687	7951	8243	8539	8783
6343	6637	6883	7159	7477	7691	7963	8263	8543	8803
6353	6653	6889	7177	7481	7699	7993	8269	8563	8807
6359	6659	6907	7187	7487	7703	8009	8273	8573	8819
6361	6661	6911	7193	7489	7717	8011	8287	8581	8821
6367	6673	6917	7207	7499	7723	8017	8291	8597	8831

8837	9013	9203	9391	9539	9739	9901
8839	9029	9209	9397	9547	9743	9907
8849	9041	9221	9403	9551	9749	9923
8861	9043	9227	9413	9587	9767	9929
8863	9049	9239	9419	9601	9769	9931
8867	9059	9241	9421	9613	9781	9941
8887	9067	9257	9431	9619	9787	9949
8893	9091	9277	9433	9623	9791	9967
8923	9103	9281	9437	9629	9803	9973
8929	9109	9283	9439	9631	9811	
8933	9127	293	9461	9643	9817	
8941	9133	9311	9463	9649	9829	
8951	9137	9319	9467	9661	9833	
8963	9151	9323	9473	9677	9839	
8969	9157	9337	9479	9679	9851	
8971	9161	9341	9491	9689	9857	
8999	9173	9343	9497	9697	9859	
9001	9181	9349	9511	9719	9871	
9007	9187	9371	9521	9721	9883	
9011	9199	9377	9533	9733	9887	

Table 2. The Smallest Primitive Root g

p	g	p	g	p	g	p	g	p	g	p	g	p	g	p	g
2	1	79	3	191	19	311	17	439	15	577	5	709	2	857	3
3	2	83	2	193	5	313	10	443	2	587	2	719	11	859	2
5	2	89	3	197	2	317	2	449	3	593	3	727	5	863	5
7	3	97	5	199	3	331	3	457	13	599	7	733	6	877	2
11	2	101	2	211	2	337	10	461	2	601	7	739	3	881	3
13	2	103	5	223	3	347	2	463	3	607	3	743	5	883	2
17	3	107	2	227	2	349	2	467	2	613	2	751	3	887	5
19	2	109	6	229	6	353	3	479	13	617	3	757	2	907	2
23	5	113	3	233	3	359	7	487	3	619	2	761	6	911	17
29	2	127	3	239	7	367	6	491	2	631	3	769	11	919	7
31	3	131	2	241	7	373	2	499	7	641	3	773	2	929	3
37	2	137	3	251	6	379	2	503	5	643	11	787	2	937	5
41	6	139	2	257	3	383	5	509	2	647	5	797	2	941	2
43	3	149	2	263	5	389	2	521	3	653	2	809	3	947	2
47	5	151	6	269	2	397	5	523	2	659	2	811	3	953	3
53	2	157	5	271	6	401	3	541	2	661	2	821	2	967	5
59	2	163	2	277	5	409	21	547	2	673	5	823	3	971	6
61	2	167	5	281	3	419	2	557	2	677	2	827	2	977	3
67	2	173	2	283	3	421	2	563	2	683	5	829	2	983	5
71	7	179	2	293	2	431	7	569	3	691	3	839	11	991	6
73	5	181	2	307	5	433	5	571	3	701	2	853	2	997	7

Table 3. The First 50 Fibonacci and Lucas Numbers

n	F_n	L_n	n	F_n	L_n
1	1	1	26	121393	271443
2	1	3	27	196418	439204
3	2	4	28	317811	710647
4	3	7	29	514229	1149851
5	5	11	30	832040	1860498
6	8	18	31	1346269	3010349
7	13	29	32	2178309	4870847
8	21	47	33	3524578	7881196
9	34	76	34	5702887	12752043
10	55	123	35	9227465	20633239
11	89	199	36	14930352	33385282
12	144	322	37	24157817	54018521
13	233	521	38	39088169	87403803
14	377	843	39	63245986	141422324
15	610	1364	40	102334155	228826127
16	987	2207	41	165580141	370248451
17	1597	3571	42	267914296	599074578
18	2584	5778	43	433494437	969323029
19	4191	9349	44	701408733	1568397607
20	6765	15127	45	1134903170	2537720636
21	10946	24476	46	1836311903	4106118243
22	17711	39603	47	2971215073	6643838879
23	28657	64079	48	4807526976	10749957122
24	46368	103682	49	7778742049	17393796001
25	75025	167761	50	12586269025	28143753123

Bibliography

Andrews, G.   Some Formulae for the Fibonacci sequence, with generalizations. *Fibonacci Quart.* 7 (1969): 113–130.

Apostol, T.   *Introduction to Analytic Number Theory.* New York: Springer-Verlag, 1976.

Brezinski, C.   *History of Continued Fractions and Pade Approximants.* New York: Springer-Verlag, 1991.

Dickson, L. E.   *History of the Theory of Numbers.* Washington: Carnegie Institute, 1919; New York: Chelsea Publishing, 1952.

Galois, E.   Demonstration d'un theoreme sur les fractions continues. *Ann. Math. Pures Appliques* 19 (1828): 294–301.

Gauss, C. F.   *Disquisitiones Arithmeticae.* 1801; English edition: New York: Springer-Verlag, 1966.

Guy, R.K. *Unsolved Problems in Number Theory.* New York: Springer-Verlag, 1981.

Hardy, G. H., and Wright, E. M.   *An Introduction to the Theory of Numbers*, 5th Ed. Oxford, England: Oxford University Press, 1979.

Hoggatt, V. E., Jr.   *Fibonacci and Lucas Numbers.* Boston: Houghton-Mifflin, 1969.

Hua Loo Keng.   *Introduction to Number Theory.* Berlin: Springer-Verlag, 1982.

Kahn, D.   *The Codebreakers.* New York: Macmillan, 1967.

Knuth, D.   *The Art of Computer Programming.* Reading, Mass.: Addison-Wesley, 1973.

Lagrange, J. L.   Addition au memoire sur la resolution des equations numeriques. *Mem. Acad. Sci. Berl.* 24 (1770): 311–352.

Lucas, E.   Theorie des fonctions numeriques simplement periodiques. *Am. J. Math.* 1 (1878): 184–240; 289–321.

Niven, I., Zuckerman, H. S., and Montgomery, H. L.   *An Introduction to the Theory of Numbers*, 5th Ed. New York: Wiley, 1991.

Ore, O.   *Number Theory and Its History.* New York: McGraw-Hill, 1948.

Perron, O.   *Die Lehre von den Kettenbruchen.* Stuttgart: Teubner, 1913.

Ribenboim, P.   *13 Lectures on Fermat's Last Theorem.* New York: Springer-Verlag, 1979.

Ribenboim, P.   *The Book of Prime Number Records.* New York: Springer-Verlag, 1988.

Riesel, H.   *Prime Numbers and Computer Methods of Factorization.* Boston: Birkhauser, 1985.

Robbins, N.   Calculating a primitive root $(\bmod\, p^n)$. *Math. Gazette* 59 (1975): 195.

Robbins, N.   A new formula for Lucas numbers. *Fibonacci Quart.* 29 (1991): 362–363.

Rosen, K. H.   *Elementary Number Theory and Its Applications*, Reading, Mass.: Addison-Wesley, 1988.

Shanks, D.   Five number theoretical algorithms. *Proc. Second Manitoba Conference on Numerical Methods.* 1972, pp. 51–70.

Sierpinski, W.   *Theory of Numbers.* Warsaw: Panstwowe Wydawnicto Naukowe, 1964.

Trost, E.   *Primzahlen.* Boston: Birkhauser, 1953.

Vajda, S.   *Fibonacci and Lucas Numbers and the Golden Section.* New York: Halsted Press, 1989.

Vorob'yev, N. N.   *Fibonacci Numbers.* New York: Blaisdell, 1961.

INDEX